What processes fixed the designs etched on the cosmic background radiation (CBR)? And what can they tell us about the early Universe and the origin and evolution of cosmic structure? This timely review covers all aspects of three decades of study of this ghostly remnant of the Hot Big Bang origin of the Universe, and examines the consequences in astrophysics, cosmology and theories of the evolution of the large-scale cosmic structure.

The observational techniques used to measure the spectrum of CBR and its angular distribution on the sky are examined in clear but critical detail; from the work of Penzias and Wilson in 1964 to the latest results from NASA's Cosmic Background Explorer (COBE) satellite. This review takes these observations and shows how they have shaped our current understanding of the early history of the Universe and of the origin and evolution of the large-scale structures in it.

As a comprehensive and up-to-date reference this book is suitable for researchers, with introductory chapters in cosmology and radio astronomy provided for graduates in physics and astronomy entering into cosmology or CBR research.

T0237069

3 K: The Cosmic Microwave Background Radiation

Cambridge astrophysics series

Series editors
Andrew King, Douglas Lin, Stephen Maran, Jim Pringle and Martin Ward.

3 K: THE COSMIC MICROWAVE BACKGROUND RADIATION

R. B. PARTRIDGE

Haverford College, Pennsylvania

CAMBRIDGE
UNIVERSITY PRESS

CAMBRIDGE UNIVERSITY PRESS
Cambridge, New York, Melbourne, Madrid, Cape Town, Singapore, São Paulo

Cambridge University Press
The Edinburgh Building, Cambridge CB2 2RU, UK

Published in the United States of America by Cambridge University Press, New York

www.cambridge.org
Information on this title: www.cambridge.org/9780521352543

First published 1995
This digitally printed first paperback version 2006

A catalogue record for this publication is available from the British Library

Library of Congress Cataloguing in Publication data

Partridge, R. B.
   3K: the cosmic microwave background radiation / R.B. Partridge.
      p. cm. – (Cambridge astrophysics series; 25)
   Includes index.
   ISBN 0 521 35254 1
   1. Cosmic background radiation.   I. Title.   II. Title: Three K.   III. Series.
QB991.C64P37   1995
523.01'875344–dc20   94-14980 CIP

ISBN-13 978-0-521-35254-3 hardback
ISBN-10 0-521-35254-1 hardback

ISBN-13 978-0-521-35808-8 paperback
ISBN-10 0-521-35808-6 paperback

To my first teachers, Robert B. and Laura L. Partridge

Contents

Preface

Humankind has made stories about the origin of the world since prehistoric times. These creation stories often have a grand beauty and are sometimes richly detailed. It is only in the present century that such myths and images have been supplanted by a well-established scientific description of the origin of the world. 'World' is now understood to mean the Universe as a whole, not just the Earth or the solar system, and the modern picture of its origin and evolution is the Hot Big Bang model. This book describes one crucial piece of astronomical evidence supporting the Big Bang model, namely the cosmic microwave background radiation, heat radiation left over from a hot and dense phase early in the history of the Universe.

The cosmic background radiation (CBR) was discovered, by accident as it happens, a quarter of a century ago. Within a few years, the basic properties of the radiation had been established. Those properties, especially the thermal 3 K spectrum and the very uniform distribution of the CBR across the sky, have convinced virtually all astrophysicists that the radiation is a relic of the Hot Big Bang, and that it comes to us from a very early time in the history of the Universe. It thus provides information about the early history of the Universe obtainable in no other way. In particular, studies of the CBR have provided important constraints on theories for the origin of large-scale structure within the Universe. Less expectedly, perhaps, related CBR observations have helped to constrain theories of elementary particle physics. In addition, almost every corner of modern cosmology has been touched in some way by observations of this background radiation. One small measure of the importance of the CBR is the number of scientific papers written about it – a number that has grown from few per year twenty years ago to well over a hundred per year by 1990.

My aim in this first monograph on the CBR is to describe the observations of the CBR and to assess their consequences. I am by training and outlook an observer, not a theorist. Hence the emphasis in this book will be on the observations themselves – on experimental techniques, frequently encountered sources of errors in the observations and analysis of the measurements. I will treat in detail measurements of the spectrum of the radiation and searches for angular variations in its intensity across the sky (that is, searches for anisotropy in the CBR). After presenting each set of observations, I will summarize what the observations have told us. These chapters on the consequences of the observational results could have been much longer, but I have elected to emphasize the observations instead. In this regard, it should be mentioned that there are now available a number of excellent reviews of the theoretical side; review papers on the observations are rarer. Hence the theory chapters here are intended more as an introduction to the literature than as an exhaustive treatment.

This book is aimed at students and at colleagues in astronomy and physics who find themselves interested in this new subfield of cosmology. It is not directed at those already active in CBR studies, though some may find it useful. Nor is it likely to be easy for non-scientists, though I would be delighted if they tried it. Undergraduates in their final years of study, and more particularly graduate students, may find it a useful supplement to texts assigned in advanced courses in cosmology or radio astronomy. To make the book more accessible to both students and colleagues outside astronomy, I have included, in Chapters 1 and 3, brief reviews of modern cosmology and of radio astronomical techniques. I have also endeavored to provide substantial and useful lists of references. In the chapters on the observations, I am confident that these lists of references are close to complete up to early 1992, and also fairly represent the experimental side of the field. Some references to more recent work appear in Appendix C. As this manuscript was nearing completion, the findings of NASA's Cosmic Background Explorer satellite became available. These results have been included in Chapters 4, 6, and 7. On the theoretical side, I have to say that I am much less confident that I have fairly sampled the vast and rapidly growing literature on the implications of measurements of the CBR.

Why am I writing this book, and why now? It has been a quarter century since the CBR was discovered, so the time seems ripe for a brief, unified presentation, which includes new and exciting results from the recent NASA experiment. In addition, that same quarter century coincides almost exactly with my scientific career, much of which has been involved with the CBR. I also love teaching, and I look forward to writing down what I have lectured about so often, with the hope that this book will provide a new way to share my enthusiasm for the subject.

A good part of this book was written while I had the good fortune to be a John Simon Guggenheim Foundation Fellow; I thank the Foundation for their support (and indeed their patience). My work on the CBR, and much of the theoretical and observational work described here, has been and continues to be supported by the U.S. National Science Foundation. Finally, preparation of this volume was supported in part by a Faculty Research grant from Haverford College, and funds deriving from the Bettye and Howard Marshall Professorship of Natural Science. The manuscript was typed (and retyped and retyped) with great care and patience by Lillian Dietrich at Haverford. For this book and much else, I owe her many thanks.

If this book is successful, it will owe much to my students and to peers who have heard me spin out these ideas before. I owe a huge debt of gratitude to my colleagues in the field of microwave astronomy, many of whom have helped me and encouraged me in the writing of this book. In particular, Ralph Alpher, Dick Bond, Steve Boughn, Marc Davis, Gianfranco De Zotti, Tom Gaisser, Robert Herman, Craig Hogan, Bernard Jones, Al Kogut, Charles Lawrence, Phil Lubin, John Mather, Jim Peebles, Paul Richards, Giorgio Sironi, George Smoot, Al Stebbins, Michael Strauss, Mike Turner, Nicolà Vittorio and Dave Wilkinson have all read chunks of my draft or helped me in other ways. They have contributed a great deal of sense and clarity to this book, but any sins of omission or of commission are mine, not theirs. Finally, not just this book but my whole career owes much to a succession of fine teachers who sparked my interest in the world and kept it alive. I have dedicated this book to my earliest teachers, but I want also to thank other teachers, friends and colleagues: Tom Carver, George Field, Bob Dicke, Jim Peebles, Dave Wilkinson and John Wheeler, all at one time or another connected with Princeton University; George Series, now at Reading University; and Martin Rees of the Institute of Astronomy in Cambridge.

1
Cosmology

The science that treats the properties and evolution of the Universe as a whole is *cosmology*. Among the sciences, it is unique in having only a single object of study – there are no other Universes for us to use as controls, nor can we readily run the whole experiment over again. As a consequence, much of the effort in modern cosmology has been to determine the best mathematical description, or 'model', of the Universe we inhabit. As we shall see, that task is not yet complete, despite the rapid advances of the past few decades. The range of possible models is presented later in this chapter. First, though, we need to look at the observational bases of modern cosmology, a set of astronomical observations which have established the Hot Big Bang theory and restricted the range of models we need to consider.

1.1 Astronomical constituents of the Universe

Since cosmology is the study of the Universe as a whole and as a single system, it is only indirectly concerned with subsystems within the Universe. Here, I will mention only two: galaxies and clusters of galaxies. The galaxies are assemblies of 10^8–10^{12} stars; many galaxies also contain appreciable amounts of interstellar gas and dust. Some of the basic physical parameters of galaxies are: radius, typically 10^3–10^4 parsecs*; luminosity, typically 10^7–10^{11} times the luminosity of the sun, or very roughly 10^{34}–10^{38} W; and mass, typically 10^8–10^{12} times the mass of the sun or about 10^{41}–10^{45} g.† The question of the mass of galaxies introduces an important issue in contemporary astronomy – the possible existence of 'dark' non-luminous matter in galaxies (see Kormendy and Knapp, 1987; Trimble, 1987; or Primack *et al.*, 1988). The mass of galaxies, especially galaxies of the spiral form shown in fig. 1.1, can be determined by measuring the speed of their rotation as a function of distance from the center, then applying a generalization of Kepler's Third Law. The masses so determined are in almost every case larger than the sum of the masses of all the stars in the galaxy, often by a factor of 3–10 or so. In addition, the measured rotation speed in the outer parts of such galaxies does not fall off as $r^{-1/2}$, but stays essentially constant (fig. 1.2), suggesting that the bulk of the mass of

---

* The parsec (1 pc = 3.08×10^{18} cm) is the standard unit of distance used in astronomy, and will be used throughout this book. A further word on units: workers in the field use a jumble of S.I. and c.g.s. units as well as some purely astronomical units like the parsec. In general, I will use the units that have become conventional in the field, giving conversions where necessary to physical units.

† In astronomy, 'solar units' are conventionally used for luminosity and mass. The luminosity of the sun is $L_\odot = 3.9 \times 10^{33}$ erg s^{-1}; the solar mass is $M_\odot = 1.99 \times 10^{33}$ g.

Fig. 1.1   A typical spiral galaxy (M81). Photograph from the Palomar and Mt. Wilson Observatories.

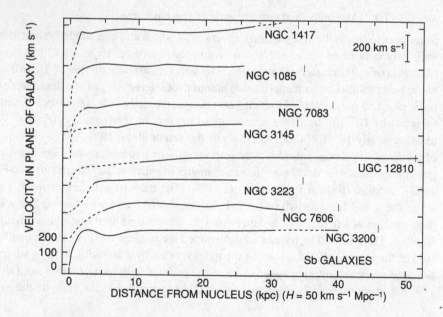

Fig. 1.2   A set of rotation curves (plots of rotation velocity as a function of distance from the center) for several spiral galaxies (from Rubin *et al.,* 1982, with permission). The velocity stays approximately constant well beyond the visible limit of the galaxy, rather than dropping as $r^{-1/2}$, suggesting an extended halo of 'dark matter' in these galaxies.

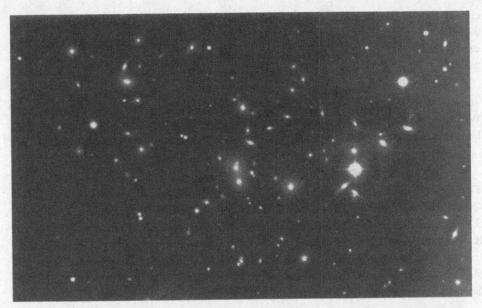

Fig. 1.3   A cluster of galaxies. A photograph from the Palomar and Mt. Wilson Observatories made with the 200-inch telescope.

galaxies is less concentrated than the luminous matter such as stars or gas. This is the 'dark matter' that we will have occasion to refer to here and in Chapters 7 and 8.

All galaxies emit radio waves as well as optical radiation at some level. In a minority of galaxies, however, the radio luminosity exceeds the optical luminosity; these are the *radio galaxies*, which, together with quasi-stellar objects, make up most of the extragalactic sources detectable with radio telescopes. We will deal with the radio emission from our own Milky Way Galaxy in Chapter 4, and with radio sources in general in Chapter 7.

Most galaxies are clumped together in small groups (our own Galaxy is a member of the Local Group, as is the Andromeda Galaxy, M31), or larger *clusters* of a few hundred to a few thousand galaxies. An example is shown in fig. 1.3. The clusters contain matter between their constituent galaxies. In some clusters, this matter is directly detected; it is ionized gas at a temperature of about 10^7–10^8 K, which emits detectable X-ray flux (this intergalactic plasma is discussed further in Chapter 8). In other clusters, the evidence for intergalactic matter is indirect. The total gravitational mass of a cluster required to hold it together may be derived by applying the virial theorem to the cluster, assuming that it is in equilibrium. The result is

$$M = \frac{R_c \bar{v}^2}{G}, \tag{1.1}$$

where v^2 is the mean-square velocity of the galaxies within the cluster, and R_c is the radius of the cluster (see Chapter IV of Peebles, 1971). The mass of clusters calculated in this fashion is 10^{47}–10^{48} g, in most cases an order of magnitude larger than the sum of the masses of the individual galaxies. Estimates of the mass of the hot intergalactic gas detected in some clusters show that it cannot account for the discrepancy; it is insufficient to bind clusters gravitationally. Nor do there appear to be enough intergalactic stars to bring the mass of clusters up to the value calculated from eqn. (1.1). Once again, the existence of some form of 'dark' matter is suggested, in this case lying between the galaxies.

1.2   Observational bases of Big Bang cosmology

We now turn to some astronomical observations, which establish properties of the Universe as a whole, and upon which our present cosmological theories are based.

1.2.1 Homogeneity

The presence of clusters of galaxies and more careful analysis of the counts of galaxies (Peebles, 1980) show that, on relatively small cosmological scales, $d \lesssim 30$ Mpc $\equiv 3 \times 10^7$ pc or about 10^{26} cm, the galaxies are inhomogeneously distributed (see figure 8.2). On larger scales $d \gtrsim 300$ Mpc, however, the distribution is approximately isotropic and homogeneous. We thus arrive at the first observational basis of cosmology – on sufficiently large scales, the Universe appears to be homogeneous and isotropic (see Section 8.2, however).

1.2.2 Expansion

One of the landmark discoveries of 20th century science is the recognition that the Universe is an *expanding* system. This expansion was discovered and characterized in the late 1920s by Edwin Hubble, who found that the atomic lines detected in the spectra of distant galaxies almost always appear at wavelengths slightly greater than the rest or laboratory wavelengths of those same atomic lines – that is, they are shifted to longer wavelengths or *redshifted*. The redshift, z, is defined by

$$z + 1 \equiv \lambda_{obs}/\lambda_{rest}, \tag{1.2}$$

where λ_{obs} is the observed wavelength. Hubble also found that, on the average, the magnitude of redshift observed in the spectrum of a galaxy was proportional to its distance, d, from us.

Hubble interpreted the redshifts he observed as instances of the Doppler effect; for recession velocities $v \ll c$, eqn. (1.2) gives

$$z = v/c \propto d.$$

In this interpretation, recession velocity is proportional to distance. This linear relation is just what one would expect for uniform expansion of the Universe. The measured constant of proportionality in the relation between v and d is now known as Hubble's constant, H_0, and is evidently a measure of the *rate* of expansion of the Universe:

$$v = H_0 d, \quad \text{or} \quad z = H_0 d/c. \tag{1.3}$$

Astronomical measurements of the redshift and d show that H_0 lies in the range $(1.3-3.2) \times 10^{-18}$ s^{-1}, or in more conventional astronomical units, 40–100 km s^{-1} per megaparsec. The factor of two uncertainty arises primarily from the difficulty of making reliable measurements of the distance of extragalactic objects (see Rowan-Robinson, 1985). To account for the uncertainty in H_0, we will generally write it as $100h$ km s^{-1} per megaparsec, with $0.4 \lesssim h \lesssim 1.0$.

The strictly linear relationship between redshift and distance breaks down for larger distances and higher velocities (see Weinberg, 1972). Since the redshift is a more easily measured quantity than distance itself, it is commonly used by cosmologists to parameterize the distance to a galaxy or other source, and is so used in this book.

While Hubble interpreted the redshift as a Doppler shift induced by motion of galaxies, the modern interpretation, based on ideas introduced in General Relativity, is somewhat different. In modern cosmological theory, the galaxies are taken as more or less fixed* in a geometry that is itself expanding. The apparent relative recessional velocity of an observer and a distant galaxy is then explained by the expansion of space between the two. The expansion is specified through a quantity R known as the *scale factor*, which is time dependent and increasing. The distance between any two objects in the Universe at time t may thus be written as

$$ d_{12}(t) = \frac{R(t)}{R(t_0)} d_{12}(t_0), $$

where $d_{12}(t_0)$ is the distance between those two objects at present (denoted throughout as t_0), and $R(t_0)$ is the present value of the dimensionless scale factor. $R(t_0)$ is often set equal to 1, and we will follow that convention. Since $\dot{R} > 0$, it follows that all lengths and distances measured in this expanding space were shorter in the past. That statement is true of the wavelengths of freely propagating photons as well (Weinberg, 1972). It thus follows that $\lambda_{obs} = R^{-1}(t) \lambda_{rest}$, for a photon emitted at some earlier time t. Hence

$$ R(t) = [z(t) + 1]^{-1}, \tag{1.4} $$

establishing the connection between the scale factor and redshift.

Likewise, if $R(t_0) \equiv 1$, it may easily be shown that $H_0 = \dot{R}(t_0)$, where the subscript '0' is used to show explicitly that we are concerned with the present value of both the scale factor and Hubble's 'constant,' since both may be functions of time.

1.2.3 Age of the Universe

If there are no forces to slow down the expansion of the Universe, $\dot{R}$ will remain constant. Under these conditions, a backward extrapolation of the present expansion reveals that $R = 0$ at some finite time in the past. As the scale factor R goes to zero, so do all distances. Hence the density goes to infinity and we cannot sensibly extrapolate further into the past. This moment of infinite (or at least very high) density is the Big Bang origin of the Universe. Again, assuming a constant value for $\dot{R}$, it is easy to show that the time elapsed since the Big Bang is H_0^{-1}. This interval is the present age of the Universe, t_0. For $\dot{R}$ = constant, t_0 lies in the range about $(3-7) \times 10^{17}$ s or 10–20 billion years, depending on the value assumed for H_0. As we shall see, for more realistic cosmological assumptions, this result is in fact an upper limit on t_0. Support for the Big Bang theory is provided by independent geophysical and astronomical measurements of the age of various constituent parts of the Universe. The Earth–Moon system, for instance, and by inference the solar system, is known to be 4.6 billion years old. The age of certain long-lived radio-isotopes found in meteoritic material is 11–12 billion years (see

---

* Galaxies may have small random or even systematic velocities relative to this background geometry. These peculiar velocities, as they are called, are typically a few hundred kilometers per second and are discussed further in Chapter 8.

Fowler, 1987; Weinberg, 1972; or Narlikar, 1983) in reasonable agreement with H_0^{-1}. So too is the calculated age of the oldest stars in our Galaxy (see summary by Tayler, 1986). It is important to note that no objects within the Universe have yet been found with ages clearly in excess of H_0^{-1} – thus age measurements are consistent with the Big Bang theory.

Finally, the age of the Universe establishes a very rough limit to its extent. In a Universe of age t_0, photons can have traveled at most a distance about ct_0, and hence an astronomer cannot 'see' further than about ct_0. This distance, very roughly 5×10^9 pc, is the effective radius of the Universe. It is important to note that this same argument implies that the Universe was smaller in the past, since it was younger. (The role of particle horizons will reappear in Chapter 8; see also standard cosmology texts.)

1.2.4 Evidence for a Hot Big Bang

The discovery of the cosmic microwave background radiation (henceforth abbreviated CBR) established that the early Universe was hot as well as dense. The key to this argument is the blackbody or thermal spectrum of the radiation (the observations are presented in Chapter 4). Let us ask what happens to a blackbody radiation field if we extrapolate backwards in time to an epoch when the scale factor R was smaller, so $z > 0$. The wavelength of all photons is decreased proportionally to R or $(z + 1)^{-1}$. The Planck function, however, depends only on the product λT. It follows (see Chapter 5) that the spectrum of the radiation was also blackbody in the past, but the temperature was higher by a factor $z + 1$ (see Weinberg, 1972, Section 15.5):

$$T(t) = T_0(z + 1), \tag{1.5}$$

where T_0 is the present temperature of the CBR, approximately 3 K. Knowing the present value of the temperature, we can calculate the temperature at any earlier epoch using eqn. (1.5). For instance, for redshifts greater than 1000, the temperature was > 3000 K, sufficient to ionize the major atomic constituent of the Universe, hydrogen. At still larger redshifts, corresponding to earlier times in the history of the expanding Universe, the temperature was even greater. Note, however, that the strict linear dependence of $z + 1$ and T breaks down at higher temperatures, where the number of light particle species goes up (see Kolb and Turner, 1990).

One earlier epoch is of particular interest. A few minutes after the Big Bang origin of the Universe, the temperature dropped to about 10^9 K, low enough to permit fusion of neutrons and protons present in the hot primordial plasma (see Section 1.6.4 below). The nuclei of light elements, primarily ^{4}He, were produced. This process of primordial nucleosynthesis has been extensively studied (Peebles, 1966; Wagoner, Fowler and Hoyle, 1967; Schramm and Wagoner, 1979; Audouze, 1987), and detailed predictions have been made of the abundances of the light nuclei produced in the Hot Big Bang. These predicted abundances (fig. 1.4) agree well with astronomical determinations of the abundances of these same nuclei in the oldest stars and other matter in our Galaxy (Boesgaard and Steigman, 1985; Walker *et al.*, 1991), providing additional strong support for the Hot Big Bang model.

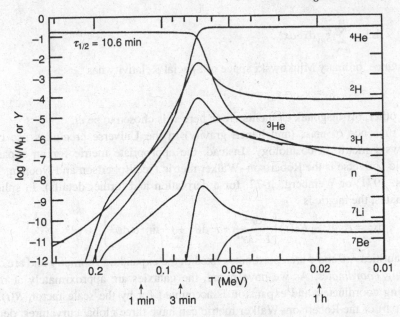

Fig. 1.4   Predicted abundances for the light nuclei produced in the first few minutes of a Hot Big Bang. Note the initial rapid rise in the abundance of deuterium (^{2}H) as the Universe cooled. Between $t = 1$ and 4 min, the deuterium was incorporated into other nuclei, especially ^{4}He. Adapted from Wagoner (1973); $n_b/n_\gamma = 3 \times 10^{-10}$ was assumed.

1.3   Cosmological models

We will now incorporate these observations into a general mathematical description for the properties of the Universe as a whole. Such descriptions are called *cosmological models*. Note the plural; as we will soon see, many mathematical models are consistent with the observational evidence now available. Of course, only a single model can *best* describe the Universe, and much of the effort in modern cosmology has been devoted to testing the models observationally, with the hope of reducing the range of possible models.

Like most models in physics, cosmological models ignore some of the details (e.g., inhomogeneities in the Universe). Most are based on the *cosmological principle*, the notion that the Universe is isotropic and homogeneous on a large scale or, more descriptively, 'the Universe is the same everywhere.'

1.3.1   The Robertson–Walker metric

If the Universe is isotropic and homogeneous on a large scale, the underlying geometry of the Universe must also be isotropic (exceptions are discussed in Section 8.3). The space–time geometry of the Universe may be completely specified by giving its metric tensor $g_{\mu\nu}$ – see texts on General Relativity; Peebles (1971), Weinberg (1972) or Narlikar (1983), for instance.

For a general set of four space–time coordinates, x^μ, the invariant interval ds^2 is given in terms of the metric tensor as

$$ds^2 = \sum_{\mu,\nu=0}^{3} g_{\mu\nu} dx^\mu dx^\nu.$$

For instance, ordinary Minkowski space of Special Relativity has

$$g_{00} = 1, \quad g_{11} = g_{22} = g_{33} = -1,$$

and all other, off-diagonal, elements are 0; here x^0 is chosen to be ct.

The presence of mass (and hence gravity) in the Universe precludes use of the Minkowski metric in cosmology. Instead, the appropriate metric for an expanding isotropic Universe is the Robertson–Walker metric (see Robertson and Noonan, 1968; Peebles, 1971; or Weinberg, 1972, for a derivation and further details). In spherical coordinates, the metric is

$$ds^2 = c^2\, dt^2 - R^2(t) \left\{ \frac{dr^2}{1-kr^2} + r^2 d\theta^2 + r^2 \sin^2\theta\, d\phi^2 \right\}. \tag{1.6}$$

The quantities r, θ and ϕ are coordinates fixed in the expanding geometry and are called *comoving coordinates*. As we have noted, the galaxies are approximately at rest in comoving coordinates, and expansion is accounted for by the scale factor, $R(t)$. The spatial part of the Robertson–Walker metric can have three global curvatures, depending on the value of the quantity k. For $k = 0$, the spatial geometry of the Universe is flat, i.e. Euclidean, so that comoving distances are given by the usual relation, $d^2 = x^2 + y^2 + z^2$. The geometry may also be positively or negatively curved (with $k \lessgtr 0$), however, in which case $d^2 \gtrless x^2 + y^2 + z^2$, respectively. A positively curved Robertson–Walker metric is a closed geometry, limited in volume but without edges, just as the two-dimensional surface of a sphere is closed, finite and without boundaries. The negatively curved case, like the flat case, is an open, infinite geometry.

1.3.2 Density and curvature

These three possible curvatures are directly linked by General Relativity to the amount of matter in the Universe. A high density produces positive curvature; in the low density case, the curvature is negative. The particular density corresponding to a flat geometry, the *critical density*, is denoted ρ_c; in the next section we will evaluate it numerically.

1.3.3 Dynamics

We know that the Universe is expanding now, so that $R(t)$ increases with time. If there were no forces* to alter the expansion, $\dot{R}$ would remain constant, as shown by curve a in fig. 1.5. Since the density of the Universe is nonzero, however, we know that at least one long-range attractive force is acting – gravity. It acts to slow the expansion. The slowing of the expansion may be represented schematically by curvature in fig. 1.5 – see curves b and c. One evident consequence of the presence of matter in the Universe is that the present age of the Universe (defined as the time since $R = 0$ at the Big Bang) is less than H_0^{-1}.

The relation governing expansion of the Universe – that is, the function $R(t)$ – may be found by solving the field equations of General Relativity (see Weinberg, 1972). Here,

---

* I recognize that, for pedagogical purposes, I am mixing Newtonian concepts like force with General Relativistic ones like space curvature.

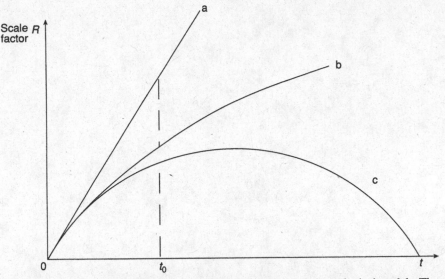

Fig. 1.5   The scale factor R as a function of time for various cosmological models. The different models are specified by their space curvature, k, and by the mean value of present density, ρ_0. The slope of the curves at the present epoch, t_0, is fixed by measurements of Hubble's constant, H_0. From the figure, it may be seen that the age of the Universe (the time since $R = 0$) is less than H_0^{-1} for models with $\rho_0 > 0$.

following McCrea and Milne (1934) and Callan, Dicke and Peebles (1965), I will take a simpler, but quite valid, approach using Newtonian physics.

Consider a sphere centered at an arbitrary point O in the expanding Universe. Let its radius at a particular time be $R(t)$; we will assume that $R(t)$ is large enough that the sphere represents a fair sample of the Universe, yet small enough that the curvature of space can be neglected. Now consider the acceleration of a unit mass on the sphere's surface towards O; it is

$$\ddot{R}(t) = -\frac{GM}{R^2(t)}, \tag{1.7}$$

where M is the mass inside the sphere.* In turn, in this Newtonian model,

$$M = \frac{4}{3}\pi R^3(t)\rho(t). \tag{1.8}$$

Here $\rho(t)$ is the density at time t.

We now make use of the conservation of mass. If the density of the Universe includes only material particles, which interact only gravitationally and exert no pressure (we specifically *exclude* radiation), then $V(t)\,\rho(t) = V(t_0)\,\rho(t_0)$, which leads to

$$\rho(t) = \frac{R^3(t_0)}{R^3(t)}\rho(t_0).$$

If, as above, we set $R(t_0) = 1$, and write the present density as ρ_0 for simplicity, we have for this simple case excluding radiation,

---

* The rest of the isotropic, homogeneous Universe outside the sphere exerts no net force on the unit mass.

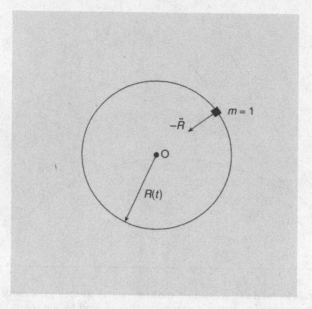

Fig. 1.6   A sphere of radius R about an arbitrary point O in a homogeneous Universe; $-\ddot{R}$ is the magnitude of the inward acceleration of the test mass, m.

$$\rho(t) = R^{-3}(t)\,\rho_0 . \tag{1.9}$$

Combining (1.7), (1.8) and (1.9),

$$\ddot{R}(t) = -\frac{4}{3}\pi G \rho_0 R^{-2}(t). \tag{1.10}$$

The first integral of eqn. (1.10) may be found by multiplying both sides by $\dot{R}$, then noting that

$$\ddot{R}\dot{R} = \frac{\mathrm{d}}{\mathrm{d}t}\left(\frac{\dot{R}^2}{2}\right), \quad \dot{R}R^{-2} = \frac{\mathrm{d}}{\mathrm{d}t}\left(\frac{-1}{R}\right).$$

Thus, after integration, we find

$$\dot{R}^2(t) = \frac{8}{3}\pi G \rho_0 R^{-1}(t) + \text{constant}. \tag{1.11}$$

In this Newtonian calculation, it may easily be shown that the constant of integration is related to the total energy per unit mass. In the full, General Relativistic solution, the connection between dynamics, density and space curvature becomes manifest; the constant of integration is found to be $-kc^2$. Hence, finally,

$$\dot{R}^2(t) = \frac{8}{3}\pi G \rho_0 R^{-1}(t) - kc^2 , \tag{1.12}$$

where $k > 0$, $= 0$ or < 0 for positively curved, flat or negatively curved space, respectively. Although we have used Newtonian concepts in this derivation, the result is the same as is found (e.g. Weinberg, 1972) from a fully relativistic calculation.

Solutions to eqn. (1.12), which is the basic equation of mathematical cosmology, are presented in texts such as those by Bondi (1960), Peebles (1971), Weinberg (1972),

Misner, Thorne and Wheeler (1973) and Narlikar (1983). Here it is sufficient to show their general form (fig. 1.5), and to comment on a few special cases. There are three general classes of solutions, depending on the value of k.

The case $k > 0$ clearly allows $\dot{R} = 0$ for a finite R; thus closed cosmological models also include the recollapse of the Universe, as shown by curve c in fig. 1.5. In Newtonian terms, these are the high density models, which have enough gravitating mass per volume to slow down *and reverse* the present expansion.

On the other hand, for $k \leqslant 0$, the Universe expands forever, as shown schematically by curve b. A specific instance is the flat-space* case, $k = 0$; then $\dot{R}^2 \propto R^{-1}$, which has as its solution $R(t) \propto t^{2/3}$. This result plus eqn. (1.4) gives $t = t_0 (z + 1)^{-3/2}$ for the Einstein–de Sitter model. Another illustrative case is the solution found as $\rho_0 \to 0$. Clearly, if the Universe is expanding, $\dot{R} > 0$; thus from eqn. (1.12), k must be < 0 in this case. As $\rho_0 \to 0$, $\dot{R} \to$ constant, as is shown by curve a.

Recall that curve a is the representation of the case when $t_0 = H_0^{-1}$; it is left as an exercise to show that $t_0 = 2/3 \, H_0^{-1}$ for the flat space case. It is also not difficult to show that $t_0 < 2/3 \, H_0^{-1}$ for $k > 0$. Measurements of both t_0 and H_0 would thus in principle allow us to determine the sign of k, and the future evolution of the Universe. Unfortunately, neither quantity has been measured with adequate precision to do so.

1.3.4 The critical density, ρ_c

Equation (1.12) also allows us to evaluate the particular value of the present density corresponding to the $k = 0$ flat space or Einstein–de Sitter model. This is the *critical density*, which we write as ρ_c. With $k = 0$, eqn. (1.12) evaluated at the present is just

$$\dot{R}^2(t_0) = \frac{8}{3}\pi G \rho_c,$$

recalling that we have taken $R(t_0) = 1$. Since $H_0 \equiv \dot{R}(t_0)$, we have

$$\rho_c \equiv \frac{3H_0^2}{8\pi G}. \tag{1.13}$$

Numerically, it may easily be shown that $\rho_c \approx (3\text{--}20) \times 10^{-30}$ g cm^{-3}, depending on the value assumed for H_0.

If the present density of the Universe exceeds ρ_c, k is greater than 0, and the spatial geometry of the Universe is closed and has positive curvature. If $\rho_0 < \rho_c$, k is less than 0 and the geometry is open and ever-expanding. The ratio ρ_0 / ρ_c is clearly a crucial cosmological parameter. It is conventionally written as Ω. For the models we have developed $\Omega \lessgtr 1 \leftrightarrow k \lessgtr 0$. Despite decades of effort, the direct measurements of Ω discussed briefly below in Section 1.4 are inconclusive, and hence do not provide a critical test of the models.

1.3.5 Models with radiation

In deriving eqn. (1.12), we explicitly excluded radiation. Yet radiation is clearly present in the Universe, both in the form of starlight and in the cosmic microwave background. Including radiation alters our Newtonian derivation in two ways. First, radiation itself

---

* This special case is called the *Einstein–de Sitter model*. Inspection of eqn. (1.12) will show that all the models discussed here approach the Einstein–de Sitter model as $R \to 0$.

acts as a source of gravity in the Universe. Both the equivalent mass density of the radiation, $\rho_\gamma = u/c^2$ g cm^{-3}, and its pressure, $P = 1/3\,u$, contribute to $\ddot{R}$. Including these effects of radiation, eqn. (1.7) and eqn. (1.8) may be combined to give

$$\ddot{R} = -\frac{4}{3}\pi G(\rho + u/c^2)R. \tag{1.14}$$

Second, there is an additional consequence of radiation pressure. The presence of pressure invalidates the simple relation (1.9) based on conservation of mass alone. We must instead look at conservation of mass–energy in the sphere of radius R, including any $P\,dV$ work done on the system as R increases or decreases. It is not hard to show that

$$\dot{\rho} = -3(\rho + P/c^2)\,\dot{R}/R \tag{1.15}$$

in this case. These two equations, (1.14) and (1.15), can be solved for $R(t)$ provided that we know the time-dependence of ρ and of P, the pressure of the radiation field.

We will soon see that the main contribution to the radiation pressure in the Universe is made by the cosmic microwave background this book addresses, with $P/c^2 = u/3c^2 \approx 3 \times 10^{-5}$ of the matter density at present. The CBR has an essentially blackbody spectrum (Chapter 4), so that its energy density and pressure are given simply by

$$u = a\,T^4 \text{ erg cm}^{-3}, \quad P = \frac{a}{3}\,T^4 \text{ dyne cm}^{-2}, \tag{1.16}$$

where constant $a = 7.565 \times 10^{-5}$ in c.g.s. units. For blackbody radiation, eqns. (1.4), (1.5) and (1.16) may be combined to show

$$\rho_\gamma \propto u \propto T^4 \propto R^{-4} \propto (z+1)^4. \tag{1.17}$$

From eqn. (1.9), we have, however,

$$\rho_{\text{matter}} \propto R^{-3} \propto (z+1)^3 \tag{1.18}$$

for the matter density alone. Thus eqn. (1.17) and eqn. (1.18) may be compared, to establish the important result that radiation energy density becomes more and more dominant as we look back to earlier and earlier epochs of the Universe, when $R(t)$ was smaller:

$$\rho_\gamma/\rho_{\text{matter}} \propto R^{-1} \propto (z+1). \tag{1.19}$$

Since $\rho_\gamma = u/c^2 \approx 10^{-4}\,\rho_{\text{matter}}$ now, we see from (1.19) that radiation, not matter, dominated the dynamics at early epochs corresponding to redshifts $z + 1 \gtrsim 10^4$. This simple conclusion profoundly simplifies our treatment of the early Universe.

In particular, we may ask how R depends on T when radiation dominates the expansion. For the radiation-dominated analog to the $k = 0$ Einstein–de Sitter model, the result is

$$R(t) \propto t^{1/2}. \tag{1.20}$$

This simple equation governs the expansion of the Universe for the first few thousand years after its Big Bang origin.

1.3.6 The cosmological constant and its effects on dynamics

A glance at eqn. (1.10) or eqn. (1.14) will show that no static ($R = $ constant) model is possible except in the uninteresting case $\rho = u = 0$. To create a static model we would need to add a long-range repulsive force to counteract gravity. Just such a notion was introduced into cosmology by Einstein in 1917 in order to obtain a static solution. It is based on the *cosmological constant* he introduced into the field equations of General Relativity. It permitted a static solution, and also appeared to Einstein to settle some other difficulties in his newly-conceived theory of General Relativity (for a historical discussion, see Chapters 2, 4 and 5 of North, 1965). In retrospect, one may question Einstein's reasons for making this addition. We now know that the Universe is expanding, not static; but in 1917, convincing observational evidence for expansion was still a decade off. The addition makes the General Theory of Relativity more 'general' by adding another parameter; to some eyes it therefore makes the theory less elegant (for such a view, see McCrea, 1971).

As it happens, suggestions for an equivalent modification of Newton's law of gravity were already in the air before 1917 (see North, 1965, Chapter 2). These may be expressed by altering eqn. (1.7) to read

$$\ddot{R} = -\frac{GM}{R^2} + \alpha R.$$

Since Newtonian gravity works very well in the solar system, we know that the coefficient α must be very small. On a cosmological scale, however, the positive term on the right-hand side of the equation can in principle have the same magnitude as the negative term, allowing $\ddot{R} \to 0$.

We will follow the usual convention and write $\alpha \equiv \Lambda/3$ where Λ is the cosmological constant with units of s^{-2} ($\alpha = \lambda c^2/3$, with λ in cm^{-2}, is sometimes used instead). From either the field equations (see Bondi, 1960; Weinberg, 1972; or Narlikar, 1983), or this Newtonian analog, we arrive at

$$\ddot{R} = -\frac{4}{3}\pi G \rho R + \frac{\Lambda}{3} R, \tag{1.21}$$

$$\dot{R}^2 = \frac{8}{3}\pi G \rho R^2 + \frac{\Lambda}{3} R^2 - kc^2. \tag{1.22}$$

Clearly, a static solution is possible if $\Lambda = 4\pi G \rho$ and $k = \Lambda R^2/c^2$; this is the solution advanced by Einstein in 1917.

Once expansion of the Universe had been firmly established, Einstein abandoned his static solution, and with it the cosmological constant, Λ. Indeed, he came to regard Λ as 'the worst mistake in my life.' Ironically, it may turn out that his dismissal of Λ was a bigger mistake. In physical terms, a value of $\Lambda > 0$ occurs because the energy density of the vacuum can be non-zero, and just such a situation is predicted by the recently developed theories of the very early Universe discussed in Section 1.5 below. In these theories, the Universe is briefly trapped in a state in which Λ is not only non-zero, it completely dominates the dynamics of the Universe. For this reason, let us solve eqns. (1.21) and (1.22) under the condition that $\Lambda \gg 4\pi G \rho$ and $\gg kc^2/R^2$. Then $\ddot{R} = \Lambda/3\ R$ and $\dot{R}^2 = \Lambda/3\ R^2$, either of which leads to the simple solution

$$R(t) = \exp\left[(\Lambda/3)^{1/2}\ t\right], \tag{1.23}$$

an *exponential* expansion. It is this exponential expansion – likened to runaway increases in prices – which gave *inflationary cosmology* its name.

One of the remaining mysteries in cosmology is why Λ is now so small or even zero, given that it was so large in the past. Many cosmologists feel that there must be some as yet undiscovered physical principle which drives Λ to 0 in the present Universe. Others consider the present value of the cosmological constant to be an open question, one to be decided by observations. If $\Lambda \neq 0$, the connection between space curvature and density is altered, and becomes

$$kc^2 - \frac{\Lambda}{3} = (\Omega - 1)H_0^2 .$$

(1.24)

This result follows from eqn. (1.22) evaluated at the present, and the substitution $\rho_0 = \Omega \rho_c = 3\Omega H_0^2/8\pi G$. Note particularly that a positive, non-zero value for Λ allows $k = 0$ for $\Omega < 1$, that is $\rho_0 < \rho_c$.

1.4   The density of the Universe

While ρ_c is very low, observational attempts to determine the actual density ρ_0 of the Universe have produced even lower values (for a recent review, see Chapter 12 of Narlikar, 1983; also Faber and Gallagher, 1979). If we add up all the mass of galaxies in a representative large volume of the Universe, for instance, we get roughly 0.01–0.1 ρ_c or $\Omega = 0.01$–0.1; the larger value includes the non-luminous matter in galaxies and clusters.

Another constraint on ρ_0 or Ω is set by nucleosynthesis early in the history of the Universe, to be discussed in more detail in Section 1.6.4 below. The amount of deuterium emerging from this epoch is a sensitive function of Ω. The observed lower limits on deuterium abundance imply *upper* limits on Ω of < 0.2. It is important to bear in mind, however, that this limit applies only to baryonic matter.

These two observations leave open the possibility that large amounts of *non-baryonic* matter could be present in intergalactic space. Since we have not detected it, this non-baryonic matter is frequently referred to as 'dark matter,' which may or may not be the same 'dark matter' needed to explain the rotation curves of galaxies (Section 1.1). Various suggestions have been made for non-baryonic 'dark matter,' including neutrinos with a non-zero rest mass of order 10 eV, and some of the particles associated with Grand Unified Theories, extensions of the standard model of particle physics (e.g., axions), or Supersymmetric Theories (photinos, etc.).

It is reasonable to ask why we invoke dark matter rather than accepting $\Omega < 1$ and the open, ever expanding Universe that value implies. One strong argument for the existence of some kind of dark matter is the need for non-luminous matter to explain the rotation curves of galaxies and the large masses of clusters of galaxies deduced from eqn. (1.1). Another, as we shall see in Chapter 8, is the extreme difficulty of reconciling theories of galaxy formation in an $\Omega < 1$, baryon-only model on the one hand with observational constraints on the small-scale anisotropy of the cosmic microwave background on the other. Neither of these arguments, of course, specifies a value for Ω. On the other hand, a value of $k = 0$ (and thus $\Omega = 1$ if $\Lambda = 0$) seems virtually required by the recent inflationary models for the very early Universe, to which we turn in the next section.

While ρ_0 is not well known, it is certainly much greater than the density of the CBR; the ratio $\rho_\gamma/\rho_0 \lesssim 10^{-4}$ at present. This result in turn may be translated into a ratio of the number density of baryons (baryons per cm^3) to the number density of photons in the

Universe; that ratio is very roughly $n_b/n_\gamma = 10^{-9}$. Note that n_b/n_γ is independent of $(z + 1)$ and therefore has remained constant as the Universe expanded, unlike the ratio of densities, $\rho_\gamma/\rho_{matter}$, which increased linearly with $(z + 1)$.*

1.5   Inflationary cosmology

Sections 1.2–1.4 present what is now conventionally called *classical cosmology* or *the standard model*. In the past decade, an influx of new ideas, mostly introduced by particle physicists rather than cosmologists or astronomers, has radically changed our picture of the very early history of the Universe. This revised version of the first fraction of a second of the history of the Universe is called *inflationary cosmology*. I will not develop it in detail here, in part because it is less well established observationally or experimentally (but see the review by Guth and Steinhardt, 1984 (at an elementary level); the compilation of Abbott and Pi, 1986; and especially Kolb and Turner, 1990). Instead I will focus on three general topics: (1) problems or 'loose ends' in classical cosmology, (2) how inflation solves those problems, and (3) the imprint that an inflationary epoch may leave in the CBR.

1.5.1   'Loose ends' in classical cosmology

The standard Hot Big Bang model sketched in sections 1.2–1.4 has some outstanding successes to its credit. Among these are the approximate agreement between the age of the Universe t_0 and the value of H_0^{-1}, and the ability of the model to predict observed ^{4}He abundances (Section 1.6.4). However, there are also observations that it does not explain, or can 'explain' only by building in specific, *ad hoc*, initial conditions. The first of these is the small value of k, the curvature of the Universe, or alternatively the fact that Ω is not grossly different from unity. Why? Classical cosmology cannot tell us why the numerical value of k is about 0 rather than 10^4. Nor does it offer an explanation of the baryon to photon ratio, $n_b/n_\gamma \simeq 10^{-9}$. Why is it not 10^{-19} or 10^9? In particular, since the Universe started in thermal equilibrium at high temperature and density, we might expect essentially all the baryons to have annihilated with their anti-baryon partners, leading to $n_b/n_\gamma \rightarrow 0$. Third, there is the large-scale homogeneity of the Universe. Observations presented in Chapter 6 show that the Universe was quite homogeneous at an early epoch corresponding to $z \simeq 1000$. The particle horizon of the Universe at that early time was far smaller than it is now. How could regions not yet causally connected 'know' to have the same density and temperature now? Classical cosmology offers no physical explanation of the approximate homogeneity of the Universe or of the isotropy of the CBR on scales larger than this horizon. Finally, classical cosmology inherently contains no mechanism for producing the inhomogeneities now observed in the Universe, such as galaxies and even larger structures. The inflationary picture ties up each of these loose ends.

1.5.2   Outline of inflationary cosmology

This new scenario for the early history of the Universe is based on two main concepts: non-conservation of baryon number and a period of exponential (or 'inflationary')

---

* In fact n_b/n_γ remained constant only after particle–antiparticle annihilations (which produce additional photons) ceased; see Chapter 3 of Kolb and Turner (1990) for details. Hence n_b/n_γ became constant only *after* the first few seconds of the history of the Universe (following e^+–e^- annihilation).

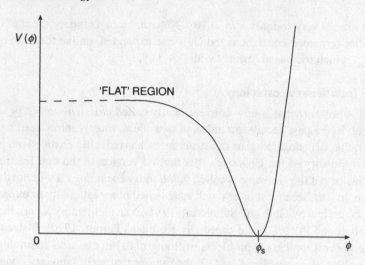

Fig. 1.7   Schematic of an inflationary potential, V. The details of the potential at small ϕ are model dependent and need not concern us here. The 'flat' region and the behavior of $V(\phi)$ near ϕ_s determine the properties of inflationary models discussed in the text.

growth in the scale factor. The first of these is intimately linked with recent developments in theories of elementary particles and forces, particularly Grand Unified Theories or other attempts to unify the four forces of physics. The second idea, introduced in 1981 by Guth, is less certainly grounded, but has proven to have great power in tying up the loose ends referred to above.

As we shall soon see, the inflationary period must have preceded the epoch at which a net baryon number was generated. Hence let us look first at the mechanism of inflation. Following Guth (see his 1981 paper and revisions; and reviews by Albrecht and Steinhardt, 1982; Linde, 1982; Turner, 1988; and Kolb and Turner, 1990), let us suppose the existence of a scalar field ϕ whose exact nature need not concern us here. Associated with this field is a potential $V(\phi)$ of the general form shown in fig. 1.7. Note in particular that V is non-zero for $\phi = 0$, and that the minimum in $V(\phi)$ occurs for a non-zero value of ϕ. It will also turn out to be important that the slope of $V(\phi)$ for $0 < \phi < \phi_s$ be small for an extended range of ϕ.

If the Universe had $\phi = 0$ as an initial condition (a variety of mechanisms have been suggested for producing this as an initial or very early condition), then $V(\phi)$ was initially non-zero, and this potential must be included in the sum of the stress-energy content of the Universe. It may be thought of as the energy density of the vacuum (called the 'false vacuum' to distinguish it from the present state where $V = 0$). The product $GV(\phi \approx 0)c^{-2}$ enters the dynamical equations of cosmology (Section 1.3) in exactly the same way as the cosmological constant discussed earlier. If $GV(\phi \approx 0)c^{-2}$ were large relative to the density term early in the history of the Universe, it would have dominated the dynamics of the expansion of the Universe, making the expansion exponential as in eqn. (1.23). This is the basis of the inflationary model.

For a potential of the shape sketched in fig. 1.7, the condition $\phi = 0$ is clearly an unstable equilibrium; the stable equilibrium is at $\phi = \phi_s$, and the Universe gradually 'rolled downhill' to reach it. How long $V(\phi)$ stayed large determines the *duration* of the inflationary phase. Hence an extended range of ϕ over which $V(\phi)$ has a small slope is required to produce an extended period of exponential expansion.

To make these ideas a bit more concrete, let us consider a particular (but not the only) mechanism for inflation, the spontaneous breaking of symmetry in some Grand Unified Theories or 'GUTs' (reviewed in Ross, 1984). This model will allow us to assign an approximate value to $V(\phi \approx 0)$ and to estimate more roughly the duration of the inflationary period. The basic premise of GUTs is that three of the fundamental forces of physics, the electromagnetic, the weak and the strong forces, are unified at sufficiently high energies (see Ross, 1984, for instance). That energy threshold is, at something like 10^{14} GeV, utterly inaccessible to our laboratory accelerators. Very early in the history of the Universe at $t \lesssim 10^{-34}$ s, however, the temperature was high enough to exceed the threshold ($kT \gtrsim 10^{14}$ GeV). Hence the strong, weak and electromagnetic forces were initially unified. We now need to follow what happened as the Universe expanded and cooled below $kT = 10^{14}$ GeV. As the Universe cooled, the symmetry between the strong force and the still-unified electro-weak force(s)* spontaneously broke down; this phase transition was associated with the $V(\phi = 0) \rightarrow V = 0$ transition. If we thus connect inflation with the spontaneous symmetry breaking in GUTs, we can express $V(\phi \approx 0)$, the energy density of the false vacuum, in terms of symmetry-breaking threshold energy, E: $V(\phi \approx 0) \propto E^4 = \eta E^4/(\hbar c)^3$ where η is a dimensionless number of order unity (see Kolb and Turner, 1990, for details). While the phase transition from $V(\phi \approx 0) \rightarrow V = 0$ was taking place, the energy density of the false vacuum dominated the dynamics of the Universe, and eqn. (1.22) becomes simply

$$\dot{R}^2 = \frac{8\pi G}{3c^2} V(\phi \approx 0)R^2 = \frac{8\pi\eta\, GE^4}{3\,\hbar^3\, c^5} R^2. \tag{1.25}$$

As already noted, this has an exponential solution:

$$R(t) = \exp\left[\left(\frac{8\pi\eta\, GE^4}{3\,\hbar^3\, c^5}\right)^{1/2} t\right]. \tag{1.26}$$

If we choose to express E in MeV, then we find to a reasonable approximation $R(t) = \exp(E^2 t)$, with t in seconds. If the symmetry-breaking scale is $E = 10^{14}$ GeV $= 10^{17}$ MeV, inflation by a factor of, say, e^{100} then requires the duration of the inflationary period to be about 10^{-32} s. It is this brief but crucial phase of exponential expansion that we will see was responsible for both large-scale homogeneity of the Universe and its flatness (curvature $k \approx 0$).

1.5.3 The effects of quantum fluctuations: reheating and structure formation

One potential flaw in this inflationary scenario is that the Universe cooled rapidly during its exponential expansion (recall that $T \propto R^{-1}$). Why then is its present temperature not extremely close to zero? In the particular inflationary scenario we are considering, the answer is supplied by quantum oscillations of the field about ϕ_s, which occurred as V

---

* The electro-weak symmetry 'broke' later as kT dropped to about 10^3 GeV.

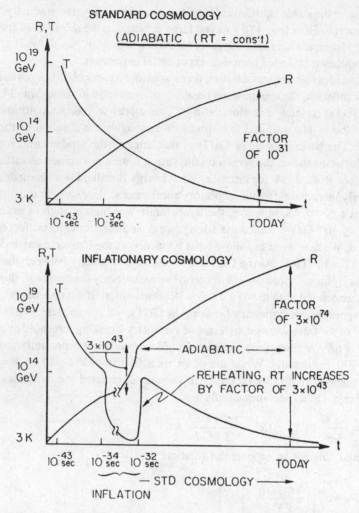

Fig. 1.8   The evolution of the temperature $T(t)$ and of the scale factor $R(t)$ in the standard model (top) and the inflationary model (bottom) as sketched by Kolb and Turner (1990). In the latter model, T dropped to ~ 0 during inflation, then rose rapidly as the Universe reheated. After $t \sim 10^{-32}$ s, the expansion was adiabatic (RT = const.) in either case, and $T \propto t^{-1/2}$ until radiation-domination ended.

dropped to its minimum at ϕ_s. These oscillations decayed into particles, including photons, which in turn thermalized and thus *reheated* the Universe. The details of this reheating process are complex and depend in detail on the coupling between the scalar field ϕ and other fields and particles. Here, we need only mention that reheating is possible, leading to a temperature history of the Universe as shown in fig. 1.8 (see Kolb and Turner, 1990, for a fuller description and additional references). Thus, in the inflationary model, the CBR is not exactly primordial, but was instead generated at $t \simeq 10^{-32}$ s after the Big Bang.

These fluctuations can introduce a small measure of inhomogeneity into an otherwise uniform distribution of matter. Thus it has been suggested that these same quantum fluctuations at $t \simeq 10^{-32}$ s were the ultimate origin of all the large-scale structure we see in the Universe today. It is therefore of considerable importance to note that the inflationary model may allow us to specify the nature, the spectrum and even the amplitude of these fluctuations.

Let us look first at the *nature* of the fluctuations. There are two main classes: those in which both the radiation and the matter were perturbed, and those in which essentially only the matter was perturbed (so that the radiation density and hence T remained essentially constant). In the first of these classes, n_γ / n_b remained the same everywhere, so this class of fluctuation is sometimes referred to as *adiabatic*. Again in the first of these classes, what fluctuated from place to place was the total density of mass–energy; consequently adiabatic fluctuations were also fluctuations in space curvature. In the second class, the curvature remained the same everywhere, so this class is referred to as *isocurvature* fluctuations. In regions in which the density of matter was higher than average, there was a decrease of much smaller amplitude in the radiation density to offset it. Thus both the total mass–energy density and the curvature remained constant. Since under these conditions the radiation was barely perturbed, isocurvature fluctuations are sometimes (loosely) called isothermal fluctuations.

As pointed out by many authors (see Kolb and Turner, 1990), inflation naturally predicts adiabatic fluctuations. Under certain conditions, however (Linde, 1985; Sekel and Turner, 1985; and references the rein), isocurvature fluctuations can also be produced. Which class predominated depends on the particle physics model assumed for the scalar field ϕ: thus there is a link between large-scale structure in the Universe and fundamental physics, which will be exploited in Chapter 8.

Inflation also predicts a particular *spectrum* for the fluctuations, that is a particular dependence of their amplitude on their scale (e.g., Press, 1980; Hawking, 1982; Bardeen *et al.*, 1983). The scale may be specified by the inverse of a wave number k if we break the fluctuating density field up into its Fourier components in comoving coordinates. Inflation predicts a *scale-invariant* spectrum of the kind first introduced into cosmology by Harrison (1970) and Zel'dovich (1972).

'Scale-invariant' has a special meaning here: in particular, it does not mean that the amplitude of fluctuations $\Delta\rho/\rho$ is independent of their present physical scale. To explore the meaning of scale-invariance, let us follow the history of a fluctuation having a particular value of the comoving wave number (corresponding to a comoving scale $l \propto k^{-1}$). At sufficiently early epochs, the physical scale of the fluctuation $l_{phys} = R(t)l$ was greater than ct, so the fluctuation was not yet within (i.e., not yet smaller than) the causal horizon of the Universe (fig. 1.9(a)). At a certain later epoch given by $t_{cr} = l_{phys}/c$, the fluctuations crossed the causal horizon.* At this particular epoch, it had an amplitude $(\Delta\rho/\rho)_{cr}$. 'Scale invariant' means simply that the value of $(\Delta\rho/\rho)_{cr}$ was the same for fluctuations of all scales as they crossed the horizon.

Thus $(\Delta\rho/\rho)_{cr}$ was independent of the mass M contained within the perturbed region of size l_{phys}. Of more direct interest, as we shall see in Chapter 8, is the dependence of $\Delta\rho/\rho$ on M *at a fixed time* (in particular, the epoch we can investigate by studying the

---

* Inflation complicates this simple argument. During exponential expansion, the causal horizon scale remained fixed (H was a constant). Thus fluctuations started out *below* the horizon scale and then recrossed it much later, as shown in fig. 1.9(b). The later history of fluctuations of various scales and types is considered further and in more detail in Chapter 8 and references therein.

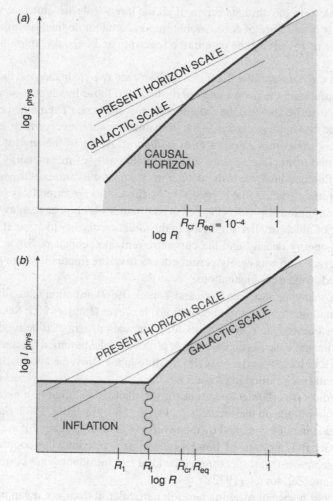

Fig. 1.9   Evolution of the physical size of the causal horizon (heavy lines) and of fluctuations of different present scales in an expanding Universe (adapted from Kolb and Turner, 1990). Only scales below the heavy line (shaded region) were causally connected. (*a*) The Standard Big Bang model without inflation. A fluctuation of a particular present scale (say the size of a galaxy) started out larger than the causal horizon and crossed it at an epoch t_{cr} corresponding to a scale factor R_{cr}. The 'knee' in the heavy line occurred as the expansion of the Universe changed from radiation dominated to matter dominated at $z_{eq} \approx 10^4$ or $R_{eq} \approx 10^{-4}$. (*b*) During inflation, at $t < t_f$, the causal horizon was fixed in size. Thus a fluctuation of a particular scale (say galactic) started in the shaded region, crossed the horizon at R_1, then re-entered the horizon at R_{cr}.

angular distribution of the CBR). Perturbations on larger scales than the horizon scale at that time had not yet reached the amplitude they would later have when they did cross the horizon. The consequence is that the spectrum of perturbations at a fixed moment in time may be written

$$\Delta\rho/\rho \propto k^2 \propto M^{-2/3}, \tag{1.27}$$

Other possible mass spectra are considered in Chapter 8.

Finally, if the functional form of $V(\phi)$ is known in detail, the *amplitude* of fluctuations emerging from the inflationary period may be calculated (see Turner, 1988, and references therein). One early model for $V(\phi)$, the Coleman–Weinberg (1973) potential, was shown (Guth and Pi, 1982; Hawking, 1982; Starobinskii, 1982; and Bardeen *et al.*, 1983) to lead to values of $\Delta\rho/\rho \simeq 10$–$100$. So large a value is incompatible with the isotropy of the CBR (Chapter 7), hence the Coleman–Weinberg potential is no longer accepted as the correct form of $V(\phi)$. I cannot help but note that purely astronomical observations have thus imposed constraints on particle physics.

The amplitude and spectrum of quantum fluctuations would have little interest unless they produced structure on an observable scale in the present Universe. Because of inflation, they do. As fig. 1.9(*b*) shows, a fluctuation of galactic size (as measured today) was for an extended period smaller than the causal horizon, so that ordinary physical processes could operate. At the moment when the scale factor reached R_1, the fluctuation expanded beyond the causal horizon; the fluctuation then remained outside the causally connected region until the Universe had expanded by a further factor R_{cr}/R_1.

Physically, the process may be understood as follows. An initially microscopic region expanded exponentially during inflation. By the end of the inflationary period, when the scale factor had reached R_f, the fluctuation had grown exponentially to a much larger size, in this case about R_f^{-1} times the present radius of a galaxy. Thus microphysical fluctuations were 'inflated' to astronomical scales during the first 10^{-32} s of the expansion of the Universe.

1.5.4   *Non-conservation of baryon number*

All of these properties and predictions follow from Guth's addition of a period of inflation to the early evolution of the Universe. We need now to consider briefly the other new piece of physics that modifies the standard picture – the non-conservation of baryon number permitted in GUTs and other particle theories. As we will soon see in more detail, non-conservation of baryon number, combined with a temporary departure from thermal equilibrium in the expanding Universe, allowed a non-zero net baryon excess to be generated in a Universe initially symmetric between matter and antimatter.

The same theories that include non-conservation of baryon number, and thus allow the Universe to generate the baryons we now see in it, also allow or even *require* the existence of additional varieties of elementary particles, some of which, like axions, photinos or neutrinos of non-zero mass, are potential 'dark matter' candidates. Since none of these exotic particles has yet been detected, we do not know which, if any, will be present in the Universe. Already, however, astronomical observations are constraining the possibilities (Kormendy and Knapp, 1987).

1.5.5   *How the inflationary model ties up the 'loose ends' of classical cosmology*

As noted, the non-conservation of baryon number allowed for in GUTs (and other unified theories) can generate a net baryon number in an expanding Universe. Hence it allows n_b/n_γ to differ from zero. Why the ratio n_b/n_γ has the particular value it does (of order 10^{-9}) may eventually be explained by particle physics. At our present stage of knowledge, however, we can only run the argument backwards: the observed ratio n_b/n_γ can be used to set constraints on theories of elementary particles.

The existence of density fluctuations responsible for structure in the Universe arises naturally from quantum fluctuations in the scalar field ϕ associated with inflation. Once again, our present understanding of fundamental physics is too slim to allow us to calculate the amplitude $\Delta\rho/\rho$ of these fluctuations unambiguously: we instead use our observations to constrain the theories (Section 1.5.3 and Chapter 8).

Two of the 'loose ends' of classical cosmology are thus tied up, at least partially. Now for the two more global questions, the flatness and the homogeneity of the Universe. It is not hard to develop an argument showing that the curvature $k = 0$. We do so by supposing that k is *not* negligibly small at present; in particular, let us assume that the two terms on the right-hand side of eqn. (1.12) are of roughly comparable magnitude today, so that

$$|kc^2| \approx \frac{8}{3}\pi G\,\rho_0\,R^{-1}(t_0).$$

Since the density term increases in magnitude as we go back to earlier epochs, the curvature term must have been negligibly small in the past:

$$|kc^2| \ll \frac{8}{3}\pi G\,\rho_0\,R^{-1}(t \ll t_0).$$

Recall that R increased by a large factor, say about e^{100}, during inflation. Thus an extraordinary fine-tuning of the initial value of k would have been required to leave the curvature term approximately equal in magnitude to the density term today. Such an *ad hoc* assumption is unpalatable to most cosmologists. It can be avoided by setting $k = 0$. Another approach leading to the same conclusion is to assume instead that the curvature and density terms were of comparable magnitude *before* inflation:

$$|kc^2| \approx \frac{8}{3}\pi G\,\rho_0\,R(t \simeq 10^{-34}\,\mathrm{s}).$$

After inflation, the curvature term would then have been about $e^{100} \simeq 10^{43}$ times larger than the density term, which completely contradicts astronomical observations. Hence many cosmologists argue that k must be rigorously zero.

A proper explanation of the large-scale homogeneity of the Universe requires analysis of the properties of particle horizons in the exponentially expanding Universe (see Harrison, 1981, Chapter 19; or, for this specific issue, Kolb and Turner, 1990, Chapter 3). With loss of some rigor, we may explain large-scale homogeneity by asking how large a causally connected region was just before inflation occurred, then asking about its size after inflation. Before inflation began, the scale of the particle horizon was given approximately by $r = ct$, with $t \simeq 10^{-34}$ s. Thus in regions of about 3×10^{-24} cm or smaller, physical processes could in principle have operated to produce homogeneity. We then allow such a region to inflate by $e^{100} = 3 \times 10^{43}$ to a size of about 10^{20} cm, far larger than the value of ct at 10^{-32} s (at the end of inflation). The *present* size of that initially causal region will be still larger, of course, since, as fig. 1.9(*b*) shows, the scale factor continued to increase between $t \simeq 10^{-32}$ s and t_0 by a further factor R_d^{-1}. The observable Universe can be approximately homogeneous today because it was, at a very early epoch, contained in a causally-connected region. Thus inflation *allows*, even if it does not *require*, large-scale homogeneity.

Since observations of the CBR provide our best check on the homogeneity of the Universe, let us go on to ask what size a region of $r \simeq 10^{20}$ cm at $t \sim 10^{-32}$ s would grow to by $t \simeq 10^{13}$ s, the earliest epoch we can observe directly by studying the CBR. To a good approximation, $R \propto t^{1/2}$ in that time interval (Section 1.3.5); hence the causally connected regions could be as large as 3×10^{42} cm at $t = 10^{13}$ s, i.e. far greater than even the present size of the Universe.

1.6   Thermal history of the Universe

As the Universe reheated at the end of its inflationary period, it reverted to radiation-dominated expansion with $R \propto t^{1/2}$. The era of standard Hot Big Bang cosmology began.

We will now sketch the history of the expanding Universe from the end of inflation at $t \simeq 10^{-32}$ s to the present ($t_0 \simeq 10^{17}$ s), with special attention devoted to those epochs in which significant changes in the properties of the Universe or its material contents took place, or in which physical processes occurred that left an imprint on the CBR. This survey of the history of the Universe will necessarily be brief and qualitative; the topic is well treated in some standard texts on cosmology (e.g., Weinberg, 1972; Narlikar, 1983) and is well reviewed by Kolb and Turner (1990).

Most noteworthy events in the history of the Universe occurred while the expansion was still dominated by the radiation we now detect as the 3 K CBR, that is at redshifts $z \gtrsim 10^4$. At epochs corresponding to $z \gtrsim 10^4$, the temperature T was proportional* to $t^{-1/2}$, and a useful (approximate) scaling to bear in mind is

$$T(t) = \frac{10^{10} \text{ K}}{\sqrt{t}} \approx \frac{1 \text{ MeV}}{\sqrt{t}}, \tag{1.28}$$

with t expressed in seconds.

1.6.1   Freeze-out

If we had sampled the contents of the Universe soon after inflation ended, what would we have found? Our sample would have consisted of particle–antiparticle pairs, including quarks and leptons, mixed with photons. To rough first order, the numbers of each kind of particle were the same, an approximate equality maintained by creation/annihilation reactions such as $e^- + e^+ \Leftrightarrow \gamma + \gamma$. Such reactions coupled the matter (electrons and positrons in this example) closely to the radiation field. As the Universe expanded, and the temperature and density dropped, particles with increasingly strong coupling can drop out of thermal contact with the radiation field. This process is appropriately called 'freeze-out' of that particular particle species. 'Freeze-out' was determined essentially by the reaction rate of the particles in question, Γ; if the reaction rate fell below the expansion rate of the Universe, the particles ceased to 'have time' to react, and dropped out of thermal contact with the other constituents of the Universe†. Thus freeze-out occurred when $\Gamma \lesssim H$, since Hubble's constant is a measure of the expansion rate of the Universe (freeze-out is explained in more detail by Narlikar, 1983, or even more fully by Kolb and Turner, 1990, Chapter 5).

---

* A simple linear proportionality did not hold exactly, because radiative decay of particles or particle–antiparticle annihilation added to the number of photons, thus increasing T; see Section 1.6.2.
† A technical point: the reaction rates in question are not just those involving coupling to photons. Reactions coupling particles to electrons and positrons, for instance, also served to maintain thermal contact because the e^+ and e^- were in turn very tightly tied to the radiation field by Compton scattering.

What happened to stable particles after they froze out? For weakly interacting particles that froze out early, when the temperature T expressed in energy units was much greater than their rest-mass energy, the answer is easy: nothing. Their number density merely decreased as $n \propto R^{-3} \propto (z + 1)^3$ following freeze-out, according to eqn (1.9). Stable particles of this sort are called 'hot,' since they were still relativistic (kT or $E \gg mc^2$) when they froze out. Neutrinos with a small non-zero mass provide an example: they froze out when the rate for the $v + \bar{v} \Leftrightarrow e^+ + e^-$ reaction fell below H, at T around a few MeV (or a few times 10^{10} K).

For stable particles that were still strongly coupled to the radiation field when $T \simeq mc^2$, the situation is more complicated (see Kolb and Turner, 1990). Here the reaction rate of interest is the annihilation rate, and the number density, n, of particles remaining after freeze-out depends on the annihilation cross-section $\bar{\sigma}$ averaged over temperature. To a first approximation (Kolb and Turner, 1990), $n \propto (m\bar{\sigma})^{-1}$, where m is the rest mass of the stable particle. Stable particles of this sort are called 'cold.' A variety of weakly interacting, massive particles (acronym WIMPs) fit this description, among them some of the particles permitted/required by the Grand Unified and Supersymmetric theories sketched in Section 1.5.

Note that we have been concerned only with *stable* particles in this section. In Section 1.6.2 we consider a specific case of freeze-out of an *unstable* particle.

1.6.2 Baryogenesis

With these general ideas in mind, let us turn to the first significant event to occur after the end of inflation: the generation of the net baryon excess $n_b/n_\gamma \simeq 10^{-9}$ by decays of some very massive particle (at very roughly $t = 10^{-25}$ s).

In an important and prescient paper, Sakharov (1967) pointed out that a net baryon excess could emerge from a Universe that was initially symmetric with respect to particles and antiparticles ($n_b = 0$) if three conditions applied:

1.   There are baryon-non-conserving reactions.
2.   The C and CP parities are violated.
3.   The expansion of the Universe provided a temporary departure from thermal equilibrium.

At the time Sakharov wrote, it was known that condition (2) held for at least one system, the K mesons. Condition (3) holds, as we have seen, and Grand Unification introduces baryon-non-conservation. If we allow for the possibility of inflation in our cosmological model, we must add a fourth condition:

4.   The baryon excess was generated after the inflationary phase.

What is the actual series of steps that resulted in the baryon excess we see today? One plausible scenario operates as follows. Suppose that there exists a massive particle (probably, but not necessarily, near the GUT scale of about 10^{14} GeV), which violated baryon conservation in its decay at an early epoch (condition 1).* As is usual, we will refer to it simply as the X boson. We next suppose that the decay rate Γ_d of the X was

---

* It is not certain what this particle is; perhaps a gauge or Higgs boson of some GUTs (Ross, 1984). No such particle has yet been detected in accelerator experiments. Note, however, that some limits on its mass can be fixed by studies of proton decay.

less than the expansion rate, H (in rough language, the X particles could not decay 'fast enough' to keep up with the expansion and cooling of the Universe). Hence they fell out of thermal equilibrium, and remained overabundant as they decayed (we have used condition 3).

Now we need to examine the decays of the X in more detail. Suppose that there were two decay modes possible, each producing a different final baryon number per decay event. For instance, one could imagine X → quark + quark, with a final baryon number $N' = 2/3$; or X → quark + lepton, with $N'' = 1/3$. Let the probability for these two processes be r and $(1 - r)$. Then the net baryon number per X decay generated by many such decays was

$$N = rN' + (1 - r)N''.$$

Now, however, we must consider that in a particle–antiparticle symmetric Universe, there must have been an $\overline{\mathrm{X}}$ for every X. Decay products of the $\overline{\mathrm{X}}$ particle must also be considered:

$$\overline{N} = \bar{r}N' + (1 - \bar{r})N''.$$

Thus the residual net baryon number δN generated as an X–$\overline{\mathrm{X}}$ pair decays was

$$\delta N \equiv N - \overline{N} = (r - \bar{r})(N' - N''). \tag{1.29}$$

Condition 1 ensures that $N' \neq N''$. Condition 2 *allows* $(r - \bar{r}) \neq 0$ (see Kolb and Turner, 1983 or 1990, for a proof). Since X and its properties are unknown, particle physics does not at present allow us to calculate $(r - \bar{r})$ and hence δN, the residual baryon number per X–$\overline{\mathrm{X}}$ decay. On the other hand, astronomical observations do permit us to evaluate δN approximately. First we write δN as a ratio of number densities of two species:

$$\delta N = \frac{n_b}{n_X + n_{\overline{X}}},$$

where n_b was the number of baryons per unit volume left at the end of baryogenesis, and n_X and $n_{\overline{X}}$ were the number densities of the X boson and its antiparticle before the decay processes began. Each X (or $\overline{\mathrm{X}}$) decay produced one or two quarks (depending on r and $\bar{r}$); therefore

$$\delta N \approx \frac{n_b}{3n_q}.$$

In turn, the number density of quarks n_q and of photons was roughly equal.* Hence

$$\delta N \approx \frac{1}{3}\left(\frac{n_b}{n_\gamma}\right),$$

and n_b/n_γ as we have seen is known to be of order 10^{-9}. It is intriguing that measurements of the temperature of the CBR (and hence n_γ) provide one of the few constraints we have on the properties of the X boson, whose rest mass energy is likely to exceed the kinetic energy of a large caliber rifle bullet in flight.

---

* Because additional photons were generated by annihilation of particles other than quarks, this result is only qualitatively correct; more detailed analyses (e.g., Kolb and Turner, 1990, Chapter 6) show $\delta N \simeq 10^{-7}$, not $\simeq 3 \times 10^{-10}$, as our crude treatment suggests.

1.6.3 The quark to hadron transition

Well after the X and $\bar{X}$ particles decayed, say at 10^{-10} s after the Big Bang, the Universe consisted of photons, leptons, and quark–antiquark pairs – with a slight excess of particles over antiparticles. When T dropped to a few hundred MeV (at $t \simeq 10^{-5}$ s), the free quarks combined to form hadrons, including particularly protons and neutrons.

The net baryon excess, with $n_b/n_\gamma \simeq 10^{-9}$, thus ended up in the form of protons and neutrons. If the fluctuations in the density of the X bosons were adiabatic, we would be left with adiabatic fluctuations in the density of protons and neutrons.

1.6.4 Big Bang nucleosynthesis

The subsequent evolution of the Universe involved particles, energies and processes that can be studied in our laboratories, and hence is better supported observationally. We can be much more confident about the evolution of the Universe from $t \simeq 10^{-5}$ s onwards than we can be about its earlier history. We can, for instance, follow the evolution of the number densities of protons and neutrons as the Universe expanded and cooled.

One of the great successes of the Hot Big Bang model is its ability to predict the light element abundances observed in primordial matter. Here we will describe briefly the synthesis of light nuclei at $t \simeq 100$ s and point out the role played by CBR photons. More detailed reviews are available in the books by Narlikar (1983), Peebles (1971), Weinberg (1972), and Kolb and Turner (1990); see also the reviews by Boesgaard and Steigman (1985) and Audouze (1987).

To form nuclei of any atom but ^{1}H, neutrons and protons must have combined. The first step in that process was formation of deuterons by the reaction $n + p \rightarrow d + \gamma$. Until the temperature of the expanding Universe dropped to about 10^9 K the abundance of deuterons was very low. After $t \simeq 100$ s, when T dropped below 10^9 K, the reaction proceeded as shown to build up deuterons. Essentially all the neutrons remaining at $t \simeq 100$ s were incorporated in nuclei and thus stabilized. Deuterons, however, are loosely bound, and subsequent nuclear reactions (such as $d + d \rightarrow {}^3\text{He} + n$; $^3\text{He} + d \rightarrow {}^4\text{He} + p$; and so on) rapidly converted most of the deuterons to much more stable nuclei of ^{4}He, as shown in fig. 1.4.

We will return to the few left-over deuterons in a moment; first let us ask what determined the amount of ^{4}He produced. It was essentially the number of neutrons available. All but a very small fraction of the neutrons available at $t \simeq 100$ s ended up in ^{4}He; if all neutrons were bound in ^{4}He nuclei, the fractional ^{4}He abundance by mass would be

$$Y = \frac{2n_n}{n_p + n_n} \, ,$$

where n_n and n_p are the number densities of neutrons and protons, respectively. Thus

$$Y = 2\left(\frac{n_n}{n_p}\right)\left(1 + \frac{n_n}{n_p}\right)^{-1} \tag{1.30}$$

can be calculated if n_n/n_p at $t \simeq 100$ s is known. That ratio in turn depended on four factors: the half-life of the free neutron (taken as 10.6 ± 0.1 min);[*] cross-sections for processes like $n + \nu \Leftrightarrow p + e$, which convert neutrons to protons and vice versa; the

---

[*] In the past few years, values more like 10.2 ± 0.1 min have been reported, e.g., Last *et al.* (1988).

present temperature of the CBR (which allows us to scale the density back to $T \simeq 10^9$ K); and the expansion rate of the Universe (which in turn depended on the number of light lepton families $\mathcal{N}_l$; see Yang *et al.*, 1984). The evolution of n_n/n_p is followed in detail by Peebles (1971) and Kolb and Turner (1990); see also the reviews by Schramm and Wagoner (1979), Yang *et al.* (1984) and Boesgaard and Steigman (1985). Following the latter, we present an equation showing how Y depends on the factors listed above:

$$Y = 0.230 + 0.011 \, ln \left(\frac{10^{10} n_b}{n_\gamma} \right) + 0.013 \, (\mathcal{N}_l - 3)$$
$$+ 0.014 \, (t_{1/2} - 10.6),$$

(1.31)

where $t_{1/2}$ is the half-life of the free neutron in minutes. Knowing T_0 and hence n_γ, we can convert n_b/n_γ to values for ρ_b, the present density of baryons in the Universe. Values derived from eqn. (1.31) are shown in graphical form in fig. 1.10. Note the very weak dependence of Y on ρ_b (or T_0). Astronomical observations of the ^{4}He abundance in the Galaxy show $Y = 0.235 \pm 0.005$ (Shaver *et al.*, 1983; Pagel, 1986). The agreement with the predictions of Hot Big Bang nucleosynthesis is gratifying. Further, we may use the good fit between astronomical observations and the calculations to show that the number of light lepton families is four or fewer (three, the e$^-$, μ and τ neutrinos, are already known).*

^{4}He and ^{1}H were not the only light nuclei to emerge from the Big Bang. Traces of ^{7}Li were produced, and traces of some of the nuclei that served as stepping stones to the formation of ^{4}He also remained. Of the latter, the remnant deuterons (or ^{2}H) are of most interest. If the nuclear reactions mentioned above had run to completion, all the deuterons would have been converted to ^{4}He, and the remaining abundance of ^{2}H would be zero. It is not zero because the nuclear reaction rates had to compete with the expansion of the Universe. The nuclear reaction rates of course depended on the number density of the reacting particles; if the number density of baryons (specifically, the ratio n_b/n_γ) had been lower, a larger fraction of deuterons would have been left over. Hence the abundance of ^{2}H emerging from the Hot Big Bang was a sharply decreasing function of baryon density, as shown in fig. 1.10. Astronomical evidence gives reliable *lower* limits on the deuterium abundance: these set a firm *upper* limit on the density of baryons $\Omega_b < 0.2$. The range $0.01 \lesssim \Omega_b h^2 \lesssim 0.025$ (or $0.01 \lesssim \Omega_b \lesssim 0.2$ for $0.4 \lesssim h \lesssim 1.0$) is currently favored by observations; see Boesgaard and Steigman (1985) or Walker *et al.* (1991). Note that this value is well below the value of $\Omega = 1$ corresponding to the $k = 0$ model strongly favored in inflationary models; on the other hand, the limit set by the deuterium abundance applies only to *baryonic* matter. If some form of dark matter is allowed, $\Omega = 1$ is possible, with $\Omega_{DM} = 1 - \Omega_b$ (Section 1.4).

Boesgaard and Steigman (1985) and also Reeves (1989) have recently emphasized that the observed abundance of ^{7}Li (e.g., Spite and Spite, 1982) provides an important independent confirmation of a low value of Ω_b. As fig. 1.10 shows, the low value of ^{7}Li abundance found by astronomers sets both upper and lower limits on Ω_b. The observed abundances of both light nuclei, taken together, are consistent with a narrow range of Ω_b, approximately $0.01 \, h^{-2} - 0.015 \, h^{-2}$. These limits were calculated with $T_0 = 2.75$ K

---

* This astrophysical constraint on particle physics has recently been confirmed by accelerator results obtained at CERN: $\mathcal{N}_l = 3.1 \pm 0.2$ (Denegri, Sadoulet and Spiro, 1990).

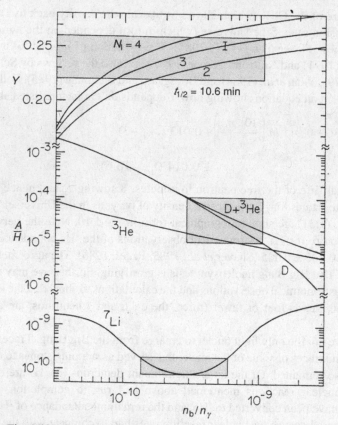

Fig. 1.10   Comparison of predicted abundances of light elements (by mass, relative to hydrogen) from primordial nucleosynthesis (lines) and the abundances derived from recent astronomical observations (shaded boxes). Note the strong dependence of deuterium (D) and ^{7}Li abundance on n_b/n_γ, which in turn is proportional to the baryon density Ω_b. For $T_0 = 2.75$ K, the relation is $\Omega_b = (3.7 \times 10^7)h^{-2}(n_b/n_\gamma)$.

(Olive *et al.*, 1990), but would change by $< 11\%$ for any value of the present CBR temperature in the range 2.65–2.85 K. As noted in Section 1.2.2, the value of H_0 is not well known but lies in the range 40–100 km s^{-1} per megaparsec, so $0.4 \leqslant h \leqslant 1.0$. In summary, then, the density of baryons is constrained by these arguments to fall in the range

$$0.01 \leqslant \Omega_b \leqslant 0.1. \tag{1.32}$$

1.6.5 Recombination: the formation of neutral atoms

Nuclei were formed at an epoch of a few hundred seconds. Neutral atoms first formed much later, when T dropped to about 3000 K (at $t \simeq 10^{13}$ s). Since $> 90\%$ of all nuclei *by number* are ^{1}H, we need consider principally the recombination of H. This was governed by the Saha equation:

$$\frac{n_p\, n_e}{n_H} = \frac{(2\pi m_e\, kT)^{3/2}}{h^3}\, e^{-\chi/kT}, \tag{1.33}$$

where n_p, n_e and n_H are the number densities of protons (ionized H), electrons, and neutral H atoms, respectively. In eqn. (1.33), m_e is the mass of the electron and χ is the ionization energy of H, 13.6 eV. The exponential dependence of n_p/n_H on T ensures that the formation of neutral atoms was prompt as the temperature of the Universe dropped below about 3000 K. Thus the Universe made a sharp transition between its earlier ionized phase and its later neutral phase (see Peebles, 1971, for details). Expressed in redshift rather than time, the duration of the transition was $\Delta z \simeq 400$, centered at $z \simeq 1400$, when $n_p/n_H = 1$ (for details, see Jones and Wyse, 1985).

This same epoch was also the earliest that we can study by observing photons of the CBR. Before recombination, at $z \gtrsim 1400$, the large number density of free electrons scattered the CBR photons frequently. The relevant scattering cross-section was the Thomson cross-section, $\sigma_T = 6.65 \times 10^{-25}$ cm^2. After recombination, the free electrons vanished. The only remaining scattering process was scattering by neutral H, which had a far lower cross-section. Since recombination was sudden, the Universe suddenly became transparent at a redshift $z = 1060^*$; at later times, the CBR photons propagated freely with no further scattering. Thus when we observe the cosmic background, we are studying a surface of last scattering at a redshift $z_s \approx 1000$ (fig. 1.11(a)) *unless* the material contents of the Universe were reionized (Section 1.7.3). The CBR thus allows us to examine the Universe at a single, early, moment in its history; a study of the distribution of the CBR across the sky is a 'snapshot' of the Universe at $z \simeq 1000$. The snapshot is not perfectly sharp, however, because of appreciable thickness of the surface of last scattering. The recombination process was rapid, as we have seen, but not instantaneous. (Again, see Jones and Wyse, 1985 who show that $\Delta z_s \simeq 100$.) The resulting thickness of the surface of last scattering smeared out detail in the CBR snapshot on angular scales below about $7\Omega^{1/2}$ arcmin if $z_s \approx 1000$. This point will reemerge as a central concern in Chapter 8.

Before continuing our thermal history of the expanding Universe, let us take note of an important physical process, which operated before recombination. It depends on the tight coupling between radiation and matter resulting from Thomson scattering (combined with the Coulomb interaction between electrons and protons). The coupling ensured that radiation will drag the matter with it, giving this process its name, 'radiation drag.' Radiation drag had a profound effect on the fluctuations in the density of *baryonic* matter. The amplitude of isocurvature (isothermal) perturbations cannot increase since this would have involved moving matter around in the (approximately) homogeneous sea of radiation. Adiabatic fluctuations in baryonic matter were actually damped out (Silk, 1968) on all scales below a critical scale of about $(10^{10}-10^{11})$ $M_\odot$, as the radiation streamed from hotter, denser regions to cooler ones, dragging matter along.† Thus the only adiabatic fluctuations in the baryon density ρ_b that survived until the epoch of recombination were those above this scale.

While radiation drag had a profound effect on baryonic matter, it had essentially *no* effect on dark matter, if the dark matter did not couple to either photons or the baryonic component. Thus the evolution of fluctuations in the density of dark matter was very

---

* I have given here the redshift at which the optical depth (Section 3.6) became unity, given the recombination history described above. This value of the redshift of last scattering depends only weakly on Ω_b.

† Since this damping involved a causal process, scales larger than the horizon were not affected. That is, fluctuations on a particular scale were damped only when that scale lay below the heavy line of fig. 1.9.

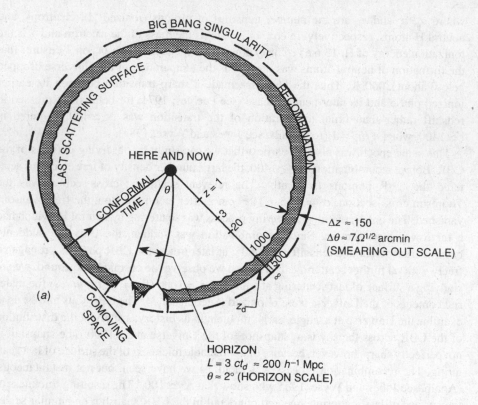

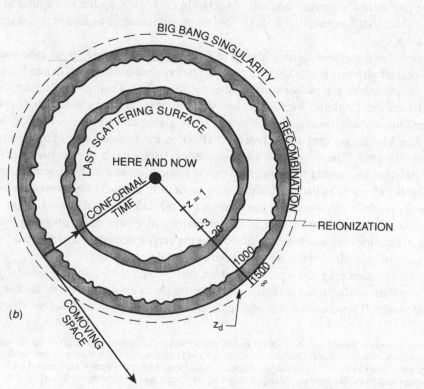

different. In particular, the damping process mentioned just above did not operate on adiabatic (or isothermal) fluctuations in the dark matter.

1.6.6 The end of radiation domination

At sufficiently early epochs, both before and after the brief interlude of inflation, the expansion of the Universe was dominated by radiation (section 1.3.5). If we neglect the possibility of a non-zero value of the cosmological constant, the expansion is now dominated by matter. The cross-over from radiation domination to matter domination occurred when $\rho_\gamma \equiv u/c^2 = \rho_{matter}$. With $u = aT^4$, and $T_0 = 2.75$ K, the corresponding redshift may easily be shown to be (from eqn. (1.19) and eqn. (1.13))

$$(z + 1) = 3.9 \times 10^4 \, \Omega \, h^2.$$

Given that $h = 0.4$–1.0, we see that the end of radiation domination will actually have occurred at a somewhat higher redshift, and therefore earlier, than recombination unless $\Omega \ll 1$.

1.7 Galaxy formation

For several reasons, the next event in the history of the expanding Universe deserves separate treatment. It was the formation of large-scale structures within the Universe, which I will refer to loosely as 'galaxy formation' even though the first objects to form may have been sub-galactic masses or even massive clusters of galaxies. Although galaxy formation is the most recent of the crucial moments that we have been considering, it is paradoxically one of the least well understood; we know neither when galaxies formed nor the primary physical mechanism responsible for their formation. In addition, galaxies have almost certainly evolved in complicated ways since their formation (unlike, say, ^{4}He atoms). Finally, while many observations bearing on the time and process of galaxy formation are now becoming available, they have not yet singled out a unique, widely accepted scenario for galaxy formation. Among the most useful constraints on models for galaxy formation are searches for small-scale structure in the CBR; hence we will return to the questions surrounding galaxy formation in Chapter 8.

In the remainder of Section 1.7, we will sketch the process of galaxy formation briefly and generally, and mention the many physical mechanisms that may be involved in establishing the structure we now see in the Universe.

1.7.1 The growth of density perturbations

One aspect of galaxy formation is widely accepted: that the presence of structure in the Universe today can be traced back to small fluctuations in the density at a much earlier epoch. As we saw in Section 1.5.3, it is plausible that these fluctuations were introduced during the period of inflation at $t \simeq 10^{-32}$ s.

We will begin our consideration of galaxy formation by examining the mechanisms that governed the *growth* of fluctuations both before and after recombination. By

---

Fig. 1.11 *(opposite)*   (a) The surface of last scattering at $z \sim 10^3$, coinciding approximately with the epoch of recombination (with permission from J. Silk). (b) Reionization can shift the surface of last scattering to a lower redshift (see Section 1.7.3), since free electrons were once again available to Thomson scatter the CBR photons.

'growth' is meant an increase in the value of $\Delta\rho/\rho$, where ρ is the average or background density and $\Delta\rho$ the excess density in a particular fluctuation. The ratio $\Delta\rho/\rho$ is the *amplitude* of the density perturbation; we will be concerned only with positive amplitudes ($\Delta\rho > 0$) and with growing modes (see, e.g., Peebles 1980; or Efstathiou, 1990).

One obvious mechanism for an increase in $\Delta\rho/\rho$ is gravity: positive density perturbations contain excess mass and hence will tend to contract. Counteracting gravity is the additional pressure of the excess matter. Hence, as Jeans (1902) showed for a static (non-expanding) fluid, there is a critical length scale, below which perturbations do not grow and above which gravity wins so that $\Delta\rho/\rho$ does increase. This scale is the Jeans length,*

$$l_{\mathrm{J}} = c_{\mathrm{s}}\left(\frac{\pi}{G\rho}\right)^{1/2} = \left(\frac{\pi k T}{G\rho m}\right)^{1/2}, \tag{1.34}$$

where c_{s} is the speed of sound in a non-relativistic gas of particles of mass m at a temperature T. Since $\gtrsim 90\%$ atoms in the Universe are $^1\mathrm{H}$, we may take $m = m_{\mathrm{p}}$, the proton mass. Corresponding to this scale is a *Jeans mass*

$$M_{\mathrm{J}} \propto l_{\mathrm{J}}^3 \, \rho \propto \rho^{-1/2} \, T^{3/2}.$$

Density perturbations of mass $< M_{\mathrm{J}}$ oscillate rather than grow. It is worth noting (Peebles and Dicke, 1968; Peebles, 1971) that the Jeans mass of baryonic matter just following recombination was about $10^5 M_\odot$ or a few times 10^{38} g, many orders of magnitude below a typical galactic mass. I will also add without proof that the Jeans length for fully relativistic matter, including radiation, is $l_{\mathrm{J}} \simeq ct$, i.e. the horizon scale.

In a *static* fluid, the growth of density perturbations of scale $> l_{\mathrm{J}}$ is exponential in time. The growth of density perturbations in an expanding Universe was treated in an important paper by Lifschitz (1946). He showed that the gravitational growth of $\Delta\rho/\rho$ was a power law in time. In fact, for matter-dominated expansion,

$$\Delta\rho/\rho \propto (z + 1)^{-1}. \tag{1.35}$$

Since $\Delta\rho/\rho$ is clearly greater than unity on galactic scales today, it follows from (1.35) that $\Delta\rho/\rho \gtrsim 10^{-3}$ at recombination on these same scales.

Gravitational growth in $\Delta\rho/\rho$ could occur only when other physical processes, such as the damping referred to in Section 1.6.5, did not prevent it. Thus all density perturbations greater than the Jeans mass of about $10^5 M_\odot$ were free to grow, or increase in amplitude, *after* recombination but not necessarily earlier. At earlier epochs no growth was possible for baryonic fluctuations of galactic scale. Fluctuations in dark matter, on the other hand, were free to grow following eqn. (1.35), provided that the dark matter was not closely coupled to the radiation. This difference is shown in fig. 1.12.

It is thus possible that $\Delta\rho/\rho$ for dark matter fluctuations was much larger than $\Delta\rho/\rho$ for fluctuations in the baryonic matter as the epoch of recombination was reached. If so, the more uniformly distributed baryons could later fall into the gravitational potential wells created by the positive perturbations in the dark matter; $\Delta\rho/\rho$ for the baryons could thus grow faster than the rate specified by eqn. (1.35), as they 'caught up' with the larger amplitude dark matter perturbations.

---

* Derived, for instance, in Peebles (1971).

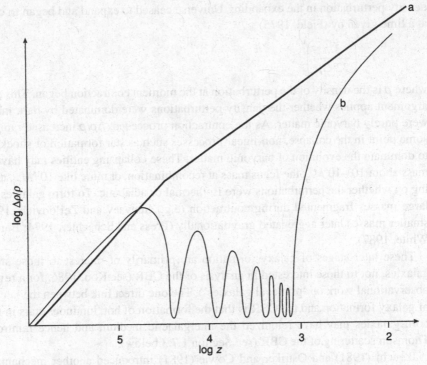

Fig. 1.12   The growth of perturbations in the dark matter density (curve a) and in the baryonic density (b) before and after the baryonic matter decoupled from the radiation at $z \sim 10^3$. Until recombination, the baryonic perturbations oscillated if their scale was less than the Jeans length; they were also damped by radiation drag (Section 1.6.5), so the amplitude of $\Delta\rho/\rho$ decreased. After decoupling the baryonic perturbations can 'catch up' with the undamped dark matter perturbations.

1.7.2   The final stages of galaxy formation

How did fluctuations in the density of matter at $z \simeq 1000$ turn into the large-scale structure we see today? The answers to this question are both uncertain and subject to debate.

A wide range of models for galaxy formation exist, but no particular model has gained the acceptance and maturity of models for formation of other components of the Universe, such as stars. In some models (Peebles and Dicke, 1968) small systems (10^5–$10^6 M_\odot$) formed first and then aggregated to form galaxies; in other models (Sunyaev and Zel'dovich, 1972) large systems ($10^{15} M_\odot$ 'pancakes') fragmented to form galaxies. Some models assume initially adiabatic density perturbations; others isothermal perturbations (the former produce 'pancakes'). In some models, galaxy formation took place (in the linear regime at least) purely by the action of gravity (Press and Schechter, 1974; White and Rees, 1978; and Peebles, 1980); in others explosive energy moved the matter around (Ikeuchi, 1981; Ostriker and Cowie, 1981; Hogan and Kaiser, 1983). Some models rely on the existence of non-baryonic dark matter (e.g., Blumenthal *et al.*, 1984; Bond, 1988).

The simplest hypothesis is that gravity alone was responsible. Essentially the same physical arguments employed in Section 1.3.3 may be used to show that a positive density perturbation in the expanding Universe ceased to expand and began to contract at a time given by (Field, 1975)

$$t = \left(\frac{3 \pi}{32 \, G\rho} \right)^{1/2},$$

where ρ is the density of the perturbation at the moment contraction began. This general argument applies whether the density perturbations were dominated by dark matter or were purely baryonic matter. As the contraction proceeded, $\Delta\rho/\rho$ increased rapidly. At some point in the collapse, non-linear processes such as star formation or shocks came to dominate the evolution of baryonic matter. These collapsing entities may have had a mass about 10^5–$10^6 M_\odot$, the Jeans mass at recombination, or more like $10^{15} M_\odot$, depending on whether the perturbations were isothermal or adiabatic. To form galaxies, either large masses fragmented during contraction (e.g., Sunyaev and Zel'dovich, 1972) or smaller masses later aggregated gravitationally (Press and Schechter, 1974; Baron and White, 1987).

These later stages of galaxy formation are primarily of interest to those studying galaxies, not to those interested in analysis of the CBR (see Koo, 1986, for a review of observational work on 'primeval galaxies'). The one direct link between the late stages of galaxy formation and the CBR is that the formation of hot, luminous stars in newly-born galaxies may have reionized the intergalactic medium and hence reintroduced Thomson scattering of the CBR (see Section 1.7.3 below).

Ikeuchi (1981) and Ostriker and Cowie (1981) introduced another mechanism for forming galaxies: explosions. In this scenario, a few massive objects formed early, before the bulk of baryonic matter had contracted. These objects exploded, and the resulting blast wave swept up intergalactic matter, forming high density shells. The surfaces of the more-or-less spherical shells thus formed then collapsed to form galaxies. Large-scale motions of ionized material are a feature of this model; these introduced temperature fluctuations into the CBR in ways to be explored in Chapter 8.

In both the explosive and the pure gravitational scenarios, *bias* may be present, that is the tendency to form galaxies only or predominantly in unusually dense regions (e.g., Bardeen, 1986; Kaiser, 1986). Such regions of particularly high density are rare. Thus adding bias to the process of galaxy formation can help explain why luminous matter, in the form of galaxies, appears less uniformly distributed than 'dark' matter, a point to be reconsidered in Chapter 8.

1.7.3 Reionization

The explosive scenario virtually requires reionization of intergalactic (baryonic) matter, and the purely gravitational model certainly allows for it. If the bulk of the matter in the Universe were reionized, we must allow for Thomson scattering of CBR photons by the free electrons produced. If the scattering optical depth exceeded unity, the surface of last scattering will shift to a much lower redshift and the photon directions will be redistributed, as shown in fig. 1.11(b). If the optical depth for Thompson scattering was > 1, fluctuations imprinted on the CBR at the epoch of recombination were erased. As Silk (1986) and Bond (1988) among others have shown, the reionization must have occurred at a large redshift (for instance, $z \gtrsim 60$ for $\Omega_b = 0.06$) to ensure a large enough optical depth to thus erase 'primordial' fluctuations.

Whether or not the baryonic matter in the Universe was reionized is a crucial question that will have to be faced when we turn, in Chapter 8, to interpretations of observational limits on the small-scale anisotropy of the CBR. It is worth noting that there is no consensus on whether reionization, if it occurred, took place at a large enough redshift to obscure our view back to $z \simeq 1000$.

Even more important is the fact that the astrophysical processes responsible for reionization could themselves have introduced fluctuations into the CBR, as Hogan (1984) and Ostriker and Vishniac (1986), among others, have noted. Both an inhomogeneous distribution of ionized matter and inhomogeneous velocity fields in the ionized matter could cause the CBR temperature to vary from point to point. Thus reionization itself can imprint fluctuations into the CBR. We will return to this issue in detail in Chapter 8.

1.7.4 Cosmic strings and galaxy formation

Cosmic strings* are massive, one-dimensional structures that are a feature of spontaneous symmetry breaking in some GUTs (Vilenkin, 1985; Kolb and Turner, 1990, Chapter 7). They are topological defects 'left over' when Grand Unification broke down; they may also be thought of as one dimensional regions of high vacuum energy density. The topological properties of cosmic strings and their origin in spontaneous symmetry breaking need not concern us here. The gravitational effects of these massive, extended structures, on the other hand, are important. To assess these effects, we need an estimate of the mass per unit length, μ, of a cosmic string. That quantity is related to the energy scale of Grand Unification: $\mu \propto E^2$. For $E \approx 10^{14}$–10^{15} GeV, $\mu \simeq 10^{18}$–10^{20} g cm^{-1}.

Because of the large negative pressure in the false vacuum, cosmic strings have a natural tension; the tension in turn causes the strings to move at relativistic speeds essentially perpendicular to their length. A massive moving string produces a 'wake', which sweeps up matter gravitationally (Silk and Vilenkin, 1984) serving to initiate galaxy formation, which then proceeded as sketched in Section 1.7.2 above. Another scenario (Turok and Brandenberger, 1986) for galaxy formation is that *loops* of string formed gravitational potential wells into which other matter fell. The loops were formed when strings intersected.

Kaiser and Stebbins (1984) and Bouchet *et al.* (1988), among others, have pointed out that the statistical properties of fluctuations in the matter density caused by strings would differ from those in a random-phase Gaussian distribution of perturbations. These non-Gaussian properties are an important signature, and will be explored further in Chapter 8.

Ostriker, Thompson and Witten (1986) have combined strings and explosions by pointing out that cosmic strings are superconducting; hence their motion may have produced electromagnetic shocks, which drove matter into collapse.

Let me conclude as I began this section by saying how unclear the present picture of galaxy formation is. Strings, for instance, may be involved, but equally possibly may be totally irrelevant. One ray of hope is that many of the suggested processes for galaxy formation may have left an imprint in the CBR. Hence the observational material this book

---

* Not to be confused with the microscopic strings of string theory. I should also note that cosmic strings are not the only topological defects possible in spontaneous symmetry breaking; monopoles are another example (see Kolb and Turner, 1990, Chapter 7).

presents may eventually help to clarify how and when the most striking objects in our Universe were formed.

1.8   Alternatives to the Hot Big Bang

There is a good deal to be said for the Hot Big Bang model outlined in the preceding sections, including its conceptual simplicity and its broad agreement with observations. It is important to bear in mind, however, that there are alternative cosmological theories. I think it is safe to say that none have gained wide acceptance, in part, I would claim, because there is no compelling observational datum forcing us to abandon the Hot Big Bang model, including its inflationary phase. In many cases, also, these alternative models conflict with astronomical data more or less seriously. In this section, I will sketch very briefly some of the alternatives, with special emphasis on what they have to say about the CBR. Not included here are variants of Big Bang cosmology, such as the anisotropic models discussed in Section 8.3, models with a time-varying gravitational constant (e.g., Dirac, 1974), or the conformal theory of Hoyle and Narlikar (see Narlikar, 1983, Chapter 10). Let us instead focus on more radical theories, which introduce new physical ideas, or deny the Big Bang origin of the Universe altogether (see Ellis, 1984, for a fuller review).

1.8.1 Is the observed expansion real?

We *observe* redshifts in the spectral lines from distant galaxies (Section 1.2); we *interpret* them as evidence for expansion. Could it be instead that light grows 'tired' – loses energy so that its wavelength increases – as it traverses large distances in a static Universe? Such a suggestion was originally made by Hubble and Tolman (1935; see also Pecker *et al.*, 1973). The 'tired light' theory encounters two difficulties: (1) the predicted relationship between redshifts and angular diameters of galaxies does not agree with observations (see Peebles, 1971, Chapter VI), and (2) it provides no explanation for the 3 K radiation which is the subject of this book.

Other theories that deny the expansion of the Universe (and hence its Big Bang origin) include the FIB cosmology of Barnothy and Forro (1944). It is a static, closed cosmology. In it, the roughly 3 K microwave background is explained as follows. Galactic starlight travels twice around the closed geometry of the Universe and is refocused on the Galaxy with an apparent redshift of about 2000 (Barnothy and Barnothy, 1972). There are two problems with this explanation: the high degree of isotropy observed in the radiation is merely fortuitous, and the spectrum of Galactic starlight is a sum of blackbody curves, which does not provide a good fit to the (almost exactly Plankian) spectrum of the CBR discussed in Chapter 4. In particular, the starlight spectrum is much broader than a blackbody spectrum.

Another static model is an inhomogeneous one with us at the center and an antipodal bright source, which generates the roughly 3 K CBR (Ellis, Maartens and Nel, 1978; see also Ellis, 1984). The observed high degree of isotropy of the radiation occurs because we are very near the center. The results of Chapter 6 may be used to show – approximately – that the Earth must fall inside a small space equivalent to about 10^{-12} of the total volume of the Universe in this model. Such geocentrism makes me uneasy.

Segal's Chronometric Theory (1976) replaces eqn. (1.3) with a quadratic law, and hence is discounted by many observers. The 3 K radiation is explained as the dispersed, scattered and redshifted radiation of all galaxies in this closed, infinite-aged, model. The Chronometric Theory does not explain the good agreement between the predicted abundances of Hot Big Bang nucleosynthesis and the observations. In addition, the magnitude of the observed dipole moment in the CBR presents problems for the theory (see Segal, 1976, p. 88).

1.8.2 The Steady State Theory

For nearly twenty years, from the late 1940s to the mid-1960s, the great rival of the Hot Big Bang model was the Steady State Theory of Bondi, Gold and Hoyle (see Bondi, 1960; and Narlikar, 1983, for details). It is based on the *perfect cosmological principle*: on a large scale, the Universe is the same everywhere *and* at all times. There is no Big Bang origin with $R = 0$; instead the expansion law is $R \propto e^{Ht}$. The exponential expansion is driven by a new field, the creation field, which also, as its name suggests, is responsible for the creation of new matter *ex nihilo*. The creation of new matter, in turn, is required to keep ρ constant as the Universe expands, in keeping with the perfect cosmological principle.

If the Universe is the same at all times, where did the thermalized 3 K background come from? Despite an ingenious attempt to produce such a background from reemitted starlight (Hoyle *et al*, 1968), the Steady State model offers no convincing explanation for the CBR. Indeed, as we will see in Chapter 2, the discovery of the CBR drove the theory out of fashion.

1.8.3 'Cool' Big Bang models

Some other cosmological models, like the 'Whimper' cosmology of Ellis (1984) and his colleagues, also eliminate the $R \to 0$ initial singularity, replacing a bang with a whimper. These models are based on different geometries than the homogeneous, isotropic Robertson–Walker metric of eqn. (1.6). By properly fixing the initial conditions, the CBR can be included.

The symmetric model of Alfvén and Klein (1962) also avoids an initial singularity. The 'symmetry' referred to is the exact symmetry between matter and antimatter, taken as an initial condition by these authors*. In this model, expansion is driven by pressure produced by the partial annihilation of particles and antiparticles (e.g., $e^+ + e^- \to 2\gamma$). To prevent complete annihilation, the overall density must always have been low (no initial high density state is permitted) and a mechanism was needed to keep regions of matter and antimatter apart. Explaining the large-scale isotropy and the blackbody spectrum of the CBR is a major difficulty for this theory. While numerous photons were produced by annihilation processes, these photons were produced at fixed energies, and hence fixed wavelengths, so the radiation spectrum is very different from a Planck curve even if cosmological redshift is included.

Another class of models (e.g. Layzer and Hively, 1973; Rees, 1978; Rowan-Robinson *et al.*, 1979; Aiello *et al.*, 1982; and Wright, 1982) include expansion and are based on ordinary physics, but generate the CBR by astrophysical processes well after the initial $R \to 0$ Big Bang. Thus the Big Bang itself was cool, and starlight, thermalized by re-

---

* There is also a Big Bang 'symmetric' model (Omnès, 1969).

emission from dust or other mechanisms at intermediate redshifts, is what we now detect as the 3 K microwave radiation. These theories must confront two major obstacles. (1) Where did the dust – which presumably must consist of heavy elements (but see Hoyle *et al.*, 1968) – come from? (2) How can starlight be thermalized to produce a precisely blackbody spectrum over the wide range of wavelengths spanned by the observations? Most absorption and reemission processes, especially when small particles are involved, are inherently wavelength dependent, so very special assumptions are necessary to produce the spectrum described in Chapter 4. Of course reemission by dust could have added to and distorted a truly cosmological spectrum; this possibility is discussed in Chapter 5.

1.9   Concluding opinions and remarks

In my view, each of the alternative models discussed in Section 1.8 has one or more serious difficulties when it confronts observations. Also, again in my view, there is no observational datum that *requires* us to abandon the Hot Big Bang model. Thus I will base the rest of the book on the Hot Big Bang model, generally including its early inflationary phase, though the evolution of the Universe at $t \lesssim 10^{-5}$ s is not as securely understood as its later history.

It must also be said that there *are* gaps and puzzles remaining in modern cosmology. Among the open questions are the small (or zero) present value of the cosmological constant (Section 1.3.6), the nature of dark matter, and the process (or processes) by which galaxies formed. In addition, while we know the present temperature of the Universe to an accuracy of 1 or 2%, other cosmological parameters like H_0, Ω and the age of the Universe are very poorly measured. As is now usual, I will continue to follow the convention of writing H_0 as $100h$ km s^{-1} per megaparsec in the remainder of the book. Since the observations (Rowan-Robinson, 1985) show $0.4 \lesssim h \lesssim 1.0$, I will use $h = 0.75$ unless otherwise stated. For the value of Ω, I will assume 1.0 unless otherwise stated, in keeping with arguments advanced in Section 1.5. In some calculations, the value of Ω has an important bearing on the result; in such instances, I will present results for both $\Omega = 1$ and $\Omega = 0.1$, say (an appropriate value if only baryonic mass is present).

This has been a brief overview of cosmology; much has been omitted to squeeze the material into a single chapter. Among the topics slighted are observational attempts to determine the crucial cosmological parameters H_0 and Ω, and the rich connection between cosmology and particle physics. Readers interested in fuller descriptions of modern cosmology should consult one of the texts by Bondi (1960), Peebles (1971, 1980), Weinberg (1972), or Kolb and Turner (1990) or similar works at a more introductory level by Sciama (1971), Berry (1976), Rowan-Robinson (1977), Harrison (1981), or Narlikar (1983).

References

Several of the papers on inflationary cosmology and related topics referred to in this chapter have been assembled and reprinted by Abbott and Pi (1986); these are indicated below by '(AP)' following the citation.

Abbott, L. F., and Pi, S.-Y. (AP) 1986, *Inflationary Cosmology*, World Scientific, Singapore.

Aiello, S., Cecchini, S., Mandolesi, N., and Melchiorri, F. 1982, in *Proceedings of the Second Marcel Grossmann Meeting on General Relativity*, ed. R. Ruffini, North-Holland, Amsterdam.

Albrecht, A., and Steinhardt, P. J. 1982, *Phys. Rev. Lett.*, **48**, 1220 (AP).

Alfvén, H., and Klein, O. 1962, *Arkiv Fysik*, **23**, 187.

Audouze, J. 1987, in *I.A.U. Symposium 124, Observational Cosmology*, eds. A. Hewitt, G. Burbidge, and L. Z. Fang, D. Reidel, Dordrecht.

Bardeen, J. 1986, in *Inner Space/Outer Space*, eds. E. W. Kolb, M. S. Turner, D. Lindley, K. Olive, and D. Seckel, University of Chicago Press, Chicago (AP).

Bardeen, J. M., Steinhardt, P. J., and Turner, M. S. 1983, *Phys. Rev.* D, **28**, 679 (AP).

Barnothy, J. M., and Barnothy, M. F. 1972, *I.A.U. Symposium 44*, ed. D. S. Evans, D. Reidel, Dordrecht.

Barnothy, J. M., and Forro, M. 1944, *Csillagaszati*, **7**, 65.

Baron, E., and White, S. D. M. 1987, *Ap. J.*, **322**, 585.

Berry, M. 1976, *Principles of Cosmology and Gravitation*, Cambridge University Press, Cambridge.

Blumenthal, G. R., Faber, S. M., Primack, J. R., and Rees, M. J. 1984, *Nature*, **311**, 517 (AP).

Boesgaard, A. M., and Steigman, G. 1985, *Ann. Rev. Astron. Astrophys.*, **23**, 319.

Bond, J. R. 1988, in *The Early Universe*, eds. W. G. Unruh and G. W. Semenoff, Reidel, Dordrecht.

Bondi, H. 1960, *Cosmology*, Cambridge University Press, Cambridge.

Bouchet, F. R., Bennett, D. P., and Stebbins, A. 1988, *Nature*, **335**, 410.

Callan, C. G., Dicke, R. H., and Peebles, P. J. E. 1965, *Amer. J. Phys.*, **33**, 105.

Coleman, S., and Weinberg, E. 1973, *Phys. Rev.* D, **7**, 1888 (AP).

Denegri, D., Sadoulet, B., and Spiro, M. 1990, *Rev. Mod. Phys.*, **62**, 1.

Dirac, P. A. M. 1974, *Proc. Roy. Soc.* A, **338**, 439.

Efstathiou, G. 1990, in *Physics of the Early Universe*, eds. J. A. Peacock, A. F. Heavens and A. T. Davies, Adam Hilger, New York.

Ellis, G. F. R., 1984, *Ann. Rev. Astron. Astrophys.*, **22**, 157.

Ellis, G. F. R., Maartens, R., and Nel, S. P. 1978, *Monthly Not. Roy. Astron. Soc.*, **184**, 439.

Faber, S. M., and Gallagher, J. S. 1979, *Ann. Rev. Astron. Astrophys.*, **17**, 135.

Field, G. B. 1975, in *Stars and Stellar Systems*, Vol. 9, University of Chicago Press, Chicago (Chap. 10).

Fowler, W. A. 1987, *Quarterly J. Roy. Astron. Soc.*, **28**, 87.

Guth, A. H. 1981, *Phys. Rev.* D, **23**, 347 (AP).

Guth, A. H., and Pi, S.-Y. 1982, *Phys. Rev. Lett.*, **49**, 1110 (AP).

Guth, A. H., and Steinhardt, P. J. 1984, *Scientific American*, May.

Harrison, E. R. 1970, *Phys. Rev.* D, **1**, 2726 (AP).

Harrison, E. R. 1981, *Cosmology: The Science of the Universe*, Cambridge University Press, Cambridge.

Hawking, S. W. 1982, *Phys. Lett.* B, **115**, 295 (AP).

Hogan, C. J. 1984, *Ap. J. (Lett.)*, **284**, L1.

Hogan, C. J., and Kaiser, N. 1983, *Ap. J.*, **274**, 7.

Hoyle, F., Wickramasinghe, N. C., and Reddish, V. C. 1968, *Nature*, **218**, 1124.

Hubble, E., and Tolman, R. C. 1935, *Ap. J.*, **82**, 307.

Ikeuchi, S. 1981, *Publ. Astron. Soc. Japan*, **33**, 211.

Jeans, J. 1902, *Philos. Trans.* A, **199**, 49.

Jones, B. J. T., and Wyse, R. F. G. 1985, *Astron. Astrophys.*, **149**, 144.

Kaiser, N. 1986, in *Inner Space/Outer Space*, eds. E. W. Kolb, M. S. Turner, D. Lindley, K. Olive, and D. Seckel, University of Chicago Press, Chicago.

Kaiser, N., and Stebbins, A. 1984, *Nature*, **310**, 391.

Kolb, E. W., and Turner, M. S. 1983, *Ann. Rev. Nuclear Particle Sci.*, **33**, 645.

Kolb, E. W., and Turner, M. S. 1990, *The Early Universe*, Addison-Wesley, New York.

Koo, D. C. 1986, in *The Spectral Evolution of Galaxies*, eds. C. Chiosi and A. Renzini, D. Reidel, Dordrecht.

Kormendy, J., and Knapp, G. A. 1987, *I.A.U. Symposium 117, Dark Matter in the Universe*, D. Reidel, Dordrecht.

Last, J., Arnold, M., Döhner, J., Dubbers, D., Freedman, S. J. 1988, *Phys. Rev. Lett.*, **60**, 995.

Layzer, D., and Hively, R. 1973, *Ap. J.*, **199**, 361.

Lifschitz, E. M. 1946, *J. Phys.*, **10**, 116.

Linde, A. 1982, *Phys. Lett.* B, **108**, 389 (AP).

Linde, A. 1985, *Phys. Lett.* B, **158**, 375.

McCrea, W. H. 1971, *Quarterly J. Roy. Astron. Soc.*, **12**, 140.

McCrea, W. H., and Milne, E. A. 1934, *Quarterly J. Math.* **5**, 73.

Misner, C. W., Thorne, K. S., and Wheeler, J. A. 1973, *Gravitation*, W. H. Freeman, San Francisco.

Narlikar, J. V. 1983, *Introduction to Cosmology*, Jones and Bartlett, Boston.

North, J. D. 1965, *The Measure of the Universe*, Oxford University Press, Oxford.

Olive, K. A., Schramm, D. N., Steigman, G., and Walker, T. P. 1990, *Phys. Lett.* B, **236**, 454.

Omnès, R. 1969, *Phys. Rev. Lett.*, **23**, 38.

Ostriker, J. P., and Cowie, L. L. 1981, *Ap. J. (Lett.)*, **243**, L127.

Ostriker, J. P. and Vishniac, E. T. 1986, *Ap. J. (Lett.)*, **306**, L51.

Ostriker, J. P., Thompson, C., and Witten, E. 1986, *Phys. Lett.* B, **180**, 231.

Pagel, B. E. J. 1986, in *Inner Space/Outer Space*, eds. E. W. Kolb, M. S. Turner, D. Lindley, K. Olive, and D. Seckel, University of Chicago Press, Chicago.

Pecker, J.-C., Tait, W., and Vigier, J. P. 1973, *Nature*, **241**, 338.

Peebles, P. J. E. 1966, *Ap. J.*, **146**, 542.

Peebles, P. J. E. 1971, *Physical Cosmology*, Princeton University Press, Princeton, New Jersey.

Peebles, P. J. E. 1980, *The Large Scale Structure of the Universe*, Princeton University Press, Princeton, New Jersey.

Peebles, P. J. E., and Dicke, R. H. 1968, *Ap. J.*, **154**, 891.

Press, W. H. 1980, *Physica Scripta*, **21**, 702.

Press, W. H. and Schechter, P. 1974, *Ap. J.*, **187**, 425.

Primack, J. R., Sekel, D., and Sadoulet, B. 1988, *Ann. Rev. Nuclear Particle Sci.*, **38**, 751.

Rees, M. J. 1978, *Nature*, **275**, 35.

Reeves, H. 1989, in *The Third ESO/CERN Symposium*, eds. M. Caffo *et al.*, Kluwer, Dordrecht.

Robertson, H. P., and Noonan, T. W. 1968, *Relativity and Cosmology*, W. B. Saunders, Philadelphia.

Ross, G. G. 1984, *Grand Unified Theories*, Benjamin Cummings, Menlo Park, California.

Rowan-Robinson, M. 1977, *Cosmology*, Oxford University Press, Oxford.

Rowan-Robinson, M. 1985, *The Cosmological Distance Ladder*, W. H. Freeman, San Francisco.

Rowan-Robinson, M., Negroponte, J., and Silk, J. 1979, *Nature*, **281**, 635.

Rubin, V. C., Ford, W. K., Thonnard, N., and Burstein, D. 1982, *Ap. J.*, **261**, 439.

Sakharov, A. 1967, *J. Exp. Theor. Phys. Lett.*, **5**, 24.

Schramm, D. N., and Wagoner, R. V. 1979, *Ann. Rev. Nuclear Particle Sci.*, **27**, 37.

Sciama, D. W. 1971, *Modern Cosmology*, Cambridge University Press, Cambridge.

Segal, I. E. 1976, *Mathematical Cosmology and Extragalactic Astronomy*, Academic Press, New York.

Sekel, D., and Turner, M. S. 1985, *Phys. Rev.* D, **32**, 3178.

Shaver, P. A., Kunth, D., and Kjar, K., eds., 1983, *ESO Workshop on Primordial Helium*, ESO, Garching bei München.

Silk, J. 1968, *Ap. J.*, **151**, 459.

Silk, J. 1986 in *Inner Space/Outer Space*, eds E. W. Kolb, M. S. Turner, D. Lindley, K. Olive, and D. Seckel, University of Chicago Press, Chicago (p. 143).

Silk, J., and Vilenkin, A. 1984, *Phys. Rev. Lett.*, **53**, 1700.

Spite, F., and Spite, M. 1982, *Astron. Astrophys.*, **115**, 357.

Starobinskii, A. A. 1982, *Phys. Lett.* B, **117**, 175 (AP).

Sunyaev, R. A. and Zel'dovich, Ya. B. 1972, *Astron. Astrophys.*, **20**, 189.

Tayler, R. J. 1986, *Quarterly J. Roy. Astron. Soc.*, **27**, 367.

Trimble, V. 1987, *Ann. Rev. Astron. Astrophys.*, **25**, 425.

Turner, M. S. 1988, in *The Early Universe*, eds. W. G. Unruh and G. W. Semenoff, D. Reidel, Dordrecht.

Turok, N., and Brandenberger, R. 1986, *Phys. Rev.* D, **33**, 2175. See also Sato, H. 1986, *Mod. Phys. Lett.* A, **1**, 9 and Stebbins, A. 1986, *Ap. J. (Lett.)*, **303**, L21.

Vilenkin, A. 1985, *Phys. Rep.*, **121**, 263.

Wagoner, R. V. 1973, *Ap. J.*, **179**, 343.

Wagoner, R. V., Fowler, W. A., and Hoyle, F. 1967, *Ap. J.*, **148**, 3.

Walker, T. P., Steigman, G., Schramm, D. N., Olive, K. A., and Kang, H.-S. 1991, *Ap. J.* **376**, 51.

Weinberg, S. 1972, *Gravitation and Cosmology*, J. Wiley, New York.

White, S. D. M., and Rees, M. J. 1978, *Monthly Not. Roy. Astron. Soc.*, **183**, 341.

Wright, E. L. 1982, *Ap. J.*, **255**, 401.

Yang, J., Turner, M. S., Steigman, G., Schramm, D. N. and Olive, K. A. 1984, *Ap. J.*, **281**, 493.

Zel'dovich, Ya. B. 1972, *Monthly Not. Roy. Astron. Soc.*, **160**, 1 (AP).

2

The early history of CBR studies

It is fair to claim that the cosmic background, announced in 1965, was discovered by telephone. As we will see, this radiation had been both *observed* and *predicted* before a fateful 1964 telephone call from Arno Penzias and Robert Wilson of the Bell Telephone Laboratories to Robert Dicke at Princeton, a few miles away. Nevertheless, that telephone conversation led to the recognition that the source of the puzzling 'excess noise' observed by Penzias and Wilson was in fact a relic of the Hot Big Bang.

In this chapter, I will start with a brief review of the state of cosmology in the early 1960s, look at earlier predictions of the radiation, particularly the prescient work of Gamow and his colleagues, and then attempt to unravel the complicated knot of circumstances that led to the telephone call of late 1964. Later in the chapter, I will describe the important early experiments which established the cosmological nature of the radiation first detected by Penzias and Wilson, and then sketch the subsequent history of the field.

2.1 Cosmology in the early 1960s

Thirty years ago, cosmology played a far less central role in physics and astronomy than it does today. For instance, cosmology is nowhere mentioned in the text* I was assigned in my first college astronomy course more than thirty years ago; the word 'Universe' appears only twice, where it is used, inappropriately, to refer to the Milky Way.

Though cosmology was then very much at the periphery of the physical sciences, there was nevertheless important work being done and there were important questions being raised in the field. A definitive monograph, which treated both cosmological models and the observations that could be used to test them, was published by Bondi in 1952 and revised in 1960. A crucial 1961 paper by Sandage on 'The ability of the 200-inch telescope to discriminate between selected World models' delineated major goals of observational cosmology in the 1960s (and after): the measurement of the rate of expansion of the Universe (H_0) and Ω_0 (or k), which, as we saw in Chapter 1, can be used to discriminate between cosmological models. The centrality of these measurements in the decade of the 1960s is reflected in the title of a later review by Sandage: 'Cosmology: a search for two numbers' (1970).

Sandage's interest lay in determining the dynamics, and hence the geometry, of the expanding Universe. In the 1960s, however, an even more fundamental, more public,

* *Astronomy*, by H. N. Russell, R. S. Dugan and J. Q. Stewart.

42

and indeed more acrimonious question dominated cosmology. It was the debate between adherents of Big Bang cosmology, formulated by Gamow and his colleagues, and those who supported the Steady State model of Bondi, Gold and Hoyle. Did the Universe have a history or not? By the early 1960s, the optical observations planned and carried out by Sandage and others were not yet able to decide the question. Counts of radio sources by Ryle (e.g., 1968) suggested that they did evolve in number or luminosity, so the Universe did change with time, but these results were controversial. In 1961, say (the year I was introduced to cosmology in a college course), scientific papers were being written in *roughly* equal numbers by adherents of the two views. The debate spilled over into the public arena (see *Scientific American*, September 1956, devoted to cosmology, and *Rival Theories of Cosmology*, a transcription by Oxford University Press of a 1959 BBC debate).

Cosmology in the 1960s was thus both starved for data (we *still* do not know H_0 to 50% accuracy) and riven by controversy about a fundamental property of the Universe. It seems entirely plausible to me that the long-running debate between proponents of the Steady State and of the Big Bang theories diminished the stature of cosmology in the eyes of physicists and astronomers in more established fields. 'Cosmology is mostly a dream of zealots who would oversimplify at the expense of deep understanding,' wrote a distinguished physicist, himself soon to make important contributions to CBR studies (W. A. Fowler in his 1968 foreword to Robertson and Noonan's *Relativity and Cosmology*).

2.2   Prediction of the CBR: Gamow's Hot Big Bang theory

In the end, of course, the Big Bang cosmology of Gamow carried the day. Here I want to look briefly at one aspect of the work of Gamow and his colleagues, specifically their recognition that a *hot* Big Bang would leave the present Universe at a non-zero temperature, first pointed out by Alpher and Herman in 1948 (and later taken up by Gamow).

Gamow introduced the Hot Big Bang in 1946 as a means to build up nuclei heavier than ^{1}H.* He recognized that temperatures of order 10^9–10^{10} K would be required (to ensure that kT was roughly equal to the binding energy of the nuclei). Originally, he hoped that Big Bang synthesis could explain the abundance of *all* the elements observed in stellar spectra. In 1949, however, Fermi and Turkevich (see, e.g., Fermi, 1949) pointed out that, since no stable nucleus of atomic mass 5 exists, a gradual building-up of heavier and heavier nuclei would end at ^{4}He. Detailed calculations by Alpher and Herman, among others, confirmed this result (see also Section 1.6.2), and produced specific predictions of the ^{4}He/^{1}H ratio in material emerging from the Big Bang. The results of these calculations appeared in both the technical literature (e.g., in a series of papers by Alpher, Gamow and Herman in volumes 74 and 75 of *The Physical Review*), and in reviews for a wider audience (e.g., Gamow's 1956 *Scientific American* article, which includes a nice diagram of Big Bang nucleosynthesis). The production of light nuclei, and the dependence of temperature on time in a radiation-dominated expanding Universe (eqn. (1.28)) are both presented in Gamow's 1952 book for lay readers, *The Creation of the Universe*, which I bought and read as a high school student.

---

* Following earlier leads by Tolman (1922) and Suzuki (1928) – see Alpher and Herman's detailed 1950 review.

Moreover, it was clear to Gamow and his group that a Hot Big Bang would leave the Universe with a calculable, non-zero temperature. In several of the papers and reviews referred to above, $T(t)$ is plotted, and it is easy to read off the present value, T_0 (typically about 10 K). On several occasions, members of this group made specific predictions of the present 'background temperature,' the phrase employed by Alpher and Herman (1949). In that paper, Alpher and Herman give $T_0 = 5$ K, and that figure appears in other papers as well.*

At first glance, it is astonishing that Alpher and Herman came within a factor of two of the presently accepted value of T_0. However, if one relies on the Hot Big Bang to produce 20–50% ^{4}He by mass, one finds T_0 about a few kelvin, independent of most cosmological details. What is more astonishing is that this discussion of a mean 'background temperature [of] ... the order of 5 K' should have dropped out of scientific sight for nearly twenty years. Why did this happen? Big Bang models remained in vogue, but Gamow's original hope of making *all* heavy elements in a Hot Big Bang was weakened by Fermi and Turkevich and by the pivotal paper of Burbidge, Burbidge, Fowler and Hoyle (1957), which showed convincingly how most heavy elements were built up in stellar interiors. Even the recognition that most elements – C, O, Fe and so on – are made in stars, however, did not entirely submerge the Hot Big Bang model. For instance, in a detailed 1965 review, Zel'dovich, a leading Soviet cosmologist, considered the Hot Big Bang model in detail. In that same year, Hoyle and Tayler noted that the large abundance of ^{4}He relative to still heavier elements was more naturally explained by a combination of Big Bang synthesis of the light nuclei like ^{4}He plus stellar nucleosynthesis than by stellar synthesis alone. In other words, a Hot Big Bang *is* needed to explain the observed abundance of some elements, especially those with atomic mass ≤ 4.

Nevertheless, most physicists and astronomers ignored the predictions of the Hot Big Bang model, perhaps because they seemed to be mere features in a 'dream of zealots.' Both the work of Gamow, Alpher and Herman and reasons for its apparent disappearance have been treated by others interested in the early history of the CBR (see Weinberg, 1972† and 1977; a more informal treatment given by Ferris, 1977; and Alpher and Herman, 1988, among others). To these analyses and to my remarks above, I would like to add a more speculative coda. It is absolutely clear that Alpher and Herman predicted a non-zero 'background temperature' for the present Universe. What is missing in these papers is the recognition that a Universe with non-zero temperature must even now be filled with more-or-less isotropic, thermal, radiation that could be detected, and indeed *had* been detected, as we will soon see. No one took up the challenge of observing the predicted background radiation.‡

* In his 1952 book (and elsewhere) Gamow quotes a much higher figure (50 K in his book) because he carelessly assumed that the expansion of the Universe remained radiation-dominated up to the present, so that $T \propto t^{-1/2}$. Thus, while he and his colleagues correctly derived eqn. (1.28), Gamow (unlike Alpher and Herman) incorrectly extrapolated that relation to the present, deriving a value for T_0 about ten times too large.

† Weinberg, in *Gravitation and Cosmology* (1972, p. 510), suggests as an explanation that, after predicting $T_0 = 5$ K, Alpher and Herman ' ... went on to express doubts as to whether this radiation would have survived until the present.' I believe Weinberg's argument misses the point; Alpher and Herman were discussing cosmic *rays* at this point, not the thermal cosmic background.

‡ Alpher and Herman have kindly informed me that they and their colleague, James Follin, did indeed explore the possibility of radio astronomical measurements, but were told by the observers that the technology of the day (the mid-1950s) would not permit them; this point may be dealt with further in a book that Alpher and Herman have in preparation.

Ironically, a radio astronomical upper limit on 'cosmic temperature' by none other than Robert Dicke and his colleagues appears in the same volume of *The Physical Review* as Gamow's first paper on the Hot Big Bang. There were false starts on the observational side as well, and it is time to turn to them.

2.3 Missed opportunities on the observational side

The existence of a universal, thermal radiation field had been predicted clearly by 1948, but, as we have seen, no attempt was made to confirm the prediction experimentally. As it happens, however, the existence of a universal radiation field had already been observationally established years earlier (Adams, 1941; McKellar, 1941)! These measurements were made by workers in a branch of astronomy quite separate from cosmology and lay forgotten until 1965. More surprisingly, radio astronomers had on several occasions before 1964 measured or set limits on an isotropic microwave background. As a complement to our examination of the theoretical side, let us look briefly at the missed opportunities and unrecognized clues on the observational side.

2.3.1 The excitation of interstellar molecules

Clouds of molecules lying between us and certain nearby stars produce narrow absorption lines in the spectra of these stars. In virtually all cases, these lines originate only from the ground states of the absorbing molecules. By 1941, the cyanogen molecule (CN) was recognized as an exception; faint absorption lines originating from the first excited rotational state, about 5×10^{-4} eV above the ground state, were detected (Adams, 1941). What was exciting the CN molecules, thus populating this state? One possibility was radiation at a wavelength of 2.64 mm. If so, the ratio of the populations in the two states leads directly to an estimate of the characteristic temperature of the radiation field (see Section 4.10). From observations of CN absorption lines, McKellar (1941) estimated a temperature of 2.3 K at 2.64 mm. Since several cases of excited CN molecules were known, in different interstellar clouds, the radiation must be widespread in the Galaxy, if not universal like the CBR. However, radiative excitation is not the only possibility, and these CN observations soon became a footnote in the study of interstellar molecules. (In his definitive monograph on molecular spectra of 1950, Herzberg writes 'From the intensity ratios of the [two CN] lines a rotational temperature of 2.3 K follows, which has of course, only a very restricted meaning,' referring, I suppose, to the possibility of collisional or other non-radiative excitation.) A brief review of these early papers is provided by Thaddeus (1972).

2.3.2 Short wavelength radar research

The important role of radar in World War II is well known. A major aim of wartime research was to produce radar equipment operating at short wavelengths, to reduce its size. In the U.S.A., one leader of this effort was Robert Dicke. As part of his work at the 'Rad Lab' at MIT, he designed and built sensitive radio/radar receivers operating at $\lambda = 1$–1.5 cm. He also introduced in 1946 the now standard convention of measuring radio intensity in terms of an equivalent temperature (see Section 3.2). More to the point, as part of his research, he made a measurement of the temperature of the sky at $\lambda = 1.25$ cm. He and his colleagues, Beringer, Kyhl and Vane (shown in fig. 2.1) determined an

Fig. 2.1   R. H. Dicke and his colleagues calibrating a microwave radiometer using an ambient temperature absorber (Dicke is holding this panel, then referred to as a 'shaggy dog'. The photo dates from the mid-1940s. At about this same time (1946) Dicke *et al.* established an upper limit of 20 K on the cosmic background at microwave frequencies using similar apparatus.

upper limit of 20 K on any isotropic 'cosmic' background radiation, which they published in 1946. The technique they employed (and the receivers available in the 1940s) were not capable of detecting an isotropic background temperature as low as 3 K. Nevertheless, Dicke's early work presaged many of the later observations; but it too was forgotten – not least by Dicke himself!

Apparently this report, like the CN results, went unrecognized by Gamow and his colleagues (even though, as noted above, the upper limit of Dicke *et al.* (1946) appeared in the same volume of the *Physical Review* as Gamow's first letter on a Hot Big Bang).

2.3.3 Early work at the Bell Telephone Laboratories

By the early 1960s, radio receivers operating at centimeter wavelengths had been improved to the point that measurements with a precision of a few tenths of a kelvin could be made. Much of this progress occurred at the Bell Telephone Laboratories, propelled by developments in telecommunications using satellites as links. At Bell Labs, these sensitive receivers were mated with carefully designed antennas and low temperature calibration sources. Years of research were invested in identifying, and then reducing or eliminating, sources of systematic error and noise.

By the early 1960s, the workers at Bell Labs (including Penzias and Wilson) began to encounter a vexing problem: excess noise entering their antennas whenever they were pointed at the sky. The first reference to this problem is in a technical paper by Ohm

(1961) on measurements at 5.3 cm; in this paper, the excess noise was effectively swallowed up in the errors assigned to other signals (Wilson, 1979, 1983). A more careful accounting for the various sources of systematic error (made after the fact by Wilson in 1983) shows that Ohm had in fact marginally detected the CBR.

Was this paper also forgotten? No; in this case the problem was not omission but misinterpretation. One contribution to the signal entering the antenna was microwave radiation from the Earth's atmosphere. Ohm referred to this as 'sky' temperature. Two alert Soviet cosmologists, Doroshkevich and Novikov (1964), noted Ohm's 1961 paper, but misinterpreted 'sky' temperature, taking it to include *both* atmospheric emission and any possible isotropic background. Since the measured 'sky' temperature closely matched the calculated atmospheric signal alone, they drew the erroneous conclusion that any possible isotropic, cosmic, background must be ≤ 1 K. This 'observation' was then used by their senior colleague, the prominent cosmologist Ya. B. Zel'dovich, to *negate* the Hot Big Bang theory: 'Radioastronomical data show that the temperature of thermal radio waves does not exceed 1 K' (Zel'dovich, 1965)! In fact there *was* a measured excess of 3.3 K in Ohm's 1961 work, even when atmospheric emission, pick-up from the ground, etc. had been accounted for.

In many ways, this strikes me as the most poignant missed opportunity of them all. Here was a group of resourceful, inventive, expert cosmologists, a group moreover familiar with the work of Gamow and his colleagues (frequently referred to in Zel'dovich's 1965 review). They appreciated the experimental advances in radio astronomy made at Bell Labs, and had even studied Ohm's 1961 paper, which appeared in a relatively obscure technical journal; but they too missed the discovery of the CBR.

I have recently learned of another unrecognized 'detection'. In the mid-1960s, Wall, Chu and Yen (see their 1970 paper) attempted to measure the absolute brightness of the radio sky at two wavelengths in order to determine the radio spectrum of the Galaxy. They expected the Galactic emission to vary with wavelength as $\lambda^{0.7}$, as is true for many radio galaxies. Their measurements at 40 and 90 cm, however, suggested a much weaker wavelength dependence, $\lambda^{0.2}$. Only later (Wall *et al.*, 1970) was the cause of this apparent discrepancy recognized – they had unknowingly measured a combination of Galactic emission and the CBR (which has a spectrum varying as λ^{-2}). This experiment too had the sensitivity to detect the CBR, but the crucial inference was not made.

2.4 The 'excess noise' of Penzias and Wilson

In the work of Ohm and his colleagues at Bell Labs, systematic errors essentially disguised the CBR signal. By 1964, a lower noise, better calibrated, radio telescope had been constructed at Bell Labs. It operated at 7.35 cm wavelength, and was coupled to a large antenna aptly described as looking like '... an alpenhorn the size of a boxcar' (quoting Ferris, 1977; see fig. 2.2). A crucial element in the new design was a switch, which allowed rapid comparison of emission from the sky and from a low temperature, stable calibrator. (Just to emphasize the intricacies of this story, this switching technique had been invented 25 years earlier by Dicke, and is widely referred to as 'Dicke switching.') With great care and diligence, Penzias and Wilson tracked down many sources of systematic error, ranging from radio signals from New York City, to the possible aftereffects of nuclear tests in the atmosphere, to pigeon dung in their antenna (for a more systematic account of their efforts, see Penzias (1979) or Wilson (1979 or 1983).

Fig. 2.2   A. A. Penzias (left) and R. W. Wilson standing in front of the horn antenna with which they detected and measured the CBR as 'excess noise' (see Appendix A here).

'Excess noise' from the sky, of intensity equivalent to 3.5 K at their wavelength of 7.35 cm, remained after all other systematic offsets had been subtracted.

What saved this work from joining the list of 'missed opportunities' was (1) the sensitivity of the receiver (the 3.5 K signal was more than ten times the statistical error in a single measurement); (2) the great care and persistence of Penzias and Wilson, who devoted months to excluding non-cosmic explanations for the 'excess noise'; and finally (3) the fateful telephone call of 1964, to which we now turn.

2.5   'Well boys, we've been scooped!'

Less than an hour's drive from Bell Labs, Robert Dicke and his Princeton colleagues were busy in 1964 reinventing the Hot Big Bang, and designing a sensitive receiver to detect the thermal background left by it. They were apparently unaware of *all* of the theoretical and observational work described above. Indeed, Dicke's motivation for a Hot Big Bang was not to build up elements heavier than hydrogen, but to destroy them. Dicke argued that a closed (recollapsing) Big Bang model might 'bounce' at the end of its collapse and then reexpand – an oscillating model, as shown in fig. 2.3. An infinitely oscillating model defines away a 'beginning,' and hence has the same philosophical tidiness as the Steady State Theory. A potential flaw in such a model is the production of heavy elements (e.g., C, N, O, Fe) in stars in each cycle – after many cycles would the Universe not be full of heavy elements? To cleanse the Universe, a high temperature state is needed at each bounce to photo-disintegrate the complex nuclei. Dicke and his colleague Jim Peebles worked out the necessary temperature, and estimated its present value about 10^{10} yrs after the most recent bounce, obtaining $T_0 \approx 10$ K. They also inde-

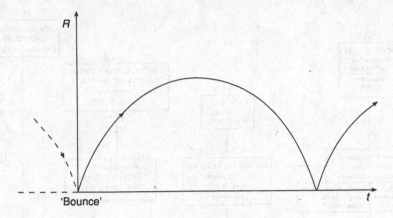

Fig. 2.3 A cyclic or 'bouncing' cosmological model. Heavy elements made in stars in each cycle need to be broken up again during the high density 'bounce' – and that requires high temperature as well as high density as $R \to 0$.

pendently rederived the synthesis of light elements in the first few minutes of expansion *after* a 'bounce.' Finally, they planned a test of the Hot Big Bang; Dicke suggested building a specialized, small radio telescope to detect the radiation left over from the hot, early phase of the Universe. Peter Roll and David Wilkinson took up that task in 1964.

In the meantime, members of the Princeton group had been presenting their ideas and calculations at scientific meetings. One of the founding fathers of radio astronomy in the U.S.A., Bernie Burke, heard (from a colleague, Ken Turner) of a talk by Jim Peebles about the Hot Big Bang, the importance of radiation in the early Universe, and the possibility of detecting that radiation now. Later, talking with Arno Penzias, Burke heard about the 3.5 K 'excess noise' problem encountered at Bell Labs, and referred to the ideas of Dicke and his colleagues. Was there a connection? Intent on following up any possible explanation of the 3.5 K 'excess,' Penzias telephoned Dicke. As it happened, the call came while the Princeton group were gathered together for a weekly 'brown bag' lunch in Dicke's office. As Dave Wilkinson relates that fateful telephone call, Dicke mostly listened, but occasionally repeated crucial phrases – 'horn antenna,' 'liquid He calibrator,' 'excess noise,' or the like – phrases familiar because the Princeton group was of course in the midst of building a system rather similar to the Bell Labs apparatus. At the end, Dicke hung up, turned to Peebles, Roll and Wilkinson, and said, 'Well boys, we've been scooped.' The CBR had been discovered.

The next day Dicke and his colleagues drove to Bell Labs, and came away convinced that the 'excess noise' was the relic of the Hot Big Bang they were seeking. The two groups agreed to publish adjacent papers in *The Astrophysical Journal Letters* (both reprinted here as appendices). The letter by Penzias and Wilson makes no reference to a Hot Big Bang origin of the radiation they detected. 'Arno [Penzias] and I were careful to exclude any discussion of the cosmological theory of the origin of background radiation from our letter because we had not been involved in any of that work. We thought, furthermore, that our measurement was independent of the theory and might outlive it. We were pleased that the mysterious noise appearing in our antenna had an explanation of any kind, especially one with such significant cosmological implications. Our mood,

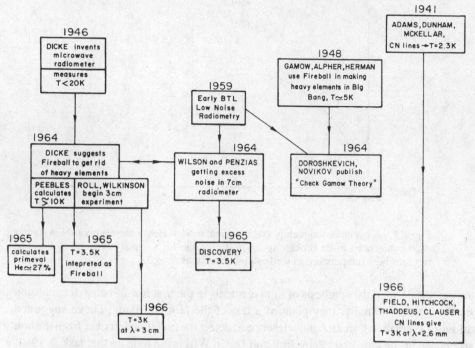

Fig. 2.4   Paths (and detours and roadblocks) to the discovery of the CBR (from Wilkinson and Peebles, 1983).

however, remained one of cautious optimism for some time' (Wilson, 1979).

This complicated tale of missed opportunities and fortuitous connections has been told several times, both by participants in the events (Wilson, 1979, 1983; Wilkinson and Peebles, 1983; Alpher and Herman, 1988) and by others (Ferris, 1977; Weinberg, 1977).

The story is summarized graphically in the crazy quilt diagram shown in fig. 2.4 (taken from Wilkinson and Peebles, 1983). How few connections there were between the physicists, the radio astronomers, and the optical spectroscopists! In saying that, I do not at all mean to diminish the achievements of those who predicted, detected and interpreted the CBR. Virtually all of classical Hot Big Bang cosmology (Sections 1.3.5 and 1.6) was presaged in the papers of Gamow, Alpher, Follin and Herman in the 1940s and on into the 1950s. In their work on CN absorption lines, optical astronomers had unknowingly measured T_0 to remarkable accuracy by 1941. Dicke and his colleagues would certainly have independently found the cosmic background radiation that they had predicted by 1965.

2.6   Confirming the Hot Big Bang

Dicke's explanation of the 'excess noise' of Penzias and Wilson was that it was a thermal cosmic background, a relic of a Hot Big Bang. Several other, non-Big-Bang, explanations were quickly proposed, however (e.g., Kaufman, 1965; Hoyle and Wickramasinghe, 1967; Layzer, 1968; and Wolfe and Burbidge, 1969).* Was the 3.5 K 'excess' truly a relic of a hot phase in the early Universe?

---

* These early, non-Big-Bang scenarios are reviewed in Partridge (1969).

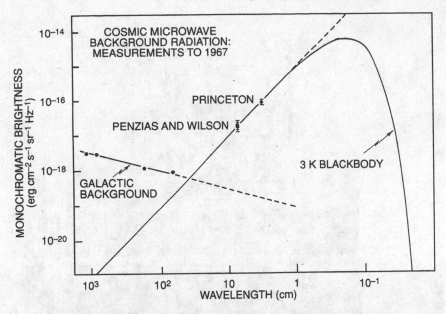

Fig. 2.5 A second measurement of the CBR at 3.0 cm (Roll and Wilkinson, 1966) confirms the discovery of a thermal background and refines the value for T_0.

It was soon recognized that there were two crucial tests of the Big Bang hypothesis. First, relic radiation from a Hot Big Bang would have a thermal (Planck) spectrum. Second, relic radiation would be more or less isotropic (equally intense in all directions). In particular, its angular distribution ought not to be correlated with local sources of radio emission, such as the solar system or the plane of our Galaxy. Within a few years of its discovery, the CBR had passed both these tests. Let us review the early observations briefly, then look at how they undermined rival explanations of the observed background.

2.6.1 Confirming the thermal spectrum

By late 1965, a few months after the letters announcing the CBR appeared in *The Astrophysical Journal*, Roll and Wilkinson had completed their measurements of the CBR intensity. Fortunately, the Princeton group had chosen a wavelength different from that used by Penzias and Wilson: $\lambda = 3$ cm, selected because waveguide components were readily available (3 cm is a radar band). Roll and Wilkinson (1966, 1967) carefully designed their instrument to allow a good absolute calibration, since there is no way to 'look away' from an isotropic background. The measured intensity at $\lambda = 3$ cm corresponded to thermal emission at 3.0 ± 0.5 K, in excellent agreement with the Bell Labs result.

These two measurements established a rough spectrum for the CBR. It was consistent with a thermal background at 3–3.5 K, but it was totally inconsistent with the possibility that the 'excess noise' was due to radio emission from our Galaxy (see fig. 2.5, taken from Roll and Wilkinson, 1966).

Before further new measurements of the CBR emerged, old ones were rediscovered: the CN measurements mentioned above. Field and Hitchcock (1966), Woolf (1966) and

Fig. 2.6   The first systematic measurements of the isotropy of the CBR were made with a 3.0-cm radiometer housed in a former pigeon coop atop the geology building at Princeton (see fig. 6.1 for a schematic of the instrument).

Thaddeus and Clauser (1966) in the U.S.A., and Shklovskii (1966) in the Soviet Union all independently recalled the excitation of these interstellar molecules. Field and Hitchcock (1966) went back to the old data and rederived limits on T_0 at $\lambda = 2.64$ mm: $2.7 \leq T_0 \leq 3.4$ K, again in agreement with a thermal spectrum. The CN results also firmly established that the CBR was not localized to the solar system; it at least filled the Galaxy.

Once the CBR was announced radio astronomers at several observatories joined in attempts to determine its spectrum, modifying existing receivers and/or devising new techniques to measure T_0. Penzias and Wilson themselves (1967) measured $T_0 = 3.2 \pm 1.0$ K at 21 cm. Astronomers in Cambridge, England, made measurements at about 21 cm and in the range 49–74 cm, obtaining 2.8 ± 0.6 K and 3.7 ± 1.2 K, respectively (Howell and Shakeshaft, 1966, 1967 – see Section 4.6). These long-wavelength results were in reasonable agreement with $T_0 \simeq 3$ K. Thus, by 1967, the observed spectrum was known to be thermal over a wavelength range of about 300. It seemed that the first test of the Big Bang hypothesis had been passed.*

---

* One false alarm was the 1968 report by Shivanandan, Houck and Harwit of a much higher flux at $\lambda \simeq 1$ mm than expected for a 3 K blackbody; subsequent measurements discussed in Section 4.8 show this result to have been in error.

2.6.2 Confirming the isotropy of the CBR

Half a year after Penzias's telephone call to Dicke, work was underway at Princeton to measure the angular distribution of the CBR across the sky. This experiment was my first involvement in CBR studies. I arrived at Princeton as a postdoc in the summer of 1965 to join Dicke's 'Gravity Group.' Four major observational programs were underway: work on a reflector to be left on the moon for laser ranging; continued effort on the Eötvös experiment (Dicke, 1961); measurements of the solar oblateness (Dicke and Goldenberg, 1967); and work on the CBR. I was first shown the solar oblateness experiment, which looked fearfully complicated. Next I was taken to see Roll and Wilkinson's instrument located in an ex-pigeon coop atop the geology building (fig. 2.6). It looked much less complex – waveguide, window screen and audio oscillators I could understand. So I chose to join Dave Wilkinson and work on the CBR, in particular on measurements of its angular distribution.

Recall that Penzias and Wilson had already established an upper limit of roughly 20% on any anisotropy in the CBR at 7.35 cm. We sought to improve this limit by roughly two orders of magnitude. We did so by building an instrument specifically designed to measure small *differences* in the CBR intensity, not its absolute value (see Section 6.2). By 1967, upper limits on large-scale anisotropy, on angular scales 15°–180°, had been set at $\Delta T/T_0 \lesssim (1\text{–}3) \times 10^{-3}$ (Partridge and Wilkinson, 1967; Wilkinson and Partridge, 1967). The second test, too, had been passed.

2.6.3 Confrontation with rival theories

By late 1965, it was thus clear that the 'excess noise' was not Galactic in origin. As we have noted, however, alternatives to a Big Bang origin certainly existed. One possibility explored in different variants by Hoyle and Wickramasinghe (1967) and Layzer (1968) was reemission by warm dust. In these models, stellar radiation (at predominantly ultraviolet or visual wavelengths) heated dust; the dust in turn reradiated with a spectrum characteristic of its temperature. If the dust lay at cosmological distances at a redshift z, its characteristic temperature would need to be about $3(z + 1)$ K to explain the observed CBR temperature of about 3 K. The major difficulty encountered by such models is the closely blackbody spectrum of the radiation. Unless the dust particles are very large, one expects their emissivity to decrease with increasing wavelength (a behavior later confirmed for far infrared emission by dust in our Galaxy). A wavelength dependent emissivity would in turn make the measured CBR temperature wavelength dependent, contrary to the observations.

Narlikar and Wickramasinghe (1967) suggested an ingenious way around this problem in an attempt to explain the microwave background in the context of Steady State Theory. If the reemission occurred primarily in *lines*, and the cosmological redshift of those lines is taken into account, a kind of sawtooth spectrum such as is shown in Fig. 2.7 would result. With a large enough number of properly chosen reemission lines, the data may be fit. As measurements at more and more wavelengths became available, however, this model looked less and less plausible.

The spectral measurements discussed in Section 2.6.1 convinced most cosmologists that the Big Bang hypothesis was to be preferred over dust models. Nevertheless, newer variants of these models continue to appear in the literature; these models are discussed

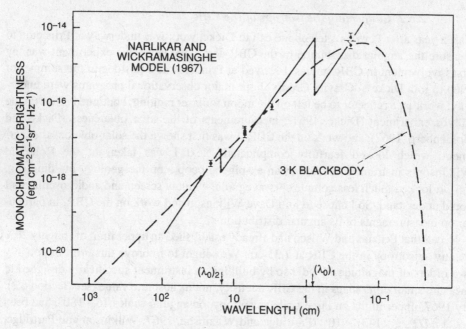

Fig. 2.7   An attempt to reconcile the Steady State Theory with early observations of the CBR. Narlikar and Wickramasinghe (1967) suggested that there might be *line* emission at rest wavelengths such as $(\lambda_0)_1$ and $(\lambda_0)_2$. The redshift then smears the lines into a spectrum which, as shown, varies as λ^{-3} for wavelengths $> \lambda_0$.

further in Section 5.3. In particular, a suggestion first made by Layzer (1968) that distant, dusty galaxies might produce the 3 K 'excess' was revived (by Bond, Carr and Hogan, 1986, among others) as an explanation for a reported submillimeter excess (see Sections 4.8 and 5.3).

In the mid-1960s it was known that many galaxies were powerful emitters of radio waves, and furthermore that the number density of radio sources appeared to increase strongly with redshift. Could the 'excess' flux discovered by Penzias and Wilson be nothing more than the summed emission of many galaxies at large distances from the Earth? Such a suggestion was made in 1969 by Wolfe and Burbidge. Given the spectral measurements, some novel mechanism for radio emission was required. Since the short-wavelength radio properties of galaxies were not well understood in the 1960s, that suggestion could not be ruled out. Another test of the radio source model was possible, however: checking the isotropy of the CBR on small angular scales (e.g., Hazard and Salpeter, 1969; Smith and Partridge, 1970). If the emitting galaxies are randomly distributed, we would expect fluctuations in their surface number density, and hence in the CBR intensity. Searches (Penzias, Schraml and Wilson, 1969; Boynton and Partridge, 1973) for small-scale isotropy specifically designed to test such a model showed that the number density of radio emitting galaxies required to explain the observed small-scale isotropy would substantially exceed the measured number density of all galaxies; galaxies cannot produce the *entire* 3 K excess. As we shall see in Chapter 8, however, emission from radio galaxies may well dominate the *fluctuations* in the CBR intensity at some wavelengths.

2.6.4 Early theoretical work

As observations supporting the Hot Big Bang hypothesis of Dicke *et al.* came in, attention turned away from attempts to explain the origin of the CBR to attempts to use it to unravel cosmological puzzles. If we accept that the CBR is indeed a relic of the Hot Big Bang, what can we learn from it? Some of the early responses to this question included the following.

1. Certain systematic distortions in the CBR spectrum would reveal energy-releasing processes at any epoch back to $z \simeq 10^6$ (Weymann, 1966; Zel'dovich and Sunyaev, 1969).

2. Measurements of the large-scale angular distribution of the CBR can determine the velocity of the Earth with respect to the bulk of the Universe (Peebles and Wilkinson, 1968) and can establish whether the expansion of the Universe is isotropic (Thorne, 1967; Novikov, 1968) and free of shear (Hawking, 1969).

3. Inhomogeneities in the distribution or velocity of matter in the Universe would introduce small-scale anisotropies or fluctuations in the CBR (Sachs and Wolfe, 1967; Ozernoi and Chernin, 1968; Silk, 1968). In particular, as Silk noted in his 1968 paper, mapping the CBR allows us to 'see' the distribution of matter at $z \simeq 1000$, just as the process of galaxy formation began.

4. If the Universe is full of thermal photons, high energy cosmic rays will interact with them; cutoffs in the spectra of γ rays and charged cosmic rays will result (e.g., Dicke and Pebbles, 1965; Felten, 1965; Jelley, 1966; Stecher, 1968; see also Section 5.5).

5. Finally, knowing T_0 allows us to calculate the fraction of ^{4}He emerging from the Big Bang (Wagoner, Fowler and Hoyle, 1967). These calculations in turn have given us another test of the Big Bang hypothesis, as we saw in Section 1.2.4.

All of these consequences – and others as well – are explored in more detail in Chapters 5 and 8, but it is worth noting how quickly the value of CBR observations was recognized (for a contemporary view, see Sciama, 1967), and how quickly they contributed to modern cosmology.

2.7 The past twenty years

A crude picture of the subsequent evolution of CBR studies is provided in fig. 2.8, in which I have plotted a rough count of the number of papers published on the CBR as a function of time. The initial burst of activity is clear; so too is a slackening of effort (or at least of published results) in the 1970s. What were the causes of this reduction in activity? In some part, I suggest, effort slackened because some of those involved in the early work reported in Section 2.6 were attracted away to other discoveries and problems of high energy astrophysics and cosmology. After all, the decade 1963–1973 was a bonanza for cosmologists – not just the CBR, but quasars, pulsars, gravity waves, nucleosynthesis, problems of galaxy formation, and faint radio source counts were all in the air. Some of us switched our allegiance to other fields permanently, some for a period of years. In larger part, we ran into technological limits. Conventional radio telescopes had been pushed to their limits in the 1960s (e.g., Conklin and Bracewell, 1967a, b; Howell and Shakeshaft, 1967); so had the first generation of small instruments built

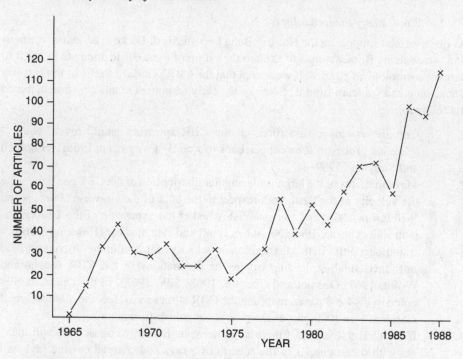

Fig. 2.8   The number of papers treating the CBR has grown rapidly, especially recently. These (approximate) numbers of CBR papers per year were culled from the *Astronomischer Jahresbericht* and *Astronomy and Astrophysics Abstracts*.

specifically to study the CBR (e.g., Partridge and Wilkinson, 1967; Wilkinson, 1967). The easy cream had been skimmed. What kept the field alive during the 1970s was in part activity on the theoretical side and in part the efforts of researchers designing *new* technology to bring to bear on the CBR. In this regard, I think particularly of those who introduced bolometric detectors to the field, Beckman, Melchiorri, Richards, Weiss and their many co-workers (see, e.g., Weiss, 1980). It was also in the early 1970s that CBR researchers began to employ balloons and high-flying aircraft to get above much of the Earth's atmosphere (e.g., Henry, 1971; Muehlner and Weiss, 1973; Smoot, Gorenstein and Muller, 1977).

By 1978 or 1979, new areas were opening up. Strong limits on large-scale polarization of the CBR had been established (Lubin and Smoot, 1979; Nanos, 1979) and the first attempt to use aperture synthesis to study the CBR on scales < 1′ was underway (Martin, Partridge, and Rood, 1980).

In the early 1980s a number of new facilities (e.g., the VLA; Section 3.5), new technologies (e.g., bolometers; Section 3.4.5), new observing techniques (e.g., the precision cold load of the 'White Mountain' collaboration; Section 4.3.1) and new experimental groups (in Japan, Italy, Canada and Great Britain as well as the U.S.A.) were beginning to have a major impact on the field. The experimental pace quickened, and progress was made on all the CBR parameters discussed in the remainder of this book. In 1990, in time for the 25th anniversary of the papers by Penzias and Wilson and Dicke *et al.*, the spectacular results of the Cosmic Background Explorer satellite (COBE) became available, confirming both the precisely thermal nature of the CBR spectrum and later its remarkable isotropy.

These experimental results will be treated in Chapters 4 and 6. First, however, for those not fully familiar with the jargon and techniques of radio astronomy (and particularly the special precautions needed when observing an isotropic background), I present a brief review of radio astronomy in Chapter 3. Other readers may wish to skip to the first of the chapters dealing directly with observations of the CBR itself, Chapter 4, which treats its spectrum.

References

Adams, S. W. 1941, *Ap. J.*, **93**, 11.
Alpher, R. A., and Herman, R. C. 1948, *Nature*, **162**, 774.
Alpher, R. A., and Herman, R. C. 1949, *Phys. Rev.*, **75**, 1089.
Alpher, R. A., and Herman, R. C. 1950, *Rev. Mod. Phys.*, **22**, 53.
Alpher, R. A., and Herman, R. C. 1988, *Phys. Today*, **41**, 24.
Bond, J. R., Carr, B. J., and Hogan, C. J. 1986, *Ap. J.*, **306**, 428.
Bondi, H. 1960, *Cosmology*, Cambridge University Press, Cambridge.
Boynton, P. E., and Partridge, R. B. 1973, *Ap. J.*, **181**, 243.
Burbidge, E. M., Burbidge, G. R., Fowler, W. A., and Hoyle, F. 1957, *Rev. Mod. Phys.*, **29**, 547.
Conklin, E. K., and Bracewell, R. N. 1967a, *Phys. Rev. Lett.*, **18**, 614.
Conklin, E. K., and Bracewell, R. N. 1967b, *Nature*, **216**, 777.
Dicke, R. H. 1961, *Scientific American*, December.
Dicke, R. H., and Goldenberg, H. M. 1967, *Phys. Rev. Lett.*, **18**, 313.
Dicke, R. H., and Peebles, P. J. E. 1965, *Space Sci. Rev.*, **4**, 419.
Dicke, R. H., Beringer, R., Kyhl, R. L., and Vane, A. V. 1946, *Phys. Rev.*, **70**, 340.
Doroshkevich, A. G., and Novikov, I. D. 1964, *Soviet Phys.*, *Doklady*, **9**, 111.
Felten, J. E. 1965, *Phys. Rev. Lett.*, **15**, 1003.
Fermi, E. 1949, *Phys. Rev.*, **75**, 1169.
Ferris, T. 1977, *The Red Limit*, W. Morrow, New York.
Field, G. B., and Hitchcock, J. L. 1966, *Phys. Rev. Lett.*, **16**, 817.
Gamow, G. 1946, *Phys. Rev.*, **70**, 572 [see erratum *ibid*, **71**, 273].
Gamow, G. 1952, *The Creation of the Universe*, Viking Press, New York.
Gamow, G. 1956, *Scientific American*, September.
Hawking, S. W. 1969, *Monthly Not. Roy. Astron. Soc.*, **142**, 129.
Hazard, C., and Salpeter, E. E. 1969, *Ap. J. (Lett.)*, **157**, L87.
Henry, P. S. 1971, *Nature*, **231**, 561.
Herzberg, G. 1950, *Molecular Spectra and Molecular Structure*, Van Nostrand, Princeton, New Jersey.
Howell, T. F., and Shakeshaft, J. R. 1966, *Nature*, **210**, 1318.
Howell, T. F., and Shakeshaft, J. R. 1967, *Nature*, **216**, 753.
Hoyle, F., and Wickramasinghe, N. C. 1967, *Nature*, **214**, 969.
Jelley, J. V. 1966, *Phys. Rev. Lett.*, **16**, 479.
Kaufman, M. 1965, *Nature*, **207**, 736.
Layzer, D. 1968, *Astrophys. Lett.*, **1**, 99.
Lubin, P. M., and Smoot, G. F. 1979, *Phys. Rev. Lett.*, **42**, 129.
Martin, H. M., Partridge, R. B., and Rood, R. T. 1980, *Ap. J. Lett.*, **240**, L79.
McKellar, A. 1941, *Publ. Dominion Astrophys. Obs.*, Victoria, **7**, 251.
Muehlner, D. J., and Weiss, R. 1973, *Phys. Rev. Lett.*, **30**, 757.
Nanos, G. P. 1979, *Ap. J.*, **232**, 341.
Narlikar, J. V., and Wickramasinghe, N. C. 1967, *Nature*, **216**, 43; *ibid*, **217**, 1235.
Novikov, I. D. 1968, *Soviet Astron. J.*, **12**, 427.
Ohm, E. A. 1961, *Bell System Technical J.*, **40**, 1065.
Ozernoi, L. M., and Chernin, A. D. 1968, *Soviet Astron. J.*, **11**, 907.
Partridge, R. B. 1969, *American Scientist*, **57**, 31.

Partridge, R. B. and Wilkinson, D. T. 1967, *Phys. Rev. Lett.*, **18**, 557.

Peebles, P. J. E., and Wilkinson, D. T. 1968, *Phys. Rev.*, **174**, 2168.

Penzias, A. A. 1979, *Science*, **205**, 549 (Nobel Lecture).

Penzias, A. A., and Wilson, R. W. 1967, *Science*, **156**, 1100.

Penzias, A. A., Schraml, J., and Wilson, R. W. 1969, *Ap. J. (Lett.)*, **157**, L49.

Robertson, H. P., and Noonan, T. W. 1968, *Relativity and Cosmology*, W. B. Saunders, Philadelphia.

Roll, P. G., and Wilkinson, D. T. 1966, *Phys. Rev. Lett.*, **16**, 405.

Roll, P. G., and Wilkinson, D. T. 1967, *Annals of Phys.*, **44**, 289.

Ryle, M. 1968, *Ann. Rev. Astron. Astrophys.*, **6**, 249.

Sachs, R. K., and Wolfe, A. M. 1967, *Ap. J.*, **147**, 73.

Sandage, A. R. 1961, *Ap. J.*, **133**, 355.

Sandage, A. R. 1970, *Physics Today*, February.

Sciama, D. W. 1967, *Scientific American*, September.

Shivanandan, K., Houck, J. R., and Harwit, M. O. 1968, *Phys. Rev. Lett.*, **21**, 1460.

Shklovskii, I. S. 1966, *Astronomicheskii Tsircular*, no. 364.

Silk, J., 1968, *Ap. J.*, **151**, 459.

Smith, M. G., and Partridge, R. B. 1970, *Ap. J.*, **159**, 737.

Smoot, G. F., Gorenstein, M. V., and Muller, R. A. 1977, *Phys. Rev. Lett.*, **39**, 898.

Stecher, F. W. 1968, *Phys. Rev. Lett.* **21**, 1016.

Suzuki, S. 1928, *Proc. Phys. Math. Soc. Japan*, **13**, 277.

Tayler, R. J. 1966, *Rep. Prog. Phys.*, **29**, 489.

Thaddeus, P. 1972, *Ann. Rev. Astron. Astrophys.*, **10**, 305.

Thaddeus, P., and Clauser, J. F. 1966, *Phys. Rev. Lett.*, **16**, 819.

Thorne, K. S. 1967, *Ap. J.*, **148**, 51.

Tolman, R. C. 1922, *J. Amer. Chem. Soc.*, **44**, 1902.

Wagoner, R. V., Fowler, W. A., and Hoyle, F. 1967, *Ap. J.*, **148**, 3.

Wall, J. V., Chu, T. Y., and Yen, J. L. 1970, *Australian J. Phys.*, **23**, 45.

Weinberg, S. 1972, *Gravitation and Cosmology*, J. Wiley, New York.

Weinberg, S. 1977, *The First Three Minutes*, Basic Books, New York.

Weiss, R. 1980, *Ann. Rev. Astron. Astrophys.*, **18**, 489.

Weymann, R. 1966, *Ap. J.*, **145**, 560.

Wilkinson, D. T. 1967, *Phys. Rev. Lett.*, **19**, 1195.

Wilkinson, D. T., and Partridge, R. B. 1967, *Nature*, **215**, 719.

Wilkinson, D. T., and Peebles, P. J. E. 1983, in *Serendipitous Discoveries in Radio Astronomy*, ed. K. Kellermann and B. Sheets, N.R.A.O., Green Bank, West Virginia.

Wilson, R. W. 1979, *Science*, **205**, 866 (Nobel Lecture).

Wilson, R. W. 1983, in *Serendipitous Discoveries in Radio Astronomy*, ed. K. Kellermann and B. Sheets, N.R.A.O., Green Bank, West Virginia.

Wolfe, A. M., and Burbidge, G. R. 1969, *Ap. J.*, **156**, 345.

Woolf, N. J. 1966, private communication.

Zel'dovich, Ya. B. 1965, *Adv. Astron. Astrophys.*, **3**, 242.

Zel'dovich, Ya. B., and Sunyaev, R. A. 1969, *Astrophys. Space Science*, **4**, 301.

3

Radio astronomy

The very first astronomical signal at radio wavelengths, by happy coincidence, was also detected at the Bell Telephone Laboratories; in 1932, Karl Jansky detected at 15 m wavelength radio emission, which he correctly identified as coming from the Galactic plane. Astronomers paid little attention. Observational radio astronomy did not really come into its own until after World War II (see, e.g., Hey, 1973, and Sullivan, 1984). It is now recognized as a powerful adjunct to optical astronomy, particularly in the study of low density cosmic matter and of energetic objects and phenomena such as quasars and the collimated jets seen in radio galaxies. We will look very briefly at radio sources later in this chapter, but most of it will be devoted to the tools and techniques of observational radio astronomy. Chapter 3 is designed to introduce the more specialized radio astronomical techniques used in studying the CBR; it is not intended to be a complete introduction to radio astronomy. For further details, readers may want to consult one or more of the following texts: Kraus (1986); Rohlfs (1986); and Christiansen and Högbom (1985). Interferometry is very fully treated by Thompson, Moran and Swenson (1986), and radio sources by Pacholczyk (1970) and Verschuur and Kellermann (1988), among others. The treatment of radio astronomy in this book is closest to the work of Rohlfs.

3.1 Fundamental radio astronomical measurements

Radio astronomy from the Earth's surface is possible because the atmosphere is transparent over a wide range of wavelengths from about 20 m to about 1 mm (the radio 'window'). At $\lambda > 20$ m, the ionosphere becomes reflective, blocking astronomical signals; at $\lambda \lesssim 1$ mm, absorption lines and bands of the constituent molecules of the atmosphere rapidly increase the opacity of the atmosphere (see Section 4.4.3). Within the radio window, we will be primarily concerned with radio emission of 10 cm $\gtrsim \lambda \gtrsim 1$ mm: the *microwave* region of the spectrum.

Radio telescopes are employed to measure the energy flux (in W m^{-2}) or the surface brightness of astronomical sources. For reasons to be explained further in Section 3.4 below, radio measurements are generally confined to a narrow frequency or wavelength band. Hence it is conventional in radio astronomy to use *flux density*, generally denoted by S, and defined as the energy flux per square meter *per hertz*. To honor the founder of radio astronomy, the unit of flux density is the Jansky (Jy) $\equiv 10^{-26}$ W m^{-2} Hz^{-1}. Since flux density, S, is a measured quantity, it clearly depends on the distance to a source as well as its luminosity and spectrum. When studying extended sources – which the CBR certainly is – surface brightness B_v is a more useful parameter. B_v is also specified in a restricted wavelength or frequency interval with the usual units being $\mathrm{W\ m^{-2}\ sr^{-1}\ Hz^{-1}}$.

Neither the CBR nor emission from radio sources is strongly polarized. Nevertheless, the polarization of their emission frequently provides useful information about the nature of radio sources, and the inputs to radio telescopes are very frequently polarized. For both these reasons, I describe briefly the notation used to describe polarized electromagnetic radiation. Here, for simplicity, I will consider only monochromatic plane waves (for further details, see Rohlfs (1986) or Thompson *et al.* (1986); note that the conventions used in radio astronomy differ from some used by optical astronomers). The polarization of a radio wave is fully specified by four *Stokes parameters*. One, I, gives the total flux; a second, V, measures the degree of circular polarization of the wave; and two others, Q and U, measure the amplitude and position angle of linear polarization. More formally, consider a wave of frequency v propagating in the $+z$ direction. At a particular value of z, the electric fields in two directions orthogonal to the direction of propagation may be written

$$E_x(t) = \varepsilon_x(t) \cos (2\pi v t + \delta_x),$$
$$E_y(t) = \varepsilon_y(t) \cos (2\pi v t + \delta_y).$$

If we define the phase difference $\delta = \delta_x - \delta_y$, we see that E_x and E_y, and hence the polarization of the wave, depend on only three parameters, the amplitudes ε_x and ε_y, and δ. Hence we expect only three of the Stokes parameters to be independent; as we shall see, $I^2 = Q^2 + U^2 + V^2$ for a monochromatic wave.

In terms of the expressions for E_x and E_y, the Stokes parameters are defined as

$$
\begin{aligned}
I &= \varepsilon_x^2 + \varepsilon_y^2 \\
Q &= \varepsilon_x^2 - \varepsilon_y^2 \\
U &= 2\,\varepsilon_x\,\varepsilon_y \cos \delta \\
V &= 2\,\varepsilon_x\,\varepsilon_y \sin \delta
\end{aligned}
\tag{3.1}
$$

For a purely circular polarized wave, $\varepsilon_x = \varepsilon_y$, and in the convention used by radio astronomers $\delta = \pi/2$ for right-hand polarization (and $-\pi/2$ for left-hand). For right- (or left-) hand polarization, then, $Q = U = 0$, and the fractional polarization $V/I = 100\%$ (or -100%). If $\varepsilon_y = 0$, the wave is 100% linearly polarized in the x direction, and so on. In general, if $\varepsilon_x \neq \varepsilon_y \neq 0$, the wave is said to be elliptically polarized.

Finally, if we drop the artificial assumption of a single purely monochromatic wave, we can allow for the possibility of *partial polarization* of flux from a source (see Rohlfs, 1986, for details). Here $I^2 > Q^2 + U^2 + V^2$, and the degree of polarization p is given by

$$p = (Q^2 + U^2 + V^2)^{1/2}/I.$$

For most astronomical sources, $p \lesssim 0.1$.

3.2   The Planck (blackbody) spectrum

As we have seen in earlier chapters, the CBR is expected to have an approximately thermal, blackbody spectrum. We reproduce here for future reference some of the characteristics of such a spectrum. First,

$$B_v = \frac{2hv^3}{c^2} \frac{1}{\exp\,(hv\,/\,kT) - 1} \ \mathrm{W\ m^{-2}\ sr^{-1}\ Hz^{-1}}, \tag{3.2}$$

where v is the frequency of observation, T the temperature and c, k and h the velocity of light, Boltzmann's constant, and Planck's constant, respectively. From this result, both the energy density u and the photon number density n may be found. For a blackbody radiation field of temperature T, they are

$$u = \frac{8\pi^5 \, k^4}{15 \; c^3 \, h^3} \; T^4 \equiv aT^4 = 7.564 \times 10^{-16} \, T^4 \; J \, m^{-3}, \tag{3.3}$$

$$n = \frac{30\zeta(3)a}{\pi^4 k} \; T^3 = 2.03 \times 10^7 \, T^3 \; m^{-3}, \tag{3.4}$$

with a as given above; $\zeta(3)$ is the Riemann Zeta function numerically equal to 1.20. Substituting $T \simeq 3$ K in (3.4) shows that each cubic meter of the Universe contains about 5×10^8 CBR photons.

The Planck spectrum (3.2) has a maximum; by differentiating the right-hand side, one may derive Wien's Law, connecting the temperature and the wavelength λ_m where B_v is a maximum:

$$\lambda_m \, T = \text{constant.}$$

For B_v expressed in the usual radio astronomical units of W m^{-2} sr^{-1} Hz^{-1}, the constant is 5.099×10^{-3} m K; if surface brightness is expressed in wavelength terms (e.g. W m^{-2} sr^{-1} Å^{-1}), the constant becomes 2.898×10^{-3} m K. For $T = 3$ K, we see that λ_m is a few millimeters.

3.2.1   The Rayleigh–Jeans law and antenna temperature

Far from the peak of the Planck function, at long wavelengths where quantum effects are small, B_v takes on a power law form. That is shown easily by expanding $e^{hv/kT}$ for small v:

$$B_v = \begin{cases} \dfrac{2 \, v^2 \, kT}{c^2} & \text{for } v \ll \dfrac{kT}{h} \\[2mm] \dfrac{2 \, kT}{\lambda^2} & \text{for } \lambda \gg \dfrac{ch}{kT} \end{cases}. \tag{3.5}$$

This λ^{-2} dependence is the Rayleigh–Jeans law, and the wavelength region in which it applies is called the Rayleigh–Jeans (R–J) region. Since the CBR itself has $T_0 \simeq 3$ K, it is reasonable to assume that all radio sources will have characteristic temperatures above 3 K. For source temperatures above 3 K, the R–J region extends to wavelengths as short as about 1 cm; thus most radio observations of thermal radio sources occur in the R–J region of their spectra, where B_v is proportional to λ^{-2} and is also directly proportional to T.

As a consequence, temperature is frequently used in radio astronomy as a measure of surface brightness: specifically, *brightness temperature* in the R–J region is defined by

$$T_B \equiv \frac{\lambda^2 \, B(v)}{2k}. \tag{3.6}$$

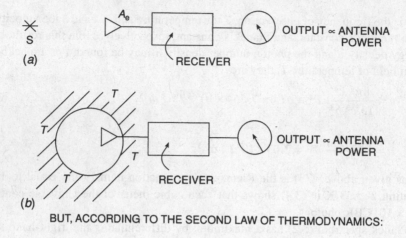

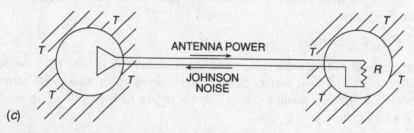

BUT, ACCORDING TO THE SECOND LAW OF THERMODYNAMICS:

Fig. 3.1 Defining antenna temperature. In (a), an antenna of effective area A_e is exposed to a source of flux density S. In (b) the same antenna is surrounded by a blackbody emitter at temperature T. The output power is the same when $T = SA_e/2k$. In (c), the second law of thermodynamics ensures that output antenna power and the Johnson noise are the same, hence establishing the proportionality of output power and T.

3.2.2 Antenna temperature and thermodynamic temperature

The use of temperature as a measure of intensity has been extended from surface brightness measurements to measurements of flux density. Consider, as shown in fig. 3.1, a simple radio antenna receiving radiation from a distant source. It will produce some energy flow p to the receiver. Now, following Dicke (1946), imagine surrounding the same antenna with a blackbody emitter, then raising its temperature T until the energy flow to the receiver is again equal to p. The temperature needed is defined as the *antenna temperature* T_A of the source. The antenna temperature of a source of flux density S is easily shown (e.g., Kraus, 1986) to be

$$T_A = \frac{SA_e}{2k},\tag{3.7}$$

where A_e is the effective area of the antenna in square meters (discussed further in Section 3.3 below). $T_A \propto A_e$ because a larger antenna collects more flux; it is important to recall that T_A, unlike flux density S, depends on the properties of the antenna as well as of the source.

Both brightness temperature and antenna temperature are derived from the Rayleigh–Jeans law, eqn. (3.5). They are thus equivalent to true thermodynamic temperature T only if $h\nu/kT \ll 1$. Since the temperature of the CBR is only about 3 K, this inequality holds only for $\nu \lesssim 30$ GHz or $\lambda \gtrsim 1$ cm. If measurements made at wavelengths $\lesssim 1$ cm are expressed in brightness or antenna temperature, they must then be corrected to give thermodynamic temperature. From (3.2) and (3.6) we see that the thermodynamic temperature T is given by

$$T = T_A \left(\frac{e^x - 1}{x} \right),$$ (3.8)

where for convenience we have written $x = h\nu/kT$; note that we need to know T itself to calculate the correction factor in parentheses.

Measurements of the CBR spectrum at short wavelengths, if expressed in antenna or brightness temperature, require large corrections to give T_0, the *thermodynamic* temperature of the CBR. For instance, if $T_0 = 3$ K, the measured brightness temperature of the background will be only 1.21 K at $\lambda = 3$ mm.

For $h\nu/kT = x$ not much larger than 1, an expansion of (3.8) gives a useful approximate relation between T and T_A:

$$T \approx T_A \left(1 + \frac{x}{2} + \frac{x^2}{6} \right).$$ (3.9)

In Chapters 6 and 7, we will be interested in measurements of fluctuations in the temperature of the CBR. Their amplitude will depend on whether we express our results in thermodynamic or antenna temperature*. By differentiating (3.8) with respect to T we find

$$\Delta T = \Delta T_A \left(\frac{(e^x - 1)^2}{x^2 e^x} \right).$$ (3.10)

Hence

$$\frac{\Delta T}{T} = \frac{\Delta T_A}{T_A} \left[\frac{e^x - 1}{x e^x} \right].$$ (3.11)

A useful approximate relation similar to eqn. (3.9) may easily be derived from eqn. (3.10); for $x < 1$,

$$\Delta T \approx \Delta T_A \left(1 + \frac{x^2}{12} \right).$$ (3.12)

3.3 Antennas and beam patterns

Radio telescopes are designed to receive radio waves from a limited solid angle of the sky. To a first approximation, that solid angle Ω is determined by diffraction theory; in this section we will explore the relation between the size of an antenna and Ω. Although some radio telescopes have cylindrical or other non-radial symmetry†, most antennas are paraboloids like the ones shown in fig. 3.2. For simplicity, we will consider only antennas with radial symmetry here.

---

* In the past this point has caused some confusion, and erroneous values of $\Delta T/T_0$ have resulted – see Chapters 6 and 7.
† One instance is the Soviet RATAN telescope mentioned briefly in Section 7.5.

Fig. 3.2   Parabolic antennas typical of the sort used in radio astronomy. In this case, they are the 25 m antennas of the Very Large Array in New Mexico (from the National Radio Astronomy Observatory, with permission). Note secondary reflectors (small convex surfaces near vertices of quadrupeds).

3.3.1 Reciprocity

Let us begin by making use of a very general result drawn from antenna theory: the reciprocity theorem (derived, for instance, by Rohlfs, 1986). This theorem establishes the equivalence between an antenna *receiving* radio waves from the sky and the same antenna *broadcasting* into the sky. Fig. 3.3 shows an antenna directing radio waves towards the sky; $P(\theta, \phi)$ specifies the angular distribution of the radiated power. The angular dependence of the response of the same antenna when used to receive radio waves from distant sources will also be given by $P(\theta, \phi)$. This equivalence gives rise to the use of the word 'beam' when referring to the response of a radio telescope, even when it is receiving, not transmitting, radiation.

3.3.2 Diffraction, beam size and effective area

Now let us see how $P(\theta, \phi)$ is determined. The purpose of the large parabolic reflectors used in radio telescopes is to gather up radio waves and bring them to a focus. Hence the reflector may be considered to function as a 'lens' of the same diameter d, as shown schematically in fig. 3.4. This imaginary lens, in turn, may be treated as a diffracting aperture. Diffraction will broaden the image of a distant point source by an angle $\theta \simeq \lambda/d$; it will also produce a diffraction pattern having secondary maxima as well as a central peak. The angular dependence of the diffracted intensity is sketched in fig. 3.5. For typical radio telescopes of 10–30 m diameter operating at $\lambda \simeq 3$–10 cm, we see that the characteristic scale of the diffraction pattern will be $\theta \simeq 3$–30′. If the antenna is radially symmetric, this pattern will be radially symmetric about the optical axis.

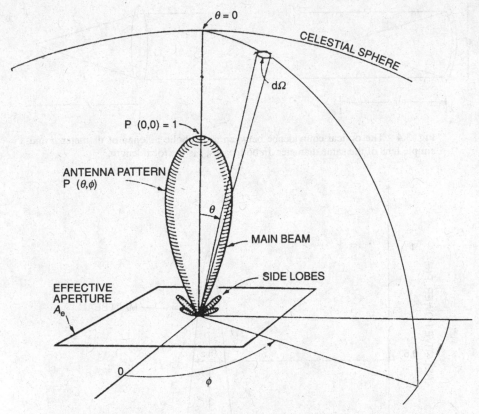

Fig. 3.3   An antenna of effective area A_e broadcasting power into the sky. The radiated power into a small solid angle $d\Omega$ at (θ, ϕ) is $\propto P(\theta, \phi)$. Note that the power is not broadcast isotropically, but is concentrated in the direction of the optical axis, coincident with $\theta = 0°$ here. (Adapted from Kraus, 1986.)

This heuristic argument has omitted many details, but it does establish why an antenna does have a well defined beam – the central diffraction maximum. It also connects the beam size, θ, with the antenna diameter. In radio astronomical jargon, this central maximum is referred to as the *main beam*; the secondary diffraction maxima are referred to as *side lobes*.

Before refining this simple heuristic argument, let us define some other terms that will be used in this book. First, it is useful to define a *normalized* power pattern as

$$P(\theta, \phi) = \frac{P(\theta, \phi)}{P(0, 0)} . \tag{3.13}$$

Then it is easy to see that the total solid angle Ω of the beam of an antenna is just

$$\Omega = \int_0^{2\pi} \int_0^{\pi} P(\theta, \phi) \sin\theta \, d\theta \, d\phi \tag{3.14}$$

This value will in general be larger than the solid angle of the main beam alone, because of the presence of side lobes. To find the main beam solid angle Ω_M alone, we need to integrate only to the first null of the function $P(\theta, \phi)$, that is over the range $0 \le \theta \le \theta_N$ (if the beam is radially symmetric):

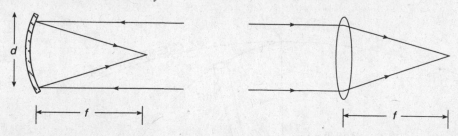

Fig. 3.4   The optical equivalence between a parabolic antenna of diameter d and a simple lens of the same diameter. In both cases, f is the focal length.

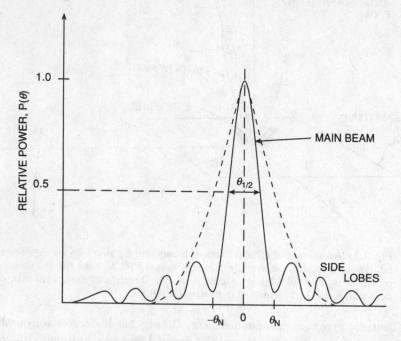

Fig. 3.5   Sketch of the diffraction pattern $P(\theta)$ of an aperture of diameter d. The central maximum, of angular width $\sim \lambda/d$ is referred to as the 'main beam,' and the secondary maxima as 'side lobes,' in radio astronomy. Also shown is $\theta_{1/2}$, the full width at half maximum, and the first nulls, θ_N. The dotted line shows the effect of taper on $P(\theta)$ – see Section 3.3.3.

$$\Omega_m = 4\pi \int_0^{\theta_N} P(\theta) \sin\theta \, d\theta.$$

(3.15)

The main beam efficiency of a telescope is then

$$\varepsilon_b = \frac{\Omega_M}{\Omega}.$$

(3.16)

Finally, we may specify the angular size of the main beam more precisely by taking the width of the pattern where $P(\theta, \phi)$ has decreased to 1/2; this is the *full width at half maximum*, often abbreviated FWHM and written here as $\theta_{1/2}$ (see fig. 3.5). If the antenna, and hence the beam, have exact radial symmetry, only a single value of the FWHM is needed; in more complicated cases, two orthogonal values of $\theta_{1/2}$ may be needed.

In actually performing the integration shown in eqn. (3.14) or eqn. (3.15), it is some-times useful to replace the spherical coordinates (θ, ϕ) with two Cartesian ones (ξ, η) in the plane of the sky and perpendicular to the optical axis of the beam, as shown in fig. 3.19.* That substitution makes it easier to find Ω_M for one simple and useful case – when $P(\theta, \phi)$ for $\theta \leq \theta_N$ can be reasonably approximated by a two-dimensional Gaussian. Treating $P(\theta, \phi)$ as a Gaussian is often not a bad approximation. Then

$$\Omega_M = \int_{-\infty}^{\infty} \int_{-\infty}^{\infty} e^{-(\xi/\alpha)^2} e^{-(\eta/\beta)^2} d\xi \, d\eta = \pi\alpha\beta.$$

Note that we have changed the limits of integration since we are not interested in side lobes. It is left as an exercise to show that, for a symmetrical Gaussian beam,

$$\Omega_M = \frac{\pi}{4 \ln 2} \theta_{1/2}^2 = 1.13 \, \theta_{1/2}^2 . \tag{3.17}$$

Let us make one final use of this rough diffraction model to relate the area of an antenna to Ω_M. Roughly, as we have seen, $\theta_{1/2} \simeq \lambda/d$, where d is a characteristic dimen-sion of the antenna. Thus $\Omega_M \propto \theta_{1/2}^2 \propto \lambda^2/d^2$ or

$$\Omega_M A \propto \lambda^2. \tag{3.18}$$

We can estimate that the constant of proportionality will be of order unity but we cannot calculate it exactly from the simple model assumed so far.

Instead, let us take a different tack, and introduce a quantity A_e, the *effective area* (or effective aperture) of the antenna. A_e is defined operationally by imagining a source of flux density S illuminating the antenna, as in fig. 3.1. If the power per unit bandwidth reaching the output of the antenna is p, then $A_e \equiv p/S$. A_e obviously can not be larger than the geometrical aperture of the antenna. In real radio telescopes, it is often 10–30% smaller, so that the *aperture efficiency* $\varepsilon_A \equiv A_e/A_{geom}$ is around 70–90%.†

A very general, thermodynamic proof developed by Pawsey and Bracewell (1954) may be used to show that

$$\Omega A_e = \lambda^2, \tag{3.19}$$

in agreement with our more approximate model, now but exact.

3.3.3 Some practical considerations

So far, we have treated radio antennas very abstractly. Let us now turn to some practi-cal details of working radio telescopes, like the ones shown in fig. 3.2. In some (e.g., the giant Arecibo dish at some wavelengths), radio waves reflected by the antenna are col-lected directly, at the *prime focus* of the antenna (fig. 3.6(a)). Usually, however, a second reflecting surface is used, as shown in fig. 3.6(b). Depending on whether the second surface is placed before or after the focal point of the primary reflector, a telescope of Cassegrain or Gregorian design results.

At the focus, in either design, a small horn antenna or other antenna is used to collect the radio waves and channel them to the receiver. At centimeter wavelengths, these small antennas frequently resemble the corrugated horns described in the following

---

* Clearly, this substitution is possible only for small θ.
† I have included the (usually small) Ohmic losses in antenna reflectivity in ε_A; not all authors do so. See, for instance, Kraus (1986) for more detail.

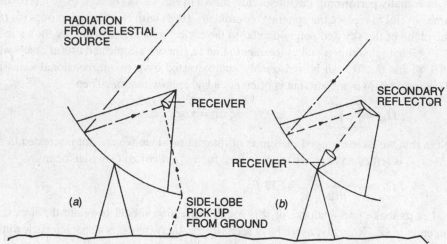

Fig. 3.6 Two conventional designs for radio telescopes: (*a*) prime focus; (*b*) Cassegrain or Gregorian. Pickup in the side lobes of the small 'feed' antenna is shown schematically for case (*a*).

section (see Christiansen and Högbom, 1985; or Rohlfs, 1986, for further details). Their role can best be understood by using the reciprocity theorem and treating the telescope as a transmitting system. The small antenna at the focus then 'feeds' or illuminates the main antenna* either directly (prime focus) or after a reflection from the secondary (Cassegrain or Gregorian). The power pattern of this small antenna is adjusted until its solid angle is approximately equal to that subtended by the primary or secondary reflector. It is usual, however, to apply a grading or taper; that is, to decrease the illumination of the edge of the primary reflector. This has the effect of reducing the amplitude of the side lobes, but at the expense of some loss in aperture efficiency. The use of taper is particularly important in prime focus telescopes, where radiation from the ground can enter the small 'feed' antenna directly around the edge of the primary reflector (as shown in fig. 3.6) unless the solid angle of the beam of the feed horn is smaller than the solid angle subtended by the primary.

A frequently used observational technique in radio (and some other branches of) astronomy is beam switching, rapid alternation between two closely adjacent patches of the sky, one containing the source of interest, the other normally nominally blank (fig. 3.7). This technique thus allows *comparative* measurements to be made. Beam switching may be accomplished by mounting two similar collecting horns on either side of the focus of the telescope, or by rocking the secondary reflector through a small angle.

3.3.4 Horn antennas

In some radio astronomical applications, where particularly low side-lobe levels are required and angular resolution is not a major concern, no reflector at all is used. Instead the radio waves are collected directly with a horn; this was the scheme used by Penzias and Wilson in their first observations of the CBR (indeed, the CBR signal was discernible in part because of the low side-lobe pickup of their horn antenna).

---

* Hence the radio astronomers' jargon for such a small antenna: a 'feed horn.'

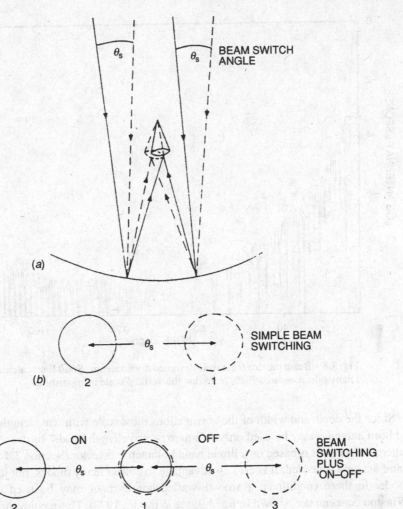

Fig. 3.7   Beam switching. In (a) beam switching is accomplished by modulating the
effective position of the receiver in the focal plane of a prime focus telescope. The
resulting pattern on the sky is shown in (b). In (c) the beam pattern resulting from
'on–off' observations is shown. This technique is described and illustrated in Section
7.2.1. In both cases, the source would be located at position 1.

The opening of a horn antenna may, as in Section 3.3.2, be treated as a diffracting
aperture. Hence side lobes will be present. These may be reduced to a very low level
($\lesssim 10^{-3}$ of P(0, 0); see for instance Otoshi and Stelzried, 1975; Mandolesi *et al.*, 1984)
by cutting grooves in the inner wall of the horn, as shown in fig. 4.6 later. These grooves
or *corrugations* decrease the current flowing along the walls of the horn, and thus lessen
its response at its rim (corrugated horns are fully described in Clarricoats and Olver,
1984). The result is to decrease the effective aperture (as with tapered illumination) and
hence to increase Ω_M. Corrugation also makes the beam more symmetrical. The main
effect of – and reason for – corrugations, however, is to reduce side lobes. Fig. 3.8 shows
how well side lobes can be suppressed in a carefully designed horn antenna.

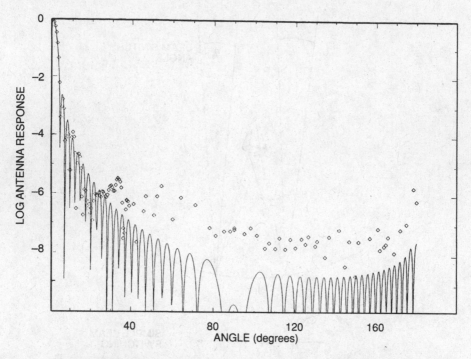

Fig. 3.8   Beam pattern for a well-designed horn antenna. Solid line: calculated pattern. Diamonds: measured $P(\theta)$. Note that the vertical scale is logarithmic.

Since the depth and width of the corrugations must scale with wavelength, corrugated horn antennas may be used only in a narrow wavelength band.* In radio telescopes where the receiver is based on a broad band bolometric detector (Section 3.4.5), a broad band antenna is needed. It is also an advantage to collect many modes, not just a single mode. In these conditions, a smooth-walled concentrator may be used (such as a Winston concentrator, shown in fig. 3.9; see Winston, 1970). These concentrators work quasi-optically; that is, they obey the laws of geometrical optics. As a consequence, the resulting beam pattern may be substantially larger than given by diffraction theory (eqn. (3.19)). Since bolometers and concentrators are used primarily at very short wavelengths, quite good angular resolution is still possible even if eqn. (3.19) breaks down.

3.4   Receivers

Once the flux from a source has been collected by the antenna of a radio telescope, the next step is to convert it to a measurable analog or digital signal. This is the role of the *receiver*. An ideal receiver would produce an amplified output signal exactly proportional to the input power $p = SA_e$ from the antenna; in other words its *gain* (amplification factor) would be linear. Radio astronomical receivers with very nearly linear gain can be constructed. It is less easy to control small *time* variations in the gain, however. Gain variations, as we shall see, are only one example of the many sources of statistical and systematic errors which afflict real receivers. Because signals from radio astronom-

---

* A technical remark: such antennas propagate only a single mode as well.

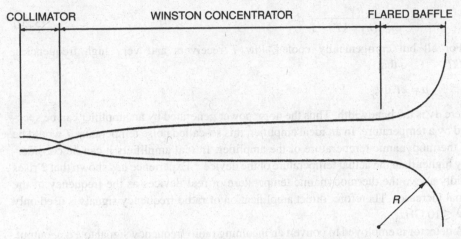

COLLIMATOR   WINSTON CONCENTRATOR   FLARED BAFFLE

Fig. 3.9   A Winston concentrator, as used in some microwave observations (including the FIRAS instrument described in Section 4.9.1). The flared baffle reduces diffraction at the antenna mouth, and hence reduces side-lobe response.

ical sources are so weak, we need to pay careful attention to sources of noise in receivers. To make this point concrete, it is useful to repeat a comparison first made, I believe, by F. Drake – all the energy ever received by all radio telescopes from all radio sources is about the same as the kinetic energy developed by a falling snowflake. Radio astronomers routinely measure powers as low as 10^{-17} W.

As a consequence, design and construction of receivers has become a painstaking, sophisticated art. I cannot hope to cover all the technical details here. Instead, I propose to outline in general the operation of various types of receivers and their major sources of systematic and statistical error. Fuller accounts may be found in the texts by Kraus (1986) and Rohlfs (1986), and in technical papers referred to in these two texts.

3.4.1 Sources of receiver noise

All receivers consist of three basic elements: (1) a device to limit the response of the receiver to the particular frequency or wavelength band of interest (loosely called a 'filter' in what follows); (2) one or more stages of amplification; and (3) a detector to convert the radio frequency signal into a d.c. (or at least much lower frequency) output.

A variety of solid state devices may be used as amplifiers at radio frequencies, including field effect transistors and high electron mobility transistors (HEMTs). Gains $\gg 10^6$ can be achieved by employing several stages of amplification. Here, I want to omit detailed description of the devices themselves, and instead to make just two points about amplifiers in general. First, as the gain G is increased, it is often the case that time variations in G increase more rapidly, so that the relative amplitude of gain fluctuations $\Delta G(t)/G$ grows larger. Second, amplifiers, like any electronic circuit element, generate noise. Even if the input power is zero, Johnson noise generated within the amplifier produces an output signal; this noise is in many cases the dominant source of statistical error in radio astronomical measurements. In an ideal amplifier, the power generated depends only on the temperature of the amplifier and the frequency: for a single mode and a single polarization,

$$p = \frac{h\nu}{\exp(h\nu / kT) - 1} \, d\nu. \tag{3.20}$$

For all but cryogenically cooled, low T receivers and very high frequencies, $h\nu/kT \ll 1$, so that

$$p = kT \, d\nu,$$

where $d\nu$ is the bandwidth. Thus the noise power generated by an amplifier can be specified by a temperature. In an ideal amplifier, this so-called noise temperature T would be the thermodynamic temperature of the amplifier. In real amplifiers it can be considerably higher than the actual temperature of the device.* Experience has shown that T rises rapidly above the thermodynamic temperature of real devices as the frequency of the signal increases. Therefore, direct amplification of radio frequency signals is used only at $\nu \lesssim 10$ GHz.

A detector is employed to convert an incoming radio frequency signal to a d.c. output. Two general classes of detectors are used in radio astronomy. The first class consists of various sorts of bolometric detectors, which produce an output signal (voltage or current) proportional to the power incident upon them, independent of its frequency; these are discussed further in Section 3.4.5. The second class are *square law detectors* (often junction devices; for a recent technical review, see White, 1988). As their name suggests, these operate by squaring the incoming signal. Let us consider the simple case of a plane polarized, monochromatic wave to illustrate their behavior. The electric field incident on the detector may then be written

$$\boldsymbol{E} = \varepsilon \cos(2\pi\nu t - \boldsymbol{k} \cdot \boldsymbol{z}) \, \hat{\boldsymbol{x}}.$$

After squaring in the detector, the resulting signal is proportional to

$$(\varepsilon^2/2) \, [\cos 4\pi\nu t + 1].$$

The high frequency signal can be filtered out, leaving a d.c. signal proportional to ε^2, i.e., directly proportional to the intensity of the incoming wave. An easy generalization to multi-frequency radiation shows that an ideal square law detector produces a d.c. output exactly proportional to the input power, p.

Real detectors, like real amplifiers, generate noise. It too may be specified by giving a noise temperature. In a working receiver, both amplifiers and a detector are employed. The noise generated in all these components may be summed to give a noise temperature for the receiver as a whole, written here T_{rec}. For the very best, cryogenically cooled receivers, a T_{rec} of 10–20 K at frequencies $\lesssim 10$ GHz can be achieved; very roughly $T_{rec} \gtrsim 10(\nu/10 \text{ GHz})$ K may be used as rule of thumb at higher frequencies.

It may be asked: 'If $T_{rec} > 10$ K, how can one hope to measure the CBR temperature of about 3 K, let alone small fluctuations in it?' The answer is that noise is a random process, and thus averages out over time, unlike the signal. Detailed arguments are given in Rohlf's (1986) book to show that the r.m.s. noise of a receiver, and hence the minimum detectable signal, is

---

* T can also be *lower* than the thermodynamic temperature of the amplifier in some special cases, such as parametric amplifiers (see Kraus, 1986; or Rohlfs, 1986).

$$T_{\text{rms}} = T_{\text{rec}} \left[\frac{1}{\Delta v \, \Delta t} + \left(\frac{\Delta G}{G} \right)^2 \right]^{1/2}, \tag{3.21}$$

where Δv is the total bandwidth of the receiver and Δt the duration of the measurement. Here, I will sketch a rough heuristic argument leading to the same result. Consider a series of n independent measurements of both a signal, specified by antenna temperature T_{A}, and noise, specified by T_{rec}. If we add up n measurements, the final signal is nT_{A}. On the other hand, if the noise is a random process, it will sum to $nT_{\text{rec}}/\sqrt{n}$. Therefore the signal to noise ratio increases as $\sqrt{n}$. In a bandwidth Δv Hz, Δv independent samples of the signal may be made per second*; thus $n = \Delta v \, \Delta t$, where Δt is the duration of the measurement in seconds. To find the minimum detectable antenna temperature, we equate nT_{A} to $nT_{\text{rec}}/\sqrt{n}$ with $n = \Delta v \, \Delta t$, yielding

$$(T_{\text{A}})_{\text{rms}} = \frac{T_{\text{rec}}}{(\Delta v \, \Delta t)^{1/2}}.$$

Time variations in the gain will also add to the statistical error in a measurement of T_{A}. If these variations are random and uncorrelated with the receiver noise, we may add the two sources of error in quadrature to recover eqn. (3.21).

A final remark on sources of noise in radio astronomy. In eqn. (3.21) only noise generated in the receiver itself is included. In a real radio telescope, there are some additional sources of noise, which add to the final noise temperature, T_{sys}, of the system as a whole. Among these are the antenna temperature of the source itself, atmospheric emission, radiation received from the side lobes of the antenna, and losses in the cable or waveguide connecting the antenna to the receiver. In most of the cases we will be considering, these additional sources of noise are less than or at worst comparable with the receiver noise. They nevertheless do affect the minimum detectable signal. Thus, for a radio telescope as a whole,

$$(T_{\text{A}})_{\text{rms}} = T_{\text{sys}} \left[\frac{1}{(\Delta v \, \Delta t)} + \left(\frac{\Delta G}{G} \right)^2 \right]^{1/2}. \tag{3.22}$$

For systems operating at centimeter wavelengths, T_{sys} can be as low as 30 K and Δv as large as 1 GHz. *Provided* that $\Delta G/G$ can be kept below 10^{-5} (see Section 3.4.4), we see from (3.22) that antenna temperatures as low as 1 mK can be detected in 1 s.

3.4.2   Simple receivers and their limitations

The simplest receivers consist of just a filter, an amplifier and a detector; two variants are shown in fig. 3.10. The direct receiver can be used only at $v \lesssim 10$ GHz; at higher frequencies, the amplifier noise becomes prohibitive.† In addition to this drawback, these simple receivers have several other limitations. One is the need for a filter to define the band pass, which can reduce the efficiency of the receiver. Another is that there is no way to control gain variations. Hence the simplest receivers, such as shown in fig. 3.10,

---

* Exactly true only for a rectangular band pass; see Rohlfs (1986) or another text for details.
† The use of high electron mobility transistors (HEMTs) is now allowing this limit to be pushed upwards to about 30 GHz (Lubin, private communication). It should also be noted that direct amplification does allow a wider bandwidth Δv than the technique to be discussed below.

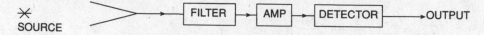

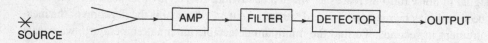

Fig. 3.10   Block diagrams of two kinds of direct receiver.

have not been used in CBR studies, and are rarely employed in any branch of radio astronomy.

3.4.3 Superheterodyne receivers

Instead, new elements are added, a *local oscillator* and a *mixer*, as shown in fig. 3.11. This is the superheterodyne receiver, the standard in radio astronomy. The local oscillator generates a radio wave at a fixed frequency ν_L. In the mixer, the local oscillator signal is multiplied by (or 'mixed with') the incoming signal. The product is then amplified. Consider again the simple case of a monochromatic, polarized input $E = \varepsilon(\cos 2\pi\nu t - k \cdot z)\hat{x}$. Let the local oscillator signal be $E_L = \varepsilon_L(\cos 2\pi\nu_L t - k \cdot z)\hat{x}$. The product will be proportional to

$$\frac{\varepsilon\varepsilon_L}{2}\,[\cos 2\pi(\nu - \nu_L)t + \cos 2\pi(\nu + \nu_L)t]. \tag{3.23}$$

If a low pass filter follows the mixer, only the signal at the difference frequency $(\nu - \nu_L)$ is passed on to the rest of the receiver. If the filter transmits only those signals with frequency $\leq \delta\nu$, only those input signals in the frequency range $\nu_L - \delta\nu \leq \nu \leq \nu_L + \delta\nu$ will reach the amplifier. Thus the low pass filter following the mixer defines the bandwidth of the receiver, and the center frequency at which the receiver operates is identical to the local oscillator frequency.

Note that the amplifier in a superheterodyne receiver is required to operate only at low frequencies, and hence is likely to be less noisy. In practical systems, the amplifier (or filter–amplifier combination) is designed to amplify signals in a band of frequencies slightly offset from 0 Hz, as shown in fig. 3.12. The center frequency of this offset band is the intermediate frequency ν_{if}. From fig. 3.12, it may be seen that the frequency response of the receiver in this case consists of two bands symmetrically placed above and below ν_L ('side bands'). When operated in this fashion, the receiver is said to be a double side band receiver; this is the norm in CBR measurements, where the precise frequency of observation is not crucial.

The amplified signal at point A (fig. 3.11) is proportional to ε only, not to the intensity ε^2. Therefore, in superheterodyne receivers, a *second detector* is employed to square the output signal from the mixer and hence to produce an output directly proportional to

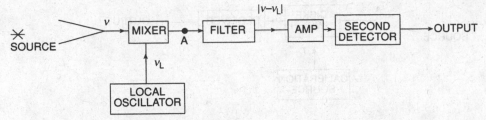

Fig. 3.11   Block diagram of a superheterodyne receiver. At point A, the output is proportional to the product of two signals at v and v_L. The low pass filter allows only the difference frequency to pass.

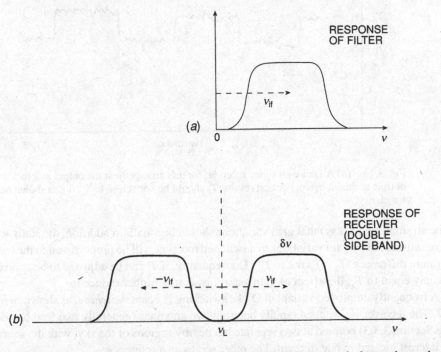

Fig. 3.12   Response of the filter–amplifier system in a practical superheterodyne receiver, (a), and the resulting response of the receiver, (b).

the input power. Note that the receiver output is also proportional to ε_L^2, so that the local oscillator power must be kept constant. In treating the system noise, time variations in ε_L^2 can be lumped in with the gain variations in the amplifiers.

While the superheterodyne receiver has many advantages, it does not solve the problem of gain variations. That requires a clever technique, first introduced by Dicke in 1946, to which we turn next.

3.4.4 Dicke switching

Dicke's idea was to switch the input to the receiver rapidly between the antenna and a standard calibration or reference source at a nominally fixed temperature T_c, as shown in fig. 3.13(a). In that way, T_A can be compared with the calibrator temperature T_c in a

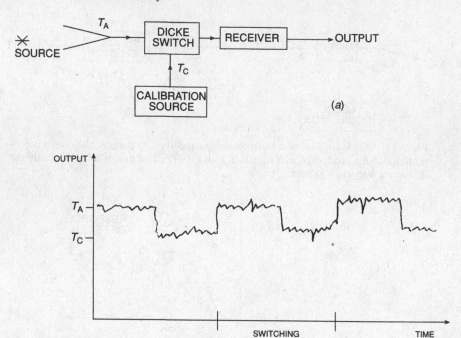

Fig. 3.13   (a) A Dicke-switched receiver; for this arrangement the output as a function of time is shown in (b). For best results, T_C should be very close to T_A, not as shown here for clarity.

time so small (< 1 s, say) that gain variations should be small. In addition, the statistical error introduced by gain variations in a switched receiver will be proportional to the temperature difference $|T_A - T_c|$ not to T_{sys}. Consequently, if T_c can be adjusted to be approximately equal to T_A, the effect of gain variations is still further reduced.

A frequently employed variant of Dicke switching is *beam switching*, as shown in fig. 3.7. The receiver is switched rapidly between two antennas (generally two feed horns – see Section 3.3.3) pointed at two separate but nearby regions of the sky, with the source of interest located in one of them. The other serves as a reference.

The output of a Dicke switched receiver is a square wave at the switching frequency, as shown in fig. 3.13(b). A final step in most radio astronomical observations is to transform the square wave to a d.c. signal proportional to the difference in temperature of the two inputs of the switch. This transformation is accomplished by phase-sensitive detection, a technique, once again, introduced by Dicke (1946). The technique is illustrated and explained in fig. 3.14.

The final output of a Dicke switched superheterodyne receiver after phase-sensitive detection is (ideally) directly proportional to the difference in antenna temperatures or powers entering the two antenna beams or two ports of the switch. Since only a difference in antenna temperature is measured, an absolute calibration is needed. A basic and fundamental way to calibrate a radio telescope is to cover the antenna (or at least the feed horn) with a blackbody source of known temperature. To discover possible non-linearities in the gain, blackbodies at two or more temperatures may be used.

A well-designed Dicke switched receiver will have gain variations $\Delta G/G < 10^{-6}$. A

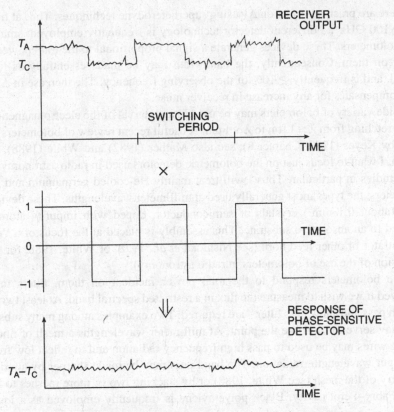

Fig. 3.14   Phase-sensitive detection. The signal coming from the receiver is multiplied (ideally) by a square wave with a period equal to the modulation of the Dicke switch. The phase is adjusted until the signal from the antenna is multiplied by +1 and the signal from the calibrator by −1. The resulting d.c. output is proportional to $(T_A - T_C)$ as shown; this signal is normally integrated over many switching periods to reduce noise.

glance at eqn. (3.22) will then show that gain variations contribute negligibly to the system noise for measurements lasting as long as hundreds of seconds, even for $\Delta \nu$ as large as 1 GHz. On the other hand, in a Dicke switched receiver the observing time is divided between the signal and the calibration or reference beam. Thus the total time spent measuring the signal of interest is only $\Delta t/2$. In addition, a temperature difference $T_A - T_C$ is measured, and both terms introduce error into the measurements. As a consequence, the minimum detectable antenna temperature in an ideal Dicke switched receiver is

$$(T_A)_{rms} = \frac{2\,T_{sys}}{(\Delta \nu\, \Delta t)^{1/2}} \ . \tag{3.24}$$

Measurements with $\Delta t \gtrsim 100$ s make it possible to reach antenna temperatures below 0.1 mK.

3.4.5 Bolometric detectors

As noted above, T_{sys} rises with frequency (with most of the increased noise arising in the mixer). Only by increasing $\Delta \nu$ can the minimum detectable temperature be kept small, and there are practical limits on Δv using superheterodyne techniques. Thus at frequencies $\gtrsim 100$ GHz*, a different detector technology is frequently employed: small solid state bolometers. These devices generate a signal proportional to the total power $\int p \, dv$ falling on them. Consequently, the bandwidth may be made essentially as large as desired, and is frequently $\gtrsim 10\%$ of the observing frequency. The increase in Δv more than compensates for any increase in receiver noise.

A wide variety of bolometers may be used in an interval of the electromagnetic spectrum stretching from $\lambda \simeq 1$ μm to $\lambda \simeq 1$ cm. A useful recent review of bolometers is provided by Keyes (1980, Chapter 8); see also Mather (1982) and White (1988). In this section, I want to focus just on the bolometric detectors used in radio astronomy and in CBR studies in particular. Thus I will treat mainly He-cooled germanium and silicon bolometers, the types most generally used at millimeter wavelengths. These devices are small (about 0.1 mm^3) crystals of semiconductor, doped with impurity atoms, and attached to an absorptive substrate. This assembly is placed at the focus of a Winston concentrator or other feed horn (see Nishioka *et al.*, 1978; or White, 1988, for further discussion of the use of bolometers in radio astronomy).

Since bolometers respond to the total power incident on them, filters must be employed if we wish to measure the flux in a restricted spectral band. At least two filters – a high pass and a low pass filter – are required. Two examples among many substances used may serve to illustrate the point. At millimeter wavelengths a mesh of fine, conducting wires may be used to pass high frequency radiation and to reflect low frequency, longer wavelength, radiation. The band pass may be made sharper by altering the geometry of the mesh (see White, 1988) or by stacking two or more meshes to form a tuned Fabry–Pérot cavity. Black polyethylene is frequently employed as a low pass filter.

Two fundamental problems are introduced by the need for filters. First, they may not be fully transparent at the frequencies of interest, lowering the receiver efficiency. More fundamentally, any material that is strongly *absorptive* at some frequency will also be *emissive* at the same frequency. To prevent radiation from the filters saturating the detector, the filters must be cooled to very low temperature. Keeping the filters nearest the bolometer cold is made easier by use of two or more stages of filter, a first stage to block most ambient radiation (e.g., the strong far infrared radiation of the atmosphere), and a second, cryogenic, stage to block the thermal radiation emitted by the first stage filters.

Let us turn now to a consideration of noise in bolometric detectors. The conventional way of specifying the noise of a bolometer is to give its noise equivalent power (NEP), the input signal power into the bolometer equal in magnitude to the noise generated internally. The units of NEP are W Hz$^{-1/2}$; values as low as about 10^{-16} W Hz$^{-1/2}$ can now be achieved at millimeter and submillimeter wavelengths.

The NEP of a bolometer is determined by six factors (Mather, 1982). Three of these are intrinsic to the bolometer (or the measurement) itself: Johnson noise, phonon noise resulting from (conductive) heat flow out of the bolometer element, and photon shot noise. We will treat all three in more detail below. Two other factors enter into the NEP

---

* Active work on SIS (superconductor–insulator–superconductor) mixers is now allowing the frequency range of superheterodynes to be extended to about 300 GHz (see Phillips and Woody, 1982; Phillips, 1988).

of realistic bolometric detectors: noise generated by the amplifier and by the load. Since these are not intrinsic to the bolometer, I ignore them here. Finally, there is 'excess noise' in real bolometers as opposed to the ideal case that we are about to consider; this 'excess' noise can be treated in the same fashion as gain fluctuations in heterodyne receivers.

Now let us consider the NEP of an *ideal* bolometer, following Mather (1982). Since the Johnson, phonon and photon noise sources are uncorrelated, the resulting NEP of an ideal detector is the quadrature sum of the three contributions. First,

$$(NEP)^2_{Johnson} = \frac{4kT\,G_t^2}{p_E\alpha^2}\,(1+\omega^2\,\tau^2), \qquad (3.25)$$

where T is the bolometer temperature, p_E is the electrical power dissipated in the bolometer, ω is the (angular) frequency, τ is the time constant of the response of the bolometer and α is the temperature co-efficient of resistance, $\alpha = (-1/R)(dR/dt)$. As Mather points out, the actual Johnson noise can be reduced below the value given above by electrothermal feedback.

Next, there is heat flow out of (or into) the bolometer to its heat sink, which generates noise (phonon noise) for which

$$(NEP)^2_{phonon} = 4\beta kT^2 G_t, \qquad (3.26)$$

where G_t is the thermal conductance (change in power per unit temperature) and β is a numerical factor of order unity, which depends on the conductivity of the detector material.

Finally, there is photon shot noise from the background radiation falling on the bolometer. Mather (1982) shows that this depends on the temperature (T_B) and emissivity (ε) of the background, and the efficiency (η) of the detector:

$$(NEP)^2_{photon} = \frac{4A_e\Omega(kT_B)^5}{c^2h^3} \int_0^\infty \frac{x^4\,dx}{e^x-1}\left(1+\frac{\varepsilon\eta}{e^x-1}\right)\varepsilon\eta, \qquad (3.27a)$$

where, as earlier, $x \equiv h\nu/kT_B$. The background temperature is frequently large enough to ensure $h\nu/kT_B \ll 1$ (consider, for instance, radiation from the sky at about 250 K or from room temperature elements in the optical path). If $x = h\nu/kT_B \ll 1$, we may simplify (3.27a) above to

$$(NEP)^2_{photon} = 2A_e\Omega kT_B(\varepsilon\eta)^2 \int_0^\infty B(T_B)\,d\nu, \qquad (3.27b)$$

where $B(T)$ is the Planck function. A more frequent, and more useful, approximation is obtained if the frequency range of the radiation is restricted to a bandwidth $\Delta\nu$ (by filters or otherwise). Then the integral in (3.27a) may be approximated by

$$\left(\frac{x^4}{e^x-1}\right)\left(1+\frac{\varepsilon\eta}{e^x-1}\right)\varepsilon\eta\,\frac{h\,\Delta\nu}{kT_B}\,.$$

If we again take $x = h\nu/kT_B \ll 1$, and expand e^x, we find

$$(NEP)^2_{photon} = \frac{4A_e\Omega kT_B\varepsilon\eta h\nu^3\,\Delta\nu}{c^2}\left(1+\frac{\varepsilon\eta kT_B}{h\nu}\right).$$

As a final step, we use eqn. (3.19) to set $A_e\Omega = c^2/\nu^2$ and obtain

$$(NEP)^2_{photon} = 4kT_B\varepsilon\eta h\nu\,\Delta\nu\left(1+\frac{\varepsilon\eta kT_B}{h\nu}\right). \qquad (3.27c)$$

The total NEP will consist of the square root of the sum of (3.25), (3.26) and (3.27). Note that the first two of these expressions depend on the bolometer temperature T; hence the drive to cool bolometric detectors to very low temperatures using liquid He, adiabatic demagnetization refrigerators, etc. Temperatures as low as 0.1 K are routinely used.

It may be asked why observers do not allow $G_t \rightarrow 0$ to reduce $(\text{NEP})_{\text{phonon}}$. The answer is that too small a thermal conductance lengthens the response time of the bolometer unacceptably.

In most astronomical applications, background photon noise (from atmospheric emission, thermal radiation from the telescope itself, etc.) usually dominates other sources of noise, provided that p_E and G_t have been chosen properly. Most observers also find that 'excess' noise in actual instruments is larger than the calculated Johnson and phonon noise.

Assume for a moment that the total NEP of a bolometric detector can be reduced to 10^{-16} W Hz$^{-1/2}$ at an operating frequency of 100 GHz. What is the corresponding value of T_{sys}? The r.m.s. power for a 1 s interval is 10^{-16} W or 10^{-9} erg s^{-1}. Using eqn. (3.20) and eqn. (3.22), and assuming that gain variations may be neglected, we have

$$10^{-9} = kT_{\text{rms}} \, \Delta v = kT_{\text{sys}} \, \Delta v^{1/2}$$

or

$$T_{\text{sys}} = 7.2 \times 10^6 \Delta v^{-1/2} = 2.3 \times 10^2 \left(\frac{1 \text{ GHz}}{\Delta v} \right)^{1/2}.$$

Since system temperatures of heterodyne receivers are around 100–200 K at 100 GHz, and $\Delta v = 1$–10 GHz can easily by achieved, we see that the two technologies have roughly equal sensitivity at 100 GHz. As the observing frequency rises beyond 100 GHz, however, bolometers are preferred, since T_{sys} for heterodyne receivers rises *roughly* as v (Section 3.4.1).

3.5   Interferometry and aperture synthesis

To achieve higher angular resolution than a single antenna can provide, radio astronomers employ a new technique, *interferometry*. Signals arriving at an array of two or more antennas are correlated; the resulting output contains information about structure on the sky with angular scales as small as about λ/D where D is the size of the array. By increasing D to 10^3–10^4 km, angular resolution finer than 10^{-3} arcsec may be obtained. In CBR studies, much smaller arrays have generally been used to probe scales of 0.1–1′.

As we shall see, an extension of this technique called aperture synthesis provides not only high angular resolution but a two-dimensional radio image, or *map*, of the sky. Interferometry also sidesteps some of the sources of systematic error encountered in observations with a single antenna. For all these reasons, interferometers have become important tools in all of radio astronomy, including now searches for fluctuations in the CBR.

This section does not pretend to be a complete treatment of interferometry in radio astronomy or of aperture synthesis. Fortunately, the subject is well treated in standard texts (e.g., Rohlfs, 1986), and in an exhaustive monograph by Thompson *et al.* (1986); interested readers may also wish to consult *Synthesis Imaging* by Perley *et al.* (1988). In

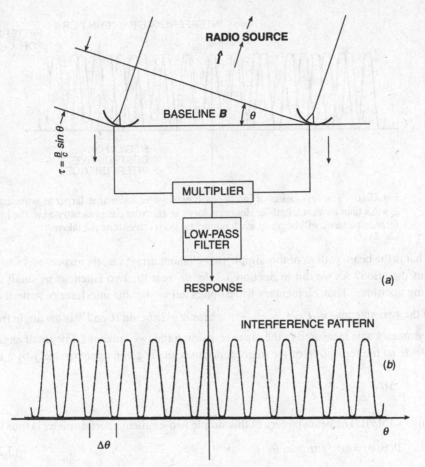

Fig. 3.15   Schematic of a simple two-element interferometer (*a*). The interference pattern from eqn. (3.28) is shown in (*b*); the separation of the interference maxima is given by $b_\lambda \sin \theta = 1$ or $\Delta\theta \sim b_\lambda^{-1}$ for small θ.

keeping with the general approach in this chapter, I will emphasize only the basic principles of aperture synthesis, and those peculiarities of the technique that have an impact on CBR observations.

3.5.1 The two-element interferometer

Let us begin with a very simple case, an interferometer consisting of just two antennas or *elements* (fig. 3.15). We will then add more details one by one until we have explored the working of a practical interferometer. For convenience we will start with the assumption that the two antennas respond equally well to radiation from any direction in the sky, that is that they are isotropic; we also assume that the position vectors of the two antennas are not changing, so that the baseline vector **B** between them is constant. The fixed separation between the two antennas may be written as $b_\lambda\lambda$, allowing us to express the baseline spacing as a multiple of the wavelength of the radiation observed.

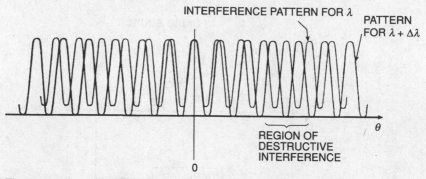

Fig. 3.16   The separation of interference maxima is somewhat larger at wavelength $\lambda + \Delta\lambda$ than at wavelength λ. Hence, at large θ, far from the symmetry axis, the interference patterns will begin to wash out unless corrective steps are taken.

What is the beam pattern of this simple two-element array, i.e., its response as a function of direction? As we did in Section 3.3, let us treat the two antennas as small diffracting apertures. Then elementary wave optics tell us that the interference pattern far from the two antennas is $I = I_0 \cos^2 \dfrac{\psi}{2}$, where $\psi = 2\pi b_\lambda \sin\theta$, and θ is the angle from the symmetry axis measured in the plane containing the two antennas. For small angles, $\sin\theta \approx \theta$, so the separation of the interference maxima is given approximately by

$$\Delta\theta = \frac{1}{b_\lambda} = \frac{\lambda}{|B|}$$

(see fig. 3.15(b)). The beam pattern of this simple two-element interferometer is thus just

$$P(\theta, \phi) = \cos^2(\pi b_\lambda \sin\theta). \tag{3.28}$$

As eqn. (3.28) shows, the spacing between interference fringes depends on the wavelength. If observations are made over a range of wavelengths $\Delta\lambda$, the fringe patterns at different wavelengths will begin to wash out as θ increases, as shown schematically in fig. 3.16; this is one aspect of *bandwidth smearing* or *fringe washing* (see Thompson *et al.*, 1986, Chapters 2 and 3). A simple dimensional argument suggests that a non-zero bandwidth will have a serious effect on the response of the interferometer at angles $\gtrsim (\lambda/b_\lambda\Delta\lambda)$ from the symmetry axis.

To permit observations to be made in directions away from the symmetry axes of the array, astronomers simply introduce a delay in one of the two signals, as shown in fig. 3.15(a). This has the effect of shifting the axis of symmetry to some direction $\hat{l}$. Providing we keep track of the relative orientation of the baseline vector B and $\hat{l}$, the appropriate delay can always be calculated and applied: it is

$$\tau = \frac{|B \times l|}{c} \, .$$

Note that applying a delay does not eliminate bandwidth smearing, it merely allows us to move the center of the fringe pattern at will. The direction $\hat{l}$ is referred to as the *fringe-stopping center* or the *phase center*.

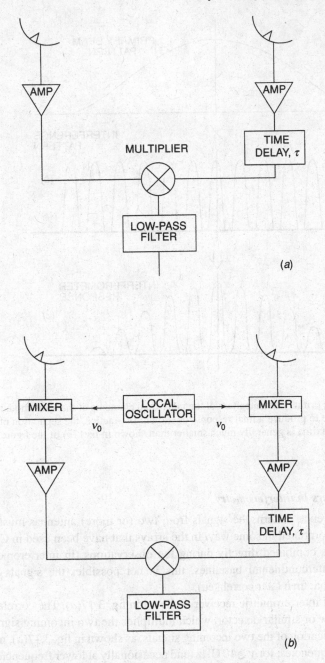

Fig. 3.17 (*a*) The simplest interferometric receiver. The two signals are combined in the multiplier. For sources not on the axis of symmetry of the antenna pair, a time delay is introduced in one of the signals. (*b*) A heterodyne receiver. Signals from the same local oscillator are fed to both mixers.

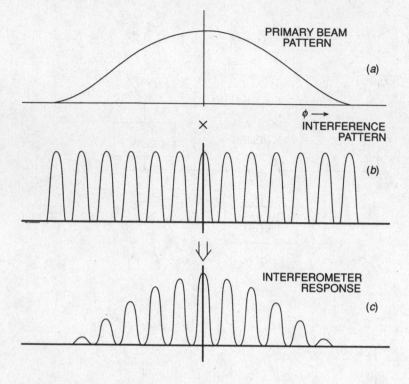

Fig. 3.18   The primary beam pattern of the individual elements of the array modulates the interference to produce a final response as shown. In practice, the separation of the interference maxima is generally much smaller than shown in part (*b*) of the figure.

3.5.2 *Receivers in interferometry*

To produce an interference pattern, the signals from two (or more) antennas must be brought together and combined in some way. In the arrays that have been used in CBR studies, the signals are combined directly during the observations (in interferometry with continental or intercontinental baselines, this is not possible; the signals are recorded at each antenna, then later correlated).

The simplest kind of interferometric receiver is shown in fig. 3.17(*a*). The correlator consists of a square law or similar detector, which multiplies the two incoming signals together. Direct amplification of the two incoming signals, as shown in fig. 3.17(*a*), may be employed at low frequencies; for $v \gtrsim 10$ GHz (and occasionally at lower frequencies), heterodyne receivers are used instead (fig. 3.17(*b*)). To preserve the phase coherence of the two signals, a single local oscillator is used.

A third kind of receiver used in interferometry is a *phase switched* device, introduced by Ryle in 1952. Here the phase of one of the incoming signals is changed regularly by 180°, by inserting an additional delay of $\lambda/2c$ s. This technique is analogous to Dicke switching, and is now used primarily to control some sources of systematic error present in interferometric observations (see Thompson *et al.*, 1986; and Section 7.8).

3.5.3 The primary beam pattern

In Section 3.5.1, we assumed that each of our two antennas was isotropic, that is equally sensitive in all directions. In reality, the antennas used in interferometric arrays are typically large paraboloids like the ones shown in fig. 3.2. Each of the elements of the array therefore has a beam pattern $P(\theta, \phi)$. This is called the *primary beam pattern*, and it modulates the interferometric response of the array, as shown schematically in fig. 3.18. Note that the angular resolution and the half width of the primary beam pattern are approximately related by $\theta/\theta_{1/2} \simeq d/b_\lambda$, for antennas of diameter d.

3.5.4 Aperture synthesis

We now want to relax our assumption that the baseline vector B is fixed in space; in an actual array, for instance, B will change as the Earth rotates. Qualitatively, as the angle between B and the direction to the phase center $\hat{l}$ changes, we would expectt to see changes in both orientation of the fringe pattern and fringe spacing. A quantitative analysis is made easier if we adopt a new set of coordinates, as shown in fig. 3.19. Two of the coordinates, u and v, are defined in the plane perpendicular to $\hat{l}$. One of these, u, is the projection of the east–west component of B into this plane; the other, v, is the corresponding north–south projection. The third component, w, is measured along $\hat{l}$. It should be clear that w can be kept very small if appropriate time delays are introduced (Section 3.5.1). It is conventional to specify distance in the u–v plane in terms of the wavelength employed in the observation, as for b_λ.

As B moves relative to $\hat{l}$, the u and v components will change. If we resolve B into east–west and north–south components B_E and B_N, we then have

$$u(h, \delta) = B_N \sin h - B_E \cos h$$
$$v(h, \delta) = - B_N \cos h \sin \delta - B_E \sin h \sin \delta \qquad (3.29)$$

for the dependence of u and v on the hour angle, h, and the declination of the phase center. During the course of an extended observation, h will slowly change as the Earth rotates, and hence B will trace a specific path in the u–v plane; it is left as an exercise to show that the trace formed by a 12 h observation with a purely east–west baseline is an ellipse in the u–v plane.

In principle, we can arrange to cover the entire u–v plane, and hence to synthesize an aperture of diameter roughly $|B|$, by combining many long observations made with different baseline orientations. This is the principle of *aperture synthesis*. More realistically, we just sample the u–v plane by selecting a few baseline orientations (see Ryle, 1975; or Thompson *et al.*, 1986, for a discussion). This sampling can be achieved by moving one element of a two-element array from place to place or by constructing a multi-element array. The latter approach, while more costly, is far more efficient and eliminates the need to move antennas in the middle of a synthesis. The Very Large Array, for instance, employs 27 antennas spaced along the arms of a symmetrical 'Y'; the resulting u–v traces for a 12 h observation of a source at $\delta = 40°$ are shown in fig. 3.20. Note that the sampling of the u–v plane, while free of large gaps, is not uniform.

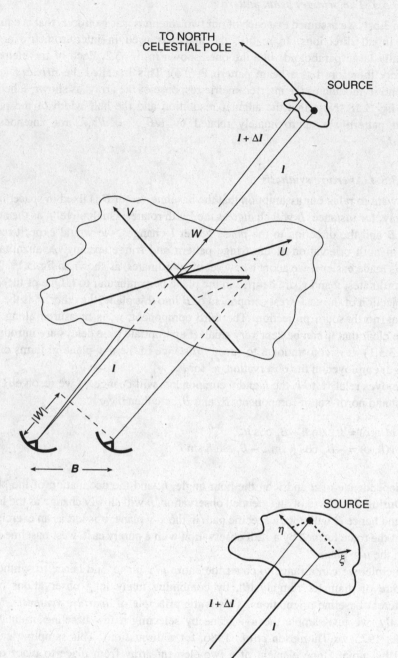

Fig. 3.19   The *u–v* coordinate system, and the projection of a baseline vector **B** into the *u–v* plane. The *u–v* plane is perpendicular to $\hat{l}$, the radius vector of (the center of) the source. The coordinate *w* is parallel to $\hat{l}$; a comparison with fig. 3.15 shows $|w| = \tau c$. The lower insert shows the Cartesian coordinates η and ξ on the plane of the sky.

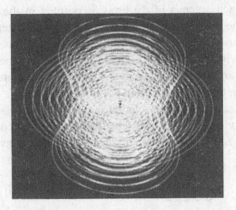

Fig. 3.20   Coverage of the u–v plane for a 12 h aperture synthesis observation made at the 27-element VLA. The source is assumed to be at declination 40°.

From fig. 3.19 and fig. 3.20, it may be seen that the resolution achievable in aperture synthesis will be about $1/u$ by $1/v$ in angular scale*. This is called the synthesized beam of the instrument; its angular scale will be represented by θ_{sy} here. Thus the goal of high angular resolution can be achieved by aperture synthesis. Let us now justify the claim that aperture synthesis also allows us to *map* the sky at this same angular scale. We will begin by considering an arbitrary direction $(\hat{l} + \Delta\hat{l})$ near the phase center. From fig. 3.19, it is clear that the distance from the phase center on the plane of the sky may be written

$$\Delta l = \Delta\delta\,\hat{\imath} + \Delta a \cos \delta\hat{\jmath}$$
$$\equiv \eta\hat{\imath} + \xi\hat{\jmath} \tag{3.30}$$

where we have introduced Cartesian† coordinates on the plane of the sky for simplicity, as in Section 3.3.2. A generalization of eqn. (3.28) then gives for the response of a two-element interferometer

$$I = I_0 \cos^2 [\pi(u\xi + v\eta)].$$

Thus the response depends on both the baseline (through u and v) and the direction (through ξ and η).

To proceed beyond this qualitative point, it is useful to introduce a quantity called the *complex visibility V*,

$$V = \int_{4\pi} P(\Delta l)\, B(\Delta l) \exp\left[-2\pi\,i\left(\frac{B\cdot\Delta l}{\lambda}\right)\right] d\Omega. \tag{3.31}$$

The choice of sign in the argument of the exponent agrees with that made by Thompson *et al.* (1986). In this expression $P(\Delta l)$ is the primary beam pattern and $B(\Delta l)$ the surface brightness at position $\hat{l} + \Delta l$, not the baseline. Reexpressed in the coordinates (ξ, η) that we have just introduced, this may be rewritten as

$$V = \int_{-\infty}^{\infty} \int_{-\infty}^{\infty} P(\xi,\,\eta)\, B(\xi,\,\eta)\, \exp\,[-2\pi i\,(u\xi + v\eta)]\, d\xi\, d\eta. \tag{3.32}$$

---

* Recall that u, v distances are measured in units of λ, the wavelength of observation.
† For small Δl, the tangent plane closely approximates the surface of a sphere.

I have assumed that each antenna of the array is pointed in the direction $\hat{l}$, so that $P(\xi, \eta)$ is centered at the phase center. A word on the limits of integration is called for. For real antennas, the primary beam pattern will fall to zero at some finite angle, so the value of the integrand will also fall to zero. Hence the integration needs to be performed only over a small solid angle of the sky including the primary beam; and thus the use of Cartesian coordinates in the tangent plane at $\hat{l}$ is justified.

The above eqn. (3.32) is the crucial relation in aperture synthesis. As may be seen, each measurement of a visibility is a measurement of one Fourier component of the convolution $P(\xi, \eta) B(\xi, \eta)$. By measuring many visibilities, that convolution may be calculated as follows:

$$P(\xi, \eta) \; B(\xi, \eta) = \int_{-\infty}^{\infty} \int_{-\infty}^{\infty} V(u, v) \; \exp\left[+2\pi i \left(u\xi + v\eta\right)\right] du \; dv. \tag{3.33}$$

If the primary beam pattern $P(\xi, \eta)$ is known, we may use eqn. (3.33) to determine the surface brightness distribution $B(\xi, \eta)$, i.e. to map the sky.

3.5.5 Some technical details

First, how are visibilities measured? In terms of V, the output of the correlator of an interferometric receiver is proportional to $|V|$:

$$I \infty |V| \cos^2\left[\pi\frac{B \cdot \Delta l}{\lambda} - \varphi\right]. \tag{3.34}$$

This may be seen by taking the real and imaginary parts of eqn. (3.32) (Thompson *et al.*, 1986, Chapter 2). Here ϕ is an arbitrary phase.

If the appropriate delay is introduced (Section 3.5.1), then the phase, ϕ, of a point source at the phase center of the map will stay constant. The phase of a point source *not* at the phase center will slowly change as the rotation of the Earth changes $B \cdot \Delta l$. By recording both the intensity and the phase of the correlator output, V may be found.

The above eqns. (3.32)–(3.34) apply to a single pair of antennas; in arrays of more than two elements, visibilities or amplitudes and phases must be recorded for all possible pairs of antennas. Thus at the Very Large Array (VLA), for instance, $(27 \times 26)/2 = 351$ visibilities must be recorded. Simultaneously, values of (u, v) must be calculated for each of the 351 pairs using eqn. (3.29). These are the raw data of an interferometric observation.

For the long observing runs used in CBR studies, 10^5–10^6 visibilities may be acquired. To perform a direct Fourier transform to obtain $P(\xi, \eta) B(\xi, \eta)$ from eqn. (3.33) would be very costly in computer time, so discrete transforms (FFT or Fast Fourier Transforms) are used instead. The measured visibilities are sorted into a $2^a \times 2^b$ grid in the u–v plane, then transformed. As a consequence, the resulting maps have $2^a \times 2^b$ resolution elements. To avoid loss of angular resolution, the resolution elements (*cells* or *pixels*) must be smaller than the synthesized beam width. It is conventional to select a cell size $\lesssim \theta_{sy}/3$.

The use of discrete Fourier transforms can produce artifacts in aperture synthesis maps. One in particular is *aliasing*: bright sources outside the boundaries of a map are in effect 'reflected' into the map. Aliasing can be minimized by making a map substantially larger than the solid angle of the primary beam. In this way, the response of the

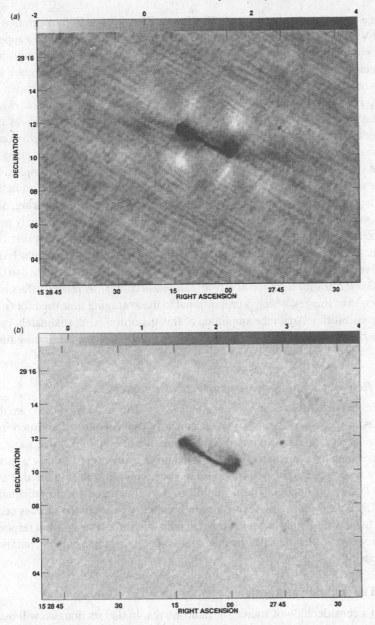

Fig. 3.21 VLA images of a radio galaxy (*a*) before and (*b*) after 'cleaning.' The radial 'spokes' in (*a*) are side lobes of the bright source. Note how faint, low surface-brightness structure emerges after careful cleaning. Images courtesy of Eric Richards.

array to sources outside the map is essentially zero. Maps of a least $6\theta_{1/2}/\theta_{sy} \approx 6\ |B|/d$ cells on a side are required.

Another artifact in aperture synthesis maps is side-lobe structure. If the coverage of the *u–v* plane is not complete, the synthesized beam will have side lobes (in the simplest case of a brief observation by a two-element interferometer, some 'side lobes' are virtually as large as the main lobe, as fig. 3.18 shows). Even with a 12 h integration at the 27-element VLA, side lobes of amplitude 1–2% are present. These side lobes appear prominently about the bright sources in figure 3.21(*a*). Since the quantities u and v are recorded for each antenna pair during an observing run, it is possible to calculate the shape of the synthesized beam, side lobes and all. With this known, it is possible to 'clean' the map of side lobes (see Högbom, 1974; or Chapter 11 of Thompson *et al.*, 1986, for a detailed description). For maps of bright sources, 'cleaning' improves the images cosmetically and may reveal features of low surface brightness, which would otherwise be masked by the side lobes (see fig. 3.21(*b*)). For aperture synthesis maps of the CBR, as we shall see in Chapter 7, the role of 'cleaning' is much more complex.

Finally, there are some features of real aperture synthesis maps that depend on radial distance from the phase center of the map. The first of these is bandwidth smearing, discussed in Section 3.5.1. The loss of visibility or fringe amplitude is approximately proportional to the radial distance. A second consequence of non-zero bandwidth is radial stretching of images far from the phase center of the map (at different wavelengths lying within the receiver bandwidth $\Delta\lambda$, the map scale varies by $\lambda - \frac{1}{2}\Delta\lambda$ to $\lambda + \frac{1}{2}\Delta\lambda$, so the stretching is $\Delta\theta = \theta\,\Delta\lambda/\lambda$ for a source at θ from the phase center). A related effect is a rotational distortion of images, which is proportional to the averaging time used for each sampling of the visibilities. Again the amplitude of this distortion is approximately proportional to the radial distance from the phase center as well as to the integrating time used.

3.5.6 *Response of interferometers to extended sources*

As we have seen, the resolution of an array is about $1/u_m \times 1/v_m$ where u_m and v_m are the maximum scales in the u–v plane. Now let us consider the response of an interferometer to sources larger than the resolution. As examination of eqn. (3.32) will show, V will have a maximum for a point source, that is when $B(\xi, \eta) = \delta(\xi, \eta)$. For a more extended source, with extent ξ or $\eta > 1/u$ or $1/v$, respectively, it is easy to see that the integral in (3.32) will be smaller. The reduction in V is illustrated for a particularly simple case in fig. 3.22. The details of this loss in sensitivity to extended sources need not concern us, but it is important to bear in mind that an interferometer does not respond at all to isotropic sources like the CBR. Such instruments can measure only *fluctuations* in the CBR intensity.

3.6   The equation of transfer

We turn now to a consideration of sources of radio waves. In this section, we will treat a general issue, the passage of electromagnetic radiation through an absorbing and emitting medium, a situation that we will encounter frequently later in the book.

Consider a source of surface brightness B_s and solid angle Ω_s viewed through an absorbing cloud, as shown in fig. 3.23. If the absorption coefficient of the material in the cloud is κ m^2 kg^{-1}, then the observed surface brightness of the source is reduced by $dB' = -B'(x')\kappa\rho\,dx'$ in a slab of thickness dx'. When the source is viewed through the entire cloud, we have

$$B(x) = B_s \exp\left(-\int_0^x \kappa\rho\,dx'\right), \tag{3.35}$$

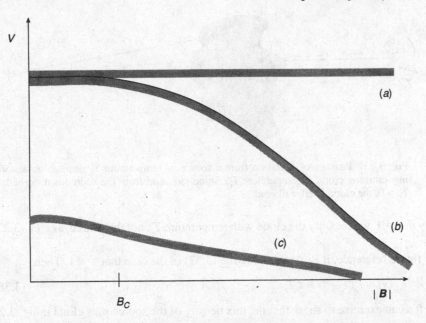

Fig. 3.22   Response of an interferometric array to sources of different angular scale. The visibility V is plotted as a function of baseline spacing $|B|$. Curve (a) is for a point source. In (b), V is shown for a slightly extended source. For baselines $> B_c$, the interference maxima are separated by an angular distance $\Delta\theta <$ the extent of the source, and the source is said to be *resolved*. In (c) the response for a heavily resolved (extended) source is shown schematically.

found by integrating $dB'/B' = - \kappa\rho \, dx'$. The quantity $\int_0^x \kappa\rho \, dx'$ is called the optical depth of the cloud and is written as τ. If the cloud is homogeneous, we have the simple result that $\tau = \kappa\rho x$, where x is the total thickness of the cloud; more frequently the density ρ will vary with position in the cloud.

It is worth remarking that, since brightness temperature is proportional to surface brightness in the R–J region, eqn. (3.35) can also be written in terms of the brightness temperature of the source:

$$T(x) = T_s \, e^{-\tau}. \tag{3.36}$$

From Kirchoff's Law, we know that a cloud that absorbs at a particular frequency will also emit at that same frequency. Thus radiation from the cloud will add to the radiation passing through it, as shown in fig. 3.23. To find the total observed surface brightness, we need to solve the equation of transfer for radiation passing through the cloud. Here I will simply write down the solution and justify it by considering three special cases (for derivation, see Chandrasekhar, 1960; Rohlfs, 1986). The solution of the equation of transfer, expressed in brightness temperature, is

$$T_{tot}(x) = T_s \, e^{-\tau} + T_c(1 - e^{-\tau}), \tag{3.37}$$

where T_c is the temperature of the cloud, assumed to be constant throughout. Now, let us justify eqn. (3.37). First, if $\tau = 0$, the cloud is totally transparent and $T_{tot}(x) = T_s$ as expected. The same will be true if $T_c = T_s$, as expected on thermodynamic grounds.

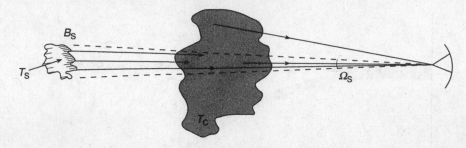

Fig. 3.23 Passage of radiation from a source of temperature T_s through an absorbing–emitting cloud at temperature T_C. Some radiation from the source is absorbed. If $T_C > 0$, the cloud itself will emit.

Finally if $\tau \gg 1$, we see only the cloud with temperature T_c, not the source, as eqn. (3.37) shows.

For future reference, it is useful to rewrite (3.37) in the case that $\tau \ll 1$. Then

$$T_{tot}(x) \approx T_s(1 - \tau) + \tau T_c. \tag{3.38}$$

It is left as an exercise to show that the flux density of the source plus cloud in fig. 3.23 is given by

$$S(x) = \frac{2kT_s}{\lambda^2} e^{-\tau} \, \Omega_s + \frac{2kT_c}{\lambda^2} \, (1 - e^{-\tau}) \, \Omega_c,$$

where Ω_c is the solid angle subtended by the entire cloud.

The passage of radio waves through the Earth's atmosphere provides one instance where (3.37) or (3.38) may be employed. As an example consider an observation of the CBR at $\lambda = 3$ cm at sea level, where $\tau \approx 0.015$. The effective temperature of the Earth's atmosphere is $T_c \simeq 250$ K. Thus the measured brightness temperature will be $T_{tot}(x) = 2.74(1 - 0.015) + 250(0.015) = 6.45$ K. The eqns. (3.38) or (3.37) may also be used to treat losses in cables or waveguides in radio telescopes; a lossy element acts just like an emitting–absorbing cloud at ambient temperature. Finally, we will make use of (3.37) when in the next section we consider radio emission from hot astrophysical plasmas.

3.7   Emission mechanisms at radio wavelengths

In radio astronomy, both spectral line emission (e.g., the 21 cm line of atomic hydrogen) and continuum emission are encountered. We will treat only the latter (for reviews of atomic and molecular line emission, see for instance Kulkarni and Heiles, 1988; and Turner, 1988).

One continuum emission process we have already treated is blackbody emission. It is encountered from warm solids (like planets) and dense gases where the optical depth, τ, at the frequency of observation is high (e.g., the Sun).

3.7.1 Bremsstrahlung

Another form of continuum emission is bremsstrahlung, radiation emitted from hot plasmas that are optically thin (i.e., $\tau < 1$; if τ is large we recover blackbody emission). The physics underlying this process is the acceleration of rapidly moving charged particles (electrons) by slower moving ions (mostly protons in astrophysical plasmas). The

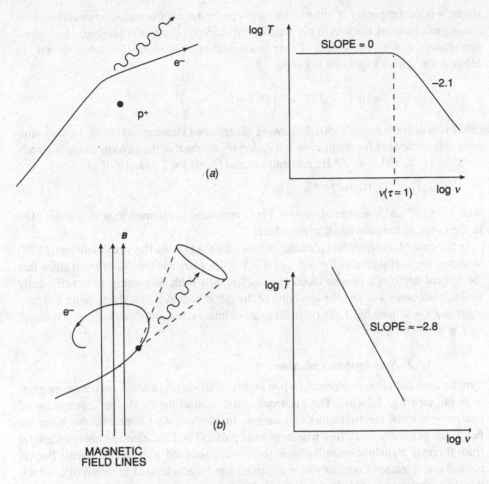

Fig. 3.24   Two emission mechanisms frequently encountered in astronomy.

In (*a*) electrons are accelerated by interactions with positive ions (e.g. protons); radiation results, called either free–free emission or thermal bremsstrahlung. The spectrum $T(v)$ is shown for a typical bremsstrahlung source; at small v, the source is opaque and radiates essentially like a blackbody. In (*b*), electrons spiral around magnetic field lines, emitting synchrotron radiation. In the case of relativistic electrons, emission is concentrated as shown in the forward core. Synchrotron sources typically have power law spectra. $S \propto v^{-0.8}$ or $T(v) \propto v^{-2.8}$.

process is shown schematically in fig. 3.24(*a*). To calculate the surface brightness temperature of an optically thin plasma, we start with its absorption coefficient κ as a function of the frequency and of the temperature of the plasma. We assume that the velocity distribution of the electrons is thermal – hence the somewhat confusing name *thermal* bremsstrahlung. In the microwave region, and for plasma temperatures low enough that the e^-/p^+ collisions may be treated nonrelativistically, we have

$$\kappa \rho = \frac{C_1 \, n_e \, n_i}{v^2 \, T^{3/2}} \ \ln \left(\frac{C_2 \, T^{3/2}}{v} \right)$$

(Shklovskii, 1960; see also Rohlfs, 1986, or Gordon, 1988), where C_1 and C_2 are constants, v is the frequency of observation, and T, n_e and n_i are the temperature and number density of electrons and ions in the plasma, respectively. For a fully ionized, pure hydrogen plasma, $n_i = n_e$; in plasmas of cosmic abundance, n_i is about 10% lower than n_e. In MKS units, with v expressed in hertz,

$$\kappa\rho = 9.8 \times 10^{-13} \, n_e^2 \, T^{-3/2} \, v^{-2} \left[19.8 + \ln \left(\frac{T^{3/2}}{v} \right) \right] \tag{3.39}$$

(Shklovskii, 1960; Lang, 1980). Following Mezger and Henderson (1967), we may simplify this somewhat for the microwave region by noting that the ln term changes slowly for $T \simeq 10^4$ K and $v \simeq 10^9$ Hz and thus rewrite (3.39) for T near 10^4 K as

$$\kappa\rho = 2.7 \times 10^{-30} n_e^2 T^{-1.35} v^{-2.1} \, \text{m}^{-1}, \tag{3.40}$$

with n_e in m^{-3} and v now *in gigahertz*. This expression is accurate to within about 10% in the range of frequencies of interest here.

In the case of optically thin plasmas, we may then substitute this result into eqn. (3.38) to obtain the useful results $T \propto v^{-2.1}$, or $S \propto v^{-0.1}$. A glance at eqn. (3.40) will show that the optical depth of a plasma cloud increases rapidly with decreasing v; at sufficiently low v, τ becomes > 1, and the spectrum of the cloud becomes blackbody, with $T = $ constant and $S \propto v^2$ (see fig. 3.24(*a*)). In all cases of interest to CBR observations, τ is small, so $T \propto v^{-2.1}$ and $S \propto v^{-0.1}$.

3.7.2 Synchrotron radiation

Synchrotron radiation is generated when highly relativistic electrons encounter magnetic fields (see fig. 3.24(*b*)). The electrons spiral around the field; the acceleration of charge generates electromagnetic radiation. In astrophysical contexts, the magnetic fields are generally weak (not true in or near pulsars) and the electron energies are far from thermal. Synchrotron radiation is therefore associated with violent events (supernovae, active galactic nuclei) where electrons can be accelerated to relativistic velocities. In such sources, it is often the case that the energy spectrum of the electrons is well described by a power law:

$$N(E) \, dE \propto E^{-a} \, dE.$$

In this case, the synchrotron radiation spectrum has the following dependence on the index a:

$$S \propto B^{(a+1)/2} v^{-(a-1)/2} \tag{3.41}$$

(see Pacholczyk, 1970; Rohlfs, 1986). Thus the spectrum of a synchrotron source depends on the electron spectrum. In a great many astronomical synchrotron sources, the flux is found to vary as $v^{-0.5}$ to $v^{-1.0}$, suggesting that a lies in the range 2–3. The observed energy spectrum of high energy cosmic ray electrons in our Galaxy has an exponent a falling in this range.

3.7.3 Inverse Compton radiation

The familiar Compton effect arises in the scattering of high energy photons off essentially stationary electrons in atoms. Given the existence of both relativistic e$^-$ and hot plasmas in astronomical settings, we must allow for the possibility of the *inverse* effect:

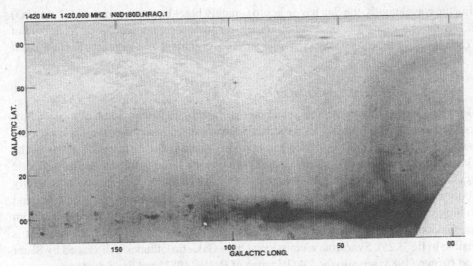

Fig. 3.25   A map of a portion of the radio sky made at 1420 MHz (Reich, 1982). Galactic coordinates are used; note the concentration of sources along the Galactic equator. Radio maps at other frequencies are shown in figs 4.8 (at 408 MHz), 6.4(*a*) (at 20 GHz) and 6.7 (at higher frequencies).

scattering of high energy e⁻ off low energy photons. The result is to transfer energy to the photon, 'boosting' its energy, E. The average increase in photon energy depends on the electron energy. For electrons in thermal equilibrium at temperature T,

$$\Delta E = h\Delta v = \frac{4kTE}{mc^2}, \text{ for } kT \gg E,$$

where m is the electron rest mass. For future reference, note that this process conserves photon *number* while increasing photon *energy*; this property is at the basis of the Sunyaev–Zel'dovich effect, which we will consider in Chapters 5 and 8.

3.8   A brief survey of radio sources

In fig. 3.25 is shown a radio wavelength 'photograph' of the sky. What are all these sources? Most of the discrete sources are extragalactic – quasars or radio galaxies. The concentration of sources along the Galactic plane, however, suggests that there are also many Galactic sources. In this section, we will discover which emission mechanism dominates in each type of source, and will also consider the role played by Galactic radio emission and by extragalactic radio sources in CBR studies. This survey will necessarily be both brief and biased – for fuller treatments, see the many review articles in Verschuur and Kellermann (1988), for instance.

3.8.1 *Blackbody sources*

We will begin close to home – with the planets and the Moon. These are blackbody sources (neglecting the very long wavelength radio emission from Jupiter). Indeed, the Moon and planets are frequently used to calibrate radio telescopes, since their temperatures and the solid angles they subtend are precisely known. From optical through millimeter wavelengths, the Sun has an approximately blackbody spectrum with $T \simeq 6000$ K. As the wavelength of observation increases, however, bremsstrahlung from the hot ($T \simeq 10^6$ K) solar corona raises the solar surface brightness as λ increases beyond about 1 cm (see Kundu, 1965).

3.8.2 The radio emission of the Galaxy

Our Galaxy emits both synchrotron and bremsstrahlung radiation. Some of the synchrotron radiation comes from localized sources, such as supernova remnants; the Crab Nebula with $S \approx 1000$ Jy at 1 GHz is an excellent example. There is also diffuse emission from the Galactic plane, caused by Galactic cosmic rays moving through the interstellar magnetic field. Synchrotron emission both from discrete sources and from cosmic rays is generally confined to the plane of the Galaxy. Regions of synchrotron emission, however, do extend to high Galactic latitudes. One example is the North Galactic Spur, which extends from the Galactic plane nearly to the North Galactic Pole along the $l = 30°$ great circle (it is visible in fig. 3.25). Synchrotron emission at high Galactic latitudes is discussed by Satter and Brown (1988; see also the 1.4 GHz map of Reich, 1982; and fig. 4.8 later).

Bremsstrahlung radiation is produced mainly by ionized regions of typical temperature 10^4 K surrounding massive, hot stars. It is sharply confined to the plane of the Galaxy, with a scale height of about 200 pc. Except near the Galactic center, the surface brightness of the bremsstrahlung emission falls to about half its central value only 2° from the Galactic plane.

Since both bremsstrahlung and synchrotron radiation are generated in the Galaxy, the radio spectrum of the Galaxy is not a simple power law, and varies with position. Hence correcting observations of the CBR for foreground Galactic emission is not a trivial task, as we shall see in the following chapter. It may be useful to bear in mind two rough rules of thumb concerning Galactic radio emission at high latitudes, where most CBR measurements are made. (1) The brightness temperature of Galactic radio emission drops to 3 K as ν rises to about 1 GHz. (2) The bremsstrahlung and synchrotron components of Galactic emission are roughly comparable at 20 GHz; clearly, synchrotron emission dominates at lower frequencies.

At wavelengths of 1 mm or less, the dominant form of emission from the Galaxy is thermal emission from small dust particles. Infrared observations made from the IRAS satellite have shown that the interstellar dust is unfortunately far from uniformly distributed (Low *et al.*, 1984). The presence of thermal emission from dust has already proved to be a major problem to those attempting measurements of the large angular scale distribution of the CBR at wavelengths of order 1 mm (see Weiss, 1980, for a review; Melchiorri *et al.*, 1981; and Chapters 6 and 7).

These two sources of emission from the Galaxy leave open a narrow wavelength window, centered at about 3 mm, which has been exploited for several of the recent CBR measurements. Even observations made in this wavelength window, however, may need corrections for Galactic emission. Such corrections are usually made by extrapolating measurements made at longer wavelengths using an appropriate combination of the two power laws given above (see Witebsky, 1978; and Fixsen *et al.*, 1983, for examples).

3.8.3 Extragalactic radio sources

Like the Milky Way, other galaxies are radio sources. The luminosity of the majority

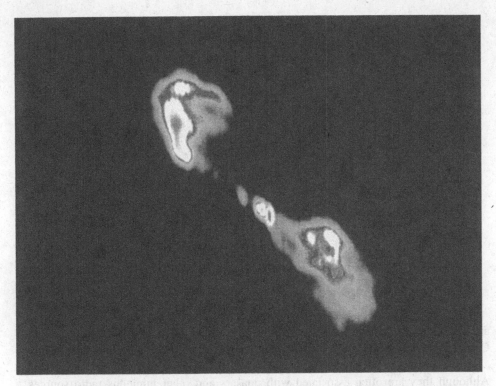

Fig. 3.26   A typical double-lobed radio source, in this case Centaurus A (National Radio Astronomy Observatory, with permission).

of galaxies lies very roughly in the range 10^{28}–10^{32} W at radio wavelengths (Hummel, 1980), and as a consequence they can be detected only if they are relatively nearby, not at cosmological distances. The situation is very different for a minority of galaxies which have a far higher radio luminosity, say 10^{34}–10^{38} W; these are known as radio galaxies. Radio galaxies have been detected out to distances corresponding to a redshift $z = 2.3$ (Chambers *et al.*, 1988). It is these high-luminosity objects, together with radio luminous quasars, which dominate the radio sky.

There are several characteristic patterns of radio emission, which can help us classify these radio sources (for a recent review, see Kellermann and Owen, 1988). In some radio sources, much or all of the emission comes from large, diffuse radio lobes located symmetrically on either side of the galaxy or quasar; fig. 3.26 shows a good example. These lobes are often far larger (10–100 times) than the optical size of the galaxy. In some cases, one or two narrow jets appear in radio images, well collimated with the lobes. These jets appear to be feeding relativistic electrons to the lobes to maintain their radio emission. The jets of galaxies that happen to be located in clusters are often bent, presumably because these galaxies are moving though an intergalactic medium (fig. 3.21(*b*) shows some evidence of this phenomenon). The mechanism responsible for radio emission from the jets and lobes is the synchrotron process. The spectral index of these radio sources is typically − 0.5 to − 1.0. Observations of such sources are reviewed by Miley (1980).

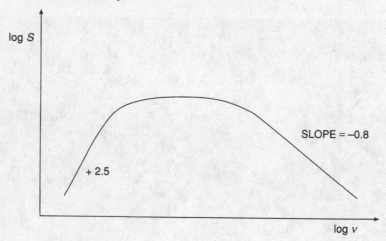

Fig. 3.27 Schematic spectrum of a compact radio source, showing the onset of synchrotron self-absorption. In the transition region, $S \sim$ constant, giving rise to the nomenclature 'flat-spectrum sources.'

Some radio sources have extremely compact structure. These radio emitting regions are often far *smaller* than the characteristic 10 kpc size of the galaxies they appear in. Compact sources are generally located at the centers of galaxies, suggesting that they are associated with galactic nuclei (for a review, see Kellermann and Pauliny-Toth, 1981). Although they are often associated with quasars and other luminous radio sources, compact sources of lower luminosity are frequently found at the centers of normal galaxies, including our own Milky Way. Finally, some radio sources have both compact sources and more extended radio emission.

The mechanism responsible for radio luminosity of compact sources is again the synchrotron process. In these sources, however, the optical depth is high, unlike the case in our Galaxy or in extended radio lobes. Thus the synchrotron radiation can be *self-absorbed* (Ginzburg and Syrovatskii, 1969). This process changes the spectral index. From arguments developed in Sections 3.2 and 3.7, we would expect the spectral index to become 2.0 as the optical depth $\tau \to \infty$. In fact, the situation is more complex in these compact synchrotron sources, and a detailed argument shows that the spectral index in the optically thick regime should be 2.5, not 2.0. In addition, in most sources the transition to high optical depth is gradual, so the spectral index rarely reaches its asymptotic value of 2.5. The spectrum of a typical compact synchrotron source is shown in fig. 3.27. Note the extended region where S is approximately constant, so that the spectrum is flat. That feature of the spectrum is frequently encountered (Kellermann and Pauliny-Toth, 1981) and has given rise to a simple classification of extragalactic radio sources: flat spectrum sources, with spectral index > -0.5, and steep spectrum sources with spectral index < -0.5. In general, as we have seen, the former are compact and the latter extended.

3.9 Radio source counts

Radio sources are interesting in their own right, not least because we do not fully understand the processes responsible for well-collimated jets or even the basic 'engine' to power the synchrotron emission. Even if we do not fully understand radio sources,

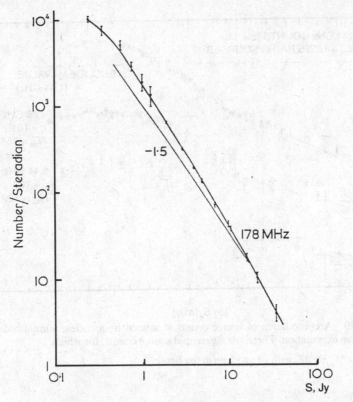

Fig. 3.28   Source counts at $\nu = 178$ MHz (from Ryle, 1968, with permission). In static Euclidean space, the integral counts would have a –1.5 slope. Note the excess counts at $S \lesssim 3$ Jy.

however, simple *counts* of sources have had an important impact on astrophysics and cosmology. Let us now set aside questions about the intrinsic properties of extragalactic radio sources and instead ask the straightforward question: 'How many sources with flux density above some level S are there per steradian on the sky?' That number is denoted $N(S)$. We can use an elementary argument to show that $N(S) \propto S^{-3/2}$ if the sources are uniformly distributed in a non-expanding Euclidean geometry. We start by assuming that all radio sources have the same intrinsic luminosity L. Such sources will have flux density $\geqslant S$ for distances $d \leqslant (L/4\pi S)^{1/2}$. Hence the number of sources per steradian with flux $\geqslant S$ is proportional to $V = 1/3\,(L/4\pi S)^{3/2}$, giving $N(S) \propto S^{-3/2}$. The dependence of N on S will be the same for sources of any L, and hence holds even if we relax our artificial assumption that all sources have the same intrinsic luminosity.

3.9.1 Evolution of radio sources

The simple 3/2 power law breaks down in non-Euclidean curved geometries, and also if cosmological expansion is included. Starting in the 1950s Ryle and others (see Ryle's 1968 review) looked for departures from the 3/2 power law as a means to discriminate among cosmological models (Section 1.3). The observed counts were surprising, and agreed with *no* plausible cosmological model (fig. 3.28; Jauncey, 1975). The rapid

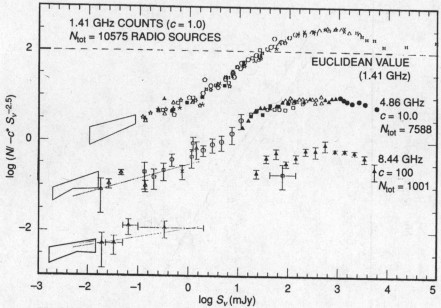

Fig. 3.29   A compilation of source counts at several frequencies, normalized to the Euclidean expectation. These are *differential* source counts, for which

$$dN = (-c^*S_\nu^{-2.5})\, dS,$$ with c^* as given in the figure.

increase in N as S decreased could not be explained by space curvature effects, but did have another natural explanation: radio sources were either more luminous or more numerous in the past. Source counts therefore demonstrate the *evolution* of radio sources (see Longair, 1966 or Condon, 1988). Further studies in the past dozen years have confirmed and enriched the situation sketched in fig. 3.28. A recent compilation of source counts made at $\lambda = 21$ cm and two shorter wavelengths is shown in fig. 3.29: here, the counts are normalized to the Euclidean 3/2 power law, shown by a dashed line. The excess counts at $S \simeq 1$ Jy are clear. So too is a sharp decrease in source counts down to $S \simeq 0.1$ mJy. Both features are explained by assuming that radio sources formed (or preexisting sources started to emit radio waves) at some epoch in the past, and that the number or luminosity of the sources then decreased following that initial burst of activity. As we count sources of lower and lower S, we are on average seeing more distant sources: as S drops below about 1 Jy, we see back to the epoch when radio sources turned on; hence the rapid decrease in their number below about 1 Jy. Condon (1988) among others presents a much more quantitative analysis of this process. For our purposes, it is sufficient to bear in mind that the normalized source counts fall off rapidly below about 1 Jy at 21 cm. At a shorter wavelength, say $\lambda = 6$ cm, the fall-off is still present, but less sharp, and occurs at $S \simeq 0.1$ Jy (see, e.g., Donnelly *et al.*, 1987).

Finally, let us focus attention on the counts of the faintest sources, at $S \lesssim 1$ mJy at 21 cm. The origin of the apparent change of slope in the $N(S)$ curve (fig. 3.29) has been controversial. Are the sources counted at $S < 1$ mJy predominantly ordinary, nearby galaxies? Is a new population of sources, such as star-forming galaxies at $z \simeq 1$ required? One working model has recently been suggested by Danese *et al*. (1987). In it, the faint source counts are ascribed to an approximately even mix of non-evolving, nearby galaxies and star-forming galaxies whose luminosity is evolving. As we will see in Section 7.3.2, the nature of these faint sources has an important bearing on attempts to measure CBR fluctuations on small angular scales: radio sources with $S \simeq 1$–100 μJy are the dominant source of foreground noise in such experiments.

Let us end this section by drawing the obvious conclusion that the evolution of radio sources makes it impossible to use radio source counts to test the geometry of the Universe.

References

Several of the papers referred to in this chapter (and other early papers in radio astronomy) have been reprinted in *Classics in Radio Astronomy* ed. W. T. Sullivan, Reidel, Dordrecht (1982); these are indicated below by '(CRA)' following the citation.

Chambers, K. C., Miley, G. K., and van Breugel, W. J. M. 1988, *Ap. J. (Lett.)*, **327**, L47.

Chandrasekhar, S. 1960, *Radiative Transfer*, Dover, New York.

Christiansen, W. N., and Högbom, J. A. 1985, *Radiotelescopes*, Cambridge University Press, Cambridge.

Clarricoats, P. J. B., and Olver, A. D. 1984, *Corrugated Horns for Microwave Antennas*, Peter Peregrinus, London.

Condon, J. J. 1988, in *Galactic and Extragalactic Radio Astronomy*, 2nd ed., eds. G. L. Verschuur and K. I. Kellermann, Springer-Verlag, Heidelberg and New York.

Danese, L., De Zotti, G., Franceschini, A., and Toffolatti, L. 1987, *Ap. J. (Lett.)*, **318**, L15.

Dicke, R. H. 1946, *Rev. Sci. Instrum.*, **17**, 268 (CRA).

Donnelly, R. H., Partridge, R. B., and Windhorst, R. A. 1987, *Ap. J.*, **321**, 94.

Fixsen, D. J., Cheng, E. S., and Wilkinson, D. T. 1983, *Phys. Rev. Lett.*, **50**, 620.

Ginzburg, V. L., and Syrovatskii, S. I. 1969, *Ann. Rev. Astron. Astrophys.*, **7**, 375.

Gordon, M. A. 1988, in *Galactic and Extragalactic Radio Astronomy*, 2nd ed., eds. G. L. Verschuur and K. I. Kellermann, Springer-Verlag, Heidelberg and New York.

Hey, J. S. 1973, *The Evolution of Radio Astronomy*, Neale Watson Academic, New York.

Högbom, J. A. 1974, *Astron. Astrophys. Suppl.*, **15**, 417.

Hummel, K. 1980, *Astron. Astrophys. Suppl.*, **41**, 151.

Jansky, K. G. 1933, *Proc. IRE*, **21**, 1387 (CRA).

Jauncey, D. L. 1975, *Ann. Rev. Astron. Astrophys.*, **13**, 23.

Kellermann, K. I., and Owen, F. N. 1988, in *Galactic and Extragalactic Radio Astronomy*, 2nd ed., eds. G. L. Verschuur and K. I. Kellermann, Springer-Verlag, Heidelberg and New York.

Kellermann, K. I., and Pauliny-Toth, I. I. K. 1981, *Ann. Rev. Astron. Astrophys.*, **19**, 373.

Keyes, R. J., ed. 1980, *Optical and Infrared Detectors*, 2nd ed., Springer-Verlag, New York.

Kraus, J. D. 1986, *Radio Astronomy*, 2nd ed., Cygnus-Quasar, Powell, Ohio.

Kulkarni, S. R., and Heiles, C. 1988, in *Galactic and Extragalactic Radio Astronomy*, 2nd ed., eds. G. L. Verschuur and K. I. Kellermann, Springer-Verlag, Heidelberg and New York.

Kundu, M. R. 1965, *Solar Radio Astronomy*, Interscience (John Wiley), New York.

Lang, K. R. 1980, *Astrophysical Formulae*, 2nd ed., Springer-Verlag, New York.

Longair, M. S. 1966, *Monthly Not. Roy. Astron. Soc.*, **133**, 421.

Low, F. J. *et al*. 1984, *Ap. J. (Lett.)*, **278**, L19.

Mandolesi, N., Calzolari, P., Cortiglioni, S., and Morigi, G. 1984, *Phys. Rev.* D, **29**, 2680.

Mather, J. C. 1982, *Applied Optics*, **21**, 1125.

Melchiorri, F., Melchiorri, B., Ceccarelli, C., and Pietranera, L. 1981, *Ap. J. (Lett.)*, **250**, L1.

Mezger, P. G., and Henderson, A. P. 1967, *Ap. J.*, **147**, 471.

Miley, G. 1980, *Ann. Rev. Astron. Astrophys.*, **18**, 165.

Nishioka, N. S., Richards, P. L., and Woody, D. P. 1978, *Applied Optics*, **17**, 1562.

Otoshi, T. Y., and Stelzried, C. T. 1975, *I.E.E.E. Trans. Instrumentation Measurement*, **24**, 174.

Pacholczyk, A. G. 1970, *Radio Astrophysics*, W. H. Freeman, San Francisco.

Pawsey, J. L., and Bracewell, R. N. 1954, *Radio Astronomy*, Oxford University Press, Oxford.

Perley, R. A., Schwab, F. R., and Bridle, A. H., eds., 1988, *Synthesis Imaging in Radio Astronomy*, Astronomical Society of the Pacific, San Francisco.

Phillips, T. G. 1988, in *Millimeter and Submillimeter Astronomy*, eds. R. D. Wolstencroft and W. B. Burton, Kluwer, Dordrecht.

Phillips, T. G., and Woody, D. P. 1982, *Ann. Rev. Astron. Astrophys.*, **20**, 285.

Reich, W. 1982, *Astron. Astrophys. Suppl.*, **48**, 219.

Rohlfs, K. 1986, *Tools of Radio Astronomy*, Springer-Verlag, Heidelberg and New York.

Ryle, M. 1952, *Proc. Roy. Soc.* A, **211**, 351 (CRA).

Ryle, M. 1968, *Ann. Rev. Astron. Astrophys.* **6**, 249.

Ryle, M. 1975, *Science*, **188**, 1071 (Nobel Lecture).

Satter, C. J., and Brown, R. L. 1988, in *Galactic and Extragalactic Radio Astronomy*, 2nd ed., eds. G. L. Verschuur and K. I. Kellermann, Springer-Verlag, Heidelberg and New York.

Shklovskii, I. S. 1960, *Cosmic Radio Waves*, Harvard University Press, Cambridge, Massachusetts.

Sullivan, W. T. 1984, *The Early Years of Radio Astronomy*, Cambridge University Press, Cambridge.

Thompson, A. R., Moran, J. M., and Swenson, G. W. 1986, *Interferometry and Synthesis in Radio Astronomy*, John Wiley, New York.

Turner, B. E. 1988, in *Galactic and Extragalactic Radio Astronomy*, 2nd ed., eds. G. L. Verschuur and K. I. Kellermann, Springer-Verlag, Heidelberg and New York.

Verschuur, G. L., and Kellermann, K. I., eds. 1988, *Galactic and Extragalactic Radio Astronomy*, 2nd ed., Springer-Verlag, Heidelberg and New York.

Weiss, R. 1980, *Annual Rev. Astron. Astrophys.*, **18**, 489.

White, G. J. 1988, in *Millimeter and Submillimeter Astronomy*, eds. R. D. Wolstencroft and W. B. Burton, Kluwer, Dordrecht.

Winston, R. 1970, *J. Optical Soc. Amer.*, **60**, 245.

Witebsky, C. 1978, COBE Report No. 5013, unpublished.

4

The spectrum of the CBR

If the microwave background radiation discovered by Penzias and Wilson is a relic of the Hot Big Bang origin of the Universe, it ought to have a thermal spectrum. The first few measurements of the spectrum of the radiation were consistent with a Planck curve with $T_0 = 3$ K, hence strengthening our belief that the microwave background was indeed cosmic in origin. In the more than 25 years that have followed the early measurements of Penzias and Wilson (1965, 1967), Roll and Wilkinson (1966), Howell and Shakeshaft (1966), Field and Hitchcock (1966) and Thaddeus and Clauser (1966), a number of increasingly precise measurements of the spectrum have been made. The range of wavelengths has been extended to 75 cm $\gtrsim \lambda \gtrsim 0.5$ mm. An early aim of these observational programs was to check the Big Bang model for the origin of the microwave radiation; another was to look for small perturbations in the spectrum of the radiation, which might have been produced by energetic processes occurring well after the Big Bang. These processes, the nature of the spectral perturbations they produce, and the limits the observations place on them, will all be discussed in Chapter 5. In this chapter, we will look at the observational techniques and the resulting values of T_0 found at different wavelengths.

We will consider two broad classes of measurement in this chapter – *direct* determinations of the temperature of the microwave background radiation (CBR), such as the work of Penzias and Wilson; and *indirect* determinations based on excitation of low-lying energy states of interstellar molecules. As we proceed, we will also examine some of the observational problems encountered in such measurements, and sources of systematic error in the results.

4.1 Direct measurements – sketch of the observational technique

The aim of such measurements is to determine the temperature (or alternatively the intensity) of the CBR at a given wavelength. The first step is to measure the intensity or temperature of radiation arriving from the zenith sky. This value is then compared with the radiation from a source of known temperature, called the cold-load calibrator (fig. 4.1). The zenith measurements must be corrected for the emission of foreground sources, such as the Earth's atmosphere. We may then express the results as follows:

$$G(S_z - S_c) = T_z - T_c + T_{offset} = T_0 + T_F - T_c + T_{offset,} \tag{4.1}$$

where S_z and S_c are the output signals from the instrument when it is directed at the zenith and at the cold-load calibrator, respectively; and G is the experimentally deter-

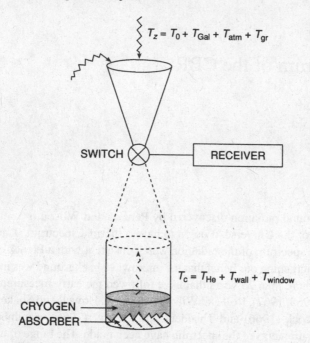

$$T_z = T_0 + T_{Gal} + T_{atm} + T_{gr}$$

SWITCH ⊗ —— RECEIVER

$$T_c = T_{He} + T_{wall} + T_{window}$$

CRYOGEN ——
ABSORBER ——

Fig. 4.1   Basis of measurements of T_0 for the CBR. T_z is the sum of T_0 and foreground emission described in this section; T_c is the cold-load comparison temperature. The receiver responds to the difference T_z-T_c.

mined conversion from the units in which S is measured to antenna temperature. T_z represents the zenith temperature, which includes the CBR signal, T_0, as well as emission from foreground sources, T_F. Finally, we include T_{offset} to allow for the possibility that the output of the instrument might change when it is moved from its zenith position to its calibrator position. Each of these terms will be treated in more detail in Sections 4.3 and 4.4.

The direct measurements described below were in most cases made with conical horn antennas, which defined beams of about 5–12° on the sky.

4.2   Determining the central wavelength and bandpass

Spectral measurements at wavelengths $\gtrsim 3$ mm have generally been made with heterodyne receivers (Section 3.4.3). These operate at a fixed wavelength, and thus respond to a relatively narrow range of wavelengths (with $\Delta\lambda/\lambda \lesssim 0.1$).

On the other hand, bolometric receivers are used for direct measurements at wavelengths less than 3 mm, and these are intrinsically broad-band devices (Section 3.4.5). In order to make a measurement of the CBR temperature at a particular wavelength, filters must be used to establish the range of frequencies over which radiation striking the antenna is allowed to reach the bolometer (fig. 4.2). The filters used must be able to block radiation at all values of λ greater than the desired wavelength and at all λ less than the desired wavelength (that is, at all higher frequencies). Such filters are not easy to construct, and rarely behave like the ideal filters shown in fig. 4.2(*a*), which have sharp cutoffs at particular wavelengths. Figure 4.2(*b*), for instance, shows the range of wavelengths passed by an actual set of filters used by Matsumoto *et al.* (1988).

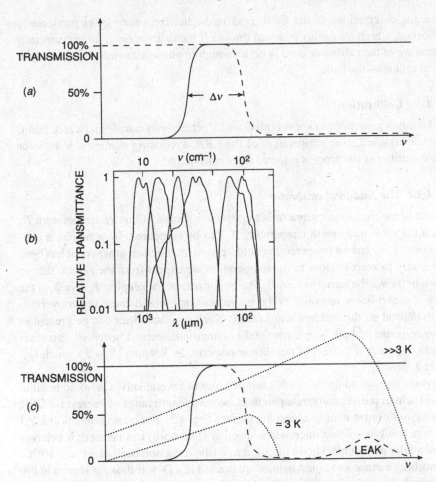

Fig. 4.2   The use of filters to define a restricted bandpass. (*a*) The solid line shows the transmission of an ideal high pass filter; the dashed line of an ideal low pass filter. (*b*) The transmission of filter sets used in actual measurements of the CBR at several short wavelengths (Matsumoto *et al.*, 1988). Note that the vertical scale is logarithmic. (*c*) Here the low pass filter has a high frequency leak (dashed line). Radiation from a high temperature ($T \gg 3$ K) calibrator (dotted line) will leak through the filter, spoiling the calibration.

The filters will also emit radiation at all wavelengths at which they absorb radiation. Hence they must be cooled to ensure that radiation from the filters does not saturate the bolometric detectors.

Finally, let me mention a more specialized point, the problem of high frequency leaks; that is, imperfections in the filters, which allow some high frequency (short wavelength) radiation to slip through to the detector, as shown schematically in fig. 4.2(*c*). What happens if the filter used to establish the short wavelength cutoff of radiation reaching the detector has such a leak? First, correcting for foreground emission becomes more difficult, because such radiation may reach the detector through the leak as well as in the desired range of wavelengths. The sources of foreground emission to be described in Section 4.4 are generally more intense at wavelengths shorter (frequencies higher) than those used for observations of the CBR, making higher frequency leaks particularly serious. Second, a high frequency leak can throw off the calibration of an instrument if the temperature of the calibrator used is high enough to allow it to emit significantly in the spectral region of the leak.

4.3   Calibration

As eqn. (4.1) shows, we need to know both G and T_c, effectively a scale and a zero point, in order to determine T_0, the temperature of the CBR. Producing numerical values for these two quantities is the process referred to as calibration.

4.3.1 The cold-load calibrator

The purpose of the cold-load calibrator is to provide a source of known temperature T_c, with which the measured zenith temperature T_z can be compared. In principle, a calibration source of *any known* temperature could be used. In practice, however, it has been found necessary to keep T_c close to the temperature we expect from the zenith, that is, several kelvin (hence the term *cold load*). The major reason for keeping T_c near T_z is the possibility of a non-linear response in the instrument that would cause the conversion factor G to depend on the incident temperature. Such non-linearities can be present in the gain or response of both heterodyne and bolometric detectors; hence all observers have sought to reduce the effect of non-linear response by keeping $(S_z - S_c)$ small, i.e., by keeping T_c near T_z.

The obvious cryogen to use to reach a temperature of several kelvin is liquid helium. Since liquid helium is essentially transparent in the wavelength range of interest for CBR observations, it is used to cool an absorbing surface (fig. 4.3). For measurements at $\lambda \gtrsim 3$ mm, a commercially available microwave absorber (Eccosorb) can be used; it behaves like an essentially perfect blackbody for $\lambda \gtrsim 3$ mm (the reflection coefficient is $\lesssim 10^{-3}$).

The absorbing surface and liquid helium are housed in a Dewar flask, as shown in fig. 4.3. The entire assembly serves as the cold-load calibrator when it is attached to an antenna of the instrument. The temperature of the absorbing surface can be measured directly by a germanium or other thermometer embedded in the absorbing material, or more accurately by measuring the pressure above the liquid helium surface.

In measurements made beneath the Earth's atmosphere, it is necessary to prevent air entering the Dewar flask and thus condensing on the cold inner walls. Hence a window is employed at the mouth of the Dewar flask, as shown in fig. 4.3.* Since the window must be at nearly ambient temperature to prevent condensation on it, it is important that the window be very transparent at the wavelength of observation. If it absorbs a fraction ε_w of incident radiation at that wavelength, it will also emit radiation with an antenna temperature $T_w \simeq \varepsilon_w T_{amb}$, where T_{amb} is the ambient temperature.† In the cold load shown in fig. 4.3, polyethylene of 23 μm thickness was used for the window, and $\varepsilon_w \lesssim 10^{-4}$ for the window. By measuring the emission from much thicker slabs of polyethylene, ε_w could be determined with sufficient accuracy to allow T_w to be calculated to an accuracy of a few millikelvin, and eventually subtracted out.

---

* In some experiments, an antenna is kept coupled to the cold load at all times, so that a window is unnecessary.

† The window acts like an emitting/absorbing cloud, as described in Section 3.6.

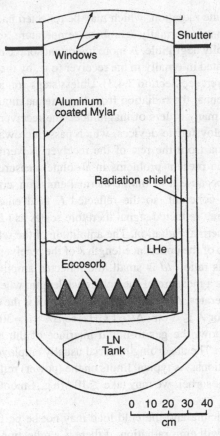

Fig. 4.3   Calibration cold load used in a series of measurements of T_0 in the wavelength range 12 cm $\geqslant \lambda \geqslant$ 0.33 cm; see Section 4.6.4 for details. The antennas used were coupled to the top of the cold load (with permission of G. Smoot).

The last paragraph illustrates the point that the cold load temperature T_c may not be identical with the temperature of the absorbing surface. Another source of spurious radiation is the inner surface of the Dewar flask above the cryogen. The inner walls will range in temperature from approximately ambient temperature at the top to the temperature of the liquid helium at the bottom. Unless they are perfectly reflective, they will emit radiation. Clean metal surfaces are, fortunately, highly reflective at microwave frequencies, so that the absorption coefficient, ε_{wall}, can be kept small (typically $\lesssim 10^{-4}$).* In addition, by using a Dewar flask with a large inner diameter, the walls can be kept far from the beam defined by the antenna coupled to it. Nevertheless, a correction for the radiation from the walls, T_{wall}, must often be made. Clearly, to make such a correction, we need to know not only ε_{wall} but also the temperature profile along the inner surface of the Dewar flask.

So far, we have been considering sources of spurious *emission* in a cold load. *Reflections* may also affect T_c. When an antenna is coupled to the cold load, it receives

---

* The inner wall of the cold load shown in fig. 4.3 was 13 μm thick aluminum coated Mylar; it was kept thin to reduce the thermal conductivity.

radiation; but the antenna also emits some radiation, which may be reflected back into it, and thus reach the detector. The antenna is normally at ambient temperature, so some thermal emission is present, though usually negligible. A more important source of radiation leaving the antenna is noise generated internally in the receiver (e.g., by the mixer or local oscillator in a heterodyne receiver; see Section 3.4.3). Unless steps are taken to isolate the local oscillator from the antenna, the radiation from the antenna mouth can exceed the purely thermal emission by many orders of magnitude. The receivers used for spectral observations therefore employ ferrite devices, which pass microwaves in only one direction to isolate the antenna from the rest of the receiver. Nevertheless, reflected emission from the receiver can present problems in absolute measurements employing superheterodyne receivers, because it is, unlike thermal emission, coherent radiation. The reflected signal is also coherent, so the reflected E field enters the receivers with a fixed phase. The resulting reflected signal therefore scales as $|E|$ and not $|E|^2$ as is the case for incoherent, thermal radiation. The amplitude of the reflected coherent signal also depends on the ratio of the coherence length l_c of the receiver to the distance to the reflecting surface. If this ratio l_c/d is small, the reflected amplitude is small. We may take $l_c = |\Delta v/v|\lambda = \Delta\lambda$ as a measure of the coherence length, where λ is the wavelength at which the receiver operates, Δv is its bandwidth, and $\Delta\lambda$ is the equivalent wavelength bandpass. Typically, for $\lambda \lesssim 3$ cm, $\Delta v \approx 1$ GHz, so $\Delta\lambda = l_c \approx 30$ cm.

Reflections can occur from the window, the gas-to-liquid interface of the helium cryogen, and the surface of the absorber. The absorbing material usually employed has a wavy surface to reduce specular reflections. As typical limits on the (power) reflection coefficients of the window, helium and absorber, we may take $\lesssim 10^{-4}$ (e.g., Smoot *et al.*, 1983; Johnson 1986).

Finally, the coupling between the antenna and the cold load may not be perfect. If some loss is involved, the imperfection will emit radiation; if there is a reflection, local oscillator power will be reflected back to the detector. These effects are generally lumped together in the offset term discussed in Section 4.3.3.

If we set aside the offset term, we may then express the cold load temperature as a sum of contributions:

$$T_c = T_{abs} + T_w + T_{wall} + T^\dagger_{reflec},\qquad(4.2)$$

where T_{abs} is the temperature of the absorbing surface immersed in liquid helium. Because of the special properties of the spurious signals resulting from reflections, their contribution is marked with a dagger.

4.3.2 Secondary calibration (determining G)

The conversion factor G is determined by measuring the output signals S_1 and S_2 when blackbodies of two different temperatures are applied to the antenna. Then

$$G = \frac{T_1 - T_2}{S_1 - S_2}.$$

Ordinarily, the higher of the two temperatures is ambient, and either the zenith sky* or the cold load is used for the other.

---

* If the sky is used, we are implicitly assuming that we already know T_0. However, an uncertainty of ± 0.3 K in T_0 will result in an uncertainty of only about 10^{-3} in G if $T_{amb} \simeq 300$ K.

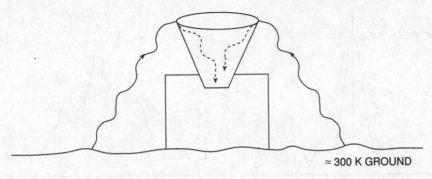

≈ 300 K GROUND

Fig. 4.4   Pickup of radiation from the ground caused by diffraction around the rim of a conical antenna.

In most of the experiments discussed in this chapter, *G* was measured frequently by placing an ambient temperature absorber over the mouth of the antenna when it was pointed at the zenith. If the response of the instrument is linear, so that *G* is a constant, then no further steps are needed. To measure any non-linearity, at least one additional measurement at a third value of temperature is needed. For instance, a liquid nitrogen load may be used. Where substantially non-linear response is suspected, a range of calibrating temperatures, preferably bracketing T_z and T_c, must be used (see, e.g., Johnson and Wilkinson, 1987; or Matsumoto *et al.*, 1988).

4.3.3 The offset term

A number of messy instrumental effects contribute to the offset term, T_{offset}, in eqn. (4.1). If the coupling between antenna and cold load is not perfect, radiation may be emitted or reflected, adding to S_c. Even when the antenna views the zenith sky, some radiation may be reflected from the antenna mouth. In addition, moving the instrument from its zenith alignment to its cold load alignment may mechanically stress the waveguide or other elements of the receiver, in turn producing a change in the output signal.

All these effects depend very sensitively on the particular instrument and observing technique used. Thus it is difficult to discuss them in general terms. As a consequence, we will move on after making the obvious remark that observers try to design instruments to keep T_{offset} small, and then try to measure the various contributions to it.

4.4   Sources of foreground emission

We turn now to contributions to the zenith temperature, T_z, other than the cosmic background. In the order in which we will treat them here, these are radiation from the ground diffracted into the antenna, radiation emitted by the Galaxy or other astronomical sources, and radiation from the Earth's atmosphere lying above the antenna.

4.4.1 Back- and side-lobe pickup of radiation from the ground

Even when an antenna is pointed at the zenith, microwave radiation from the ground can enter it by diffracting around the mouth of the antenna (fig. 4.4). The open mouth of a conical horn antenna of the type used for spectral measurements acts as a diffracting aperture with diameter typically an order of magnitude larger than the wavelength at

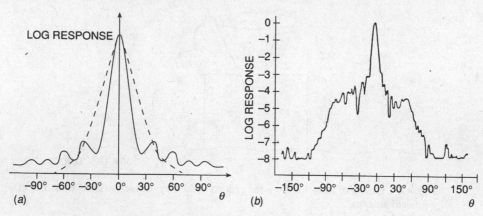

Fig. 4.5   (*a*) Schematic diffraction pattern of a horn antenna (solid line). The dashed line shows schematically the effect of employing corrugations in the walls of the horn, or of flaring the antenna: lower resolution but also much-reduced side lobes. (*b*) Actual measured beam pattern of the Soviet satellite-borne instrument described in Section 6.7.1.

which the observations are made. In addition to the central diffraction maximum (called the 'main beam' by radio astronomers), secondary diffraction maxima are present. These may extend to angles > 90° away from the optical axis, as shown schematically in fig. 4.5. Thus an instrument pointing at the zenith will respond to radiation entering these secondary diffraction maxima of the antenna pattern as well as the main beam. These secondary maxima are called 'side lobes' (for $\theta < 90°$) and 'back lobes' (for $\theta \geqslant 90°$) of the antenna.

Because the Earth's surface is about 300 K and subtends a full 2π sr, it can contribute a substantial signal to T_z even if the back lobes are much smaller in amplitude than the main beam. This unwanted contribution to the measured temperature of the zenith sky will be written as T_{gr}.

Two experimental techniques are used to reduce T_{gr} to a manageable level. First, the horn antennas are designed to have low amplitude side and back lobes. This is accomplished (see, for instance, Gorenstein *et al.*, 1978; or Bielli *et al.*, 1983; and fig. 4.6) by machining $\lambda/4$ corrugations into the walls of the antennas; these serve to prevent radiation flowing along the antenna walls. As a result, the response of the antenna is peaked at the center of its aperture, and falls away at the edges. A corrugated antenna is therefore less sensitive to radiation diffracted at its rim, so that its side- and back-lobe response is lower*. On the other hand, because the response is concentrated towards the center of the antenna, the effective aperture is smaller than the physical aperture of the antenna, and, as a consequence, the main beam has a larger solid angle than would be expected from an aperture of that physical diameter. Using a corrugated (or flared) horn antenna thus reduces side lobes and back lobes at the expense of some loss in angular resolution.

Corrugated horns can readily be constructed with all side lobes smaller than 10^{-4} of the amplitude of the main beam, and with back lobes $\lesssim 10^{-6}$ of the main beam response.

* A similar result may be achieved by flaring the antenna, like the bell of a trumpet (Mather, 1982). These techniques are similar to apodizing optical elements.

Fig. 4.6   Photograph of a horn antenna used in a measurement of the CBR at 6 cm (Mandolesi *et al.*, 1986); note the corrugations machined into the horn walls.

At that level, radiation from the 300 K surface of the Earth will contribute $\lesssim 0.05$ $(\theta/10°)^{-2}$ K, where θ is the full width of the main beam in degrees.

Even that small contribution can be reduced by an order of magnitude by a second technique – the use of reflecting screens to shield the antenna from direct radiation from the Earth (fig. 4.7). These ground screens, as they are called, can be constructed from sheet metal or even wire mesh for low frequencies. If the ground screens are arrayed as in fig. 4.7, they effectively reflect away radiation from the ground and replace it by the much less intense radiation of the sky.

Direct measurements of the temperature of the CBR made in the past decade have generally employed both corrugated horns and ground screens. While the magnitude of T_{gr} varies from experiment to experiment, a typical value for carefully designed apparatus is $\lesssim 3$ mK. In addition, by altering the configuration of the ground screens or otherwise, the observers can measure T_{gr} to within about 30% accuracy, so that it contributes very little to the uncertainty of these spectral measurements (see, e.g., Smoot *et al.*, 1985).

4.4.2 Emission from astronomical sources, especially the Galaxy

Let us begin by considering the contributions from celestial radio sources which are small relative to the size of the beams (5–12°) employed in direct measurements of the spectrum – these may be referred to as point sources, even if they can be resolved by the eye.

Radiation from the Sun and Moon into either the main beam or side lobes may be avoided by making observations when those two sources are near or below the horizon. The next brightest solar system source is Jupiter. Suppose we directed an antenna with

≈ 300 K GROUND

Fig. 4.7   The use of ground screens to reduce side- and back-lobe pickup from the ground.

a main beam of 10° full width at Jupiter. The resulting antenna temperature would be (Section 3.6)

$$T = \frac{\Omega_J}{\Omega_{beam}} T_J,$$

where $T_J \simeq 150$ K is the temperature of Jupiter at microwave wavelengths (Kraus, 1966). From the figures given above and the angular diameter of Jupiter (0.8′), we find $\Omega_J/\Omega_{beam} \simeq 1.8 \times 10^{-6}$, so Jupiter would contribute only 0.3 mK to T_z even if we looked directly at it.

The same kind of calculation may be made for other discrete radio sources in the Galaxy or beyond it (such as Taurus A or Cygnus A). From eqn. (3.7) and eqn. (3.19), we have

$$T = \frac{S\lambda^2}{2k\Omega}$$

for the antenna temperature of a source of flux density S observed at the center of a beam of solid angle Ω. If we express S in the usual units of Jansky, and again take 10° as a typical value for the full width of the antenna beam, we find numerically

$$T \approx 10^{-6} S\lambda^2 \text{ K},$$

with λ in centimeters; hence radio sources with $S \lesssim 10^4 \lambda^{-2}$ Jy make less than about 0.3% contribution to T_z*. For wavelengths less than, say, 20 cm, there are only a dozen or so sources with $S > 25$ Jy, and these can be avoided or if necessary corrected for.

Now we must turn to the extended diffuse radio emission of the Galaxy as a whole, which *can* add substantially to T_z. Two mechanisms give rise to the extended radio emission from our Galaxy.

The first of these is the synchrotron process: the acceleration of relativistic electrons as they spiral around the magnetic fields in interstellar space (see Kraus, 1966; Pacholczyk, 1970; or Section 3.7.2 for details). Both the energetic electrons and the

---

* Sources do, however, place important limits on measurements of the *isotropy* of the CBR; see Section 7.3.2.

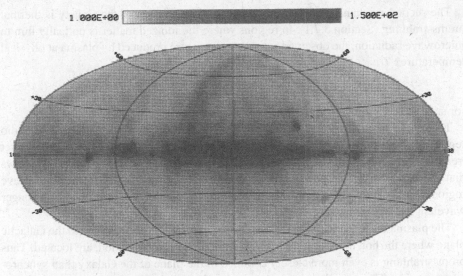

1.000E+00   1.500E+02

Data units are Degrees Kelvin (Log Scale)

Fig. 4.8 A map of the Galaxy at a frequency of 408 MHz (Haslam *et al.*, 1982). Galactic coordinates are shown. At this low radio frequency, synchrotron radiation dominates. While most emission is concentrated on the Galactic plane, there are clearly regions of emission at higher latitudes as well. Note also that the radio emission of the Galaxy at this frequency can greatly exceed T_0: the grey scale of this figure runs from 1 to 150 K.

magnetic fields are largely confined to the disk of our spiral Galaxy, so the synchrotron radiation is also confined to a band around the sky aligned with the plane of the Galaxy (fig. 4.8). Synchrotron emission is brightest towards the Galactic center and weaker in the anticenter direction. The radio frequency spectrum of synchrotron radiation depends on the energy spectrum of the relativistic electrons which produce it. If the energy spectrum is a power law

$$N(E)\, dE \propto E^{-\alpha}\, dE,$$

as often appears to be the case, then the spectrum of the emitted synchrotron radiation will also be a power law (eqn. (3.41)):*

$$S(\nu) \propto \nu^{(1-\alpha)/2}$$

or

$$T(\lambda) \propto \lambda^{(\alpha+3)/2}.$$

In our Galaxy, α is close to 2.6, so $T(\nu) \propto \nu^{-2.8}$, with a scatter of ± 0.2 in the index (Lawson *et al.*, 1987). Synchrotron radiation, then, will affect primarily low frequency (long wavelength) spectral measurements.

* I neglect here synchrotron self-absorption and other details, which have no bearing on the diffuse emission of the Galaxy at microwave frequencies (for details, consult Kraus, 1966; or Pacholczyk, 1970).

The second mechanism responsible for the radio emission of the Galaxy is thermal bremsstrahlung (Section 3.7.1). In regions where the ionized matter is optically thin to microwave radiation, the observed antenna temperature produced by plasma at physical temperature* T is

$$T = T(1 - e^{-\tau}) \approx T\tau$$

for small τ.

To a good approximation, the emission coefficient of ionized matter in the radio regime is proportional to λ^2. Thus both τ and T are also proportional to λ^2. A more exact treatment (e.g., Chapter 3 of Verschuur and Kellermann, 1974, or Section 3.7.1) shows that the exponent of the power law dependence lies closer to 2.1 in the microwave region. We see once again that Galactic emission presents a larger problem for longer wavelength measurements.

The plasma responsible for thermal bremsstrahlung is tightly confined to the Galactic plane where the hot, luminous stars, which emit the ionizing radiation, are located. Thus bremsstrahlung is even more closely confined to the plane of the Galaxy than synchrotron emission. There are, however, some isolated local clouds of plasma, which appear above or below the Galactic plane (fig. 4.8).

The fact that both sorts of emission are largely confined to the Galactic plane suggests that foreground emission from the Galaxy can be minimized by making observations near one or the other of the Galactic poles. The north Galactic pole region, at $\alpha \simeq 13\,\text{h}$ right ascension and $\delta \simeq +30°$ declination is accessible to observers in the northern hemisphere.

In addition to avoiding regions of intense emission in the Galactic plane, observers may also correct each zenith measurement for the Galactic emission $T_{gal}(\alpha, \delta)$ along that line of sight. These corrections are derived from maps made of Galactic radio emission, such as those shown in fig. 4.8 and fig. 6.4 later. If the map has been made at a frequency different from that used in the CBR measurement, it is necessary to extrapolate to the observing frequency using an appropriate combination of the two power laws described above (e.g., Witebsky, 1978). To give a sense of the magnitude of the corrections required, let us consider a pair of observations, one made near the north Galactic pole and the other in the plane at Galactic longitude 170°. For measurements made at 2.5 GHz ($\lambda = 12$ cm), the Galactic contributions are 0.12 and 0.27 K, respectively (see Sironi and Bonelli, 1986); for observations at 10 GHz ($\lambda = 3$ cm), the corresponding corrections are only about 0.003 and 0.015 K.

Synchrotron and bremsstrahlung processes dominate the Galactic emission for all wavelengths above about 3 mm. At shorter wavelengths, and especially in the submillimeter region, a new source of emission takes over, namely thermal emission from warm dust in the Galaxy. The dust is heated by starlight, and reradiates at its own characteristic temperature of about 20 K. Since the dust grains are very small ($r \ll 1$ mm), their emissivity varies strongly with wavelength, and hence frequency. As a consequence the antenna temperature of the warm dust also has a strong wavelength dependence, $T \propto \lambda^\gamma$ where γ lies in the range -1 to -2. Thus dust emission influences spectral observations of the CBR only at the shortest wavelengths, below 1 mm.

---

* Typical values of T are about 10^4 K for ionized regions in our Galaxy.

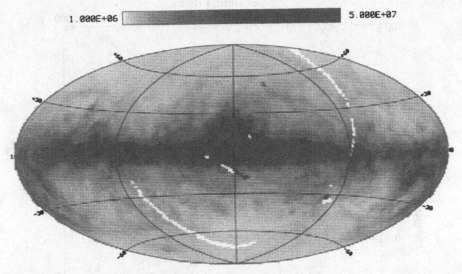

Data units are Log(Jansky/Steradian)

Fig. 4.9   A map of the Galaxy at $\lambda = 100$ µm ($\nu \approx 3000$ GHz) made by the IRAS satellite, again in Galactic coordinates. Pure white regions are areas where data are absent. Emission from warm ($T \sim 20$ K) dust dominates; note that patchy emission is present even at high Galactic latitudes, especially at $l \sim 180°$ (NASA diagram, with permission).

Thermal emission from dust is most intense in the Galactic plane, but is *not* confined to the plane. That fact is demonstrated vividly in fig. 4.9, a map of the Galaxy made at 0.1 mm (100 µm) by the IRAS satellite. Wispy emission (called 'infrared cirrus' because of its appearance) is present at all Galactic latitudes; as may be seen, its distribution is very irregular. The patchy distribution of the dust makes it harder to correct for its contribution to the temperature of the zenith sky. In addition, a large and uncertain extrapolation is required from the wavelength of the IRAS map (0.1 mm) to the wavelengths of interest for CBR measurements (generally $\lambda \gtrsim 0.5$ mm). Data from the four IRAS channels at 100, 60, 25 and 12 µm indicate that a λ^{-2} emissivity law provides the best fit to emission, but confirming work at $\lambda > 0.1$ mm is needed. The COBE satellite, to be described in Section 4.9, is beginning to provide important data on Galactic dust emission.

Our discussion of various sources of Galactic emission is summarized in fig. 4.10. The antenna temperature (vertical) scale applies to observations made near the Galactic pole. As fig. 4.10 shows, Galactic emission is $\lesssim 1\%$ of the roughly 3 K temperature of the CBR in the broad wavelength range $6 \gtrsim \lambda \gtrsim 0.07$ cm. Not surprisingly, the most accurate determinations of T_0 have been made in this wavelength range.

4.4.3   Emission from the Earth's atmosphere

The atmosphere of the Earth is not perfectly transparent at microwave frequencies, and hence behaves like an absorbing/emitting foreground cloud when we observe the CBR

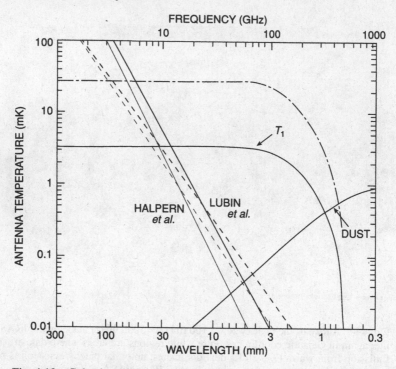

Fig. 4.10   Galactic emission, expressed in antenna temperature, measured or expected near the Galactic pole. The solid and dashed lines dropping rapidly with frequency are the synchrotron and bremsstrahlung contributions, respectively. Two models are shown, from Lubin and Villela (1985) and Halpern *et al.* (1988), with slightly different assumptions about the spectral index. Also shown (dot dash line) is 1% of the CBR spectrum and (solid line) the amplitude of the dipole component, T_1.

through it. Although it is only approximately true, let us make the assumption that the atmosphere has the same physical temperature T_{ph} at all heights; then eqn. (3.37) applied to CBR measurements through the atmosphere gives

$$T = T_0 e^{-\tau} + T_{ph}(1-e^{-\tau}) \approx T_0(1-\tau) + T_{ph}\,\tau \qquad (4.3)$$

The approximate relation holds only when the optical depth τ is much less than unity, but that has been the case for all spectral observations made to date. Since $T_{ph} \simeq 250$ K, it is easy to see that the main problem caused by the atmosphere is *emission*, not the absorption of the CBR radiation. As we shall see, the atmospheric emission, $T_{atm} = T_{ph}(1-e^{-\tau}) \approx T_{ph}\,\tau$, exceeds T_0 for all wavelengths $\lesssim 3$ cm for sea-level observations. It has proven to be the dominant source of statistical and systematic error in most measurements of the spectrum of the CBR.

At wavelengths between about 1 mm and 100 cm, where virtually all CBR spectral measurements have been made, the opacity of the atmosphere is dominated by oxygen and water vapor (we will mention the contribution of other components of the atmosphere later in this section). For both molecules, line and continuum absorption contributes to the opacity, and hence to the antenna temperature of the atmosphere.

Oxygen. Because of its symmetry, the O_2 molecule has no electric dipole moment. Hence its spectral lines result from magnetic dipole transitions, and as a consequence are

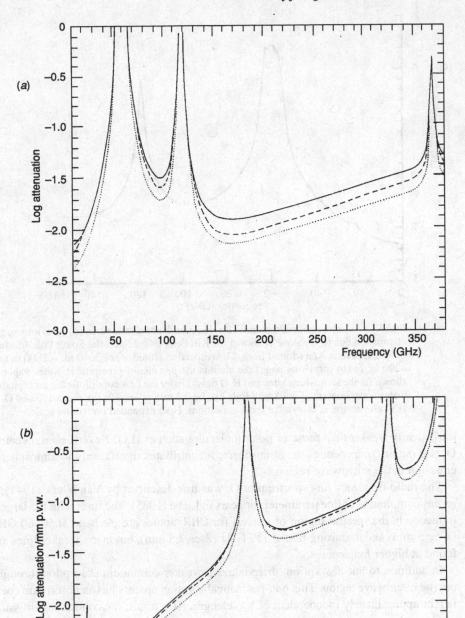

Fig. 4.11.   (*Legend on p 118.*)

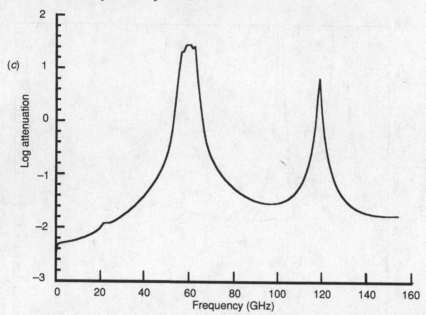

Fig. 4.11   Atmospheric attenuation for good, high altitude sites. In (*a*) (previous page) attenuation due to O_2 alone is shown for Kitt Peak (solid line), the South Pole (dashed line) and Mauna Kea (dotted line). The respective altitudes are 2040 m, ~ 2800 m and 4200 m. In (*b*) (previous page) the attenuation *per mm* of precipitable water vapor is shown for the same three sites and H_2O only. Under the best conditions the precipitable water vapor content at the South Pole can be 0.3 mm or less. In (*c*), the combined O_2 + H_2O attenuation is shown for these conditions. Note expanded horizontal scale.

intrinsically weaker than those of polar molecules such as H_2O. Nevertheless, because O_2 is a major component of the atmosphere, it contributes significantly to atmospheric emission in the microwave region.

The radio frequency *line* spectrum of O_2 was first described by Van Vleck (1947); a recent compilation of line parameters appears in Liebe (1985). The lines with the largest influence in the spectral region of interest for CBR studies are the band at 50–60 GHz ($\lambda \simeq 5$ mm) and the strong line at 119 GHz ($\lambda \simeq 2.5$ mm), but many weaker lines are found at higher frequencies.

In addition to line absorption, there is much weaker continuum absorption throughout the microwave region. This non-resonant absorption process has an absorption coefficient approximately independent of wavelength. As a result, the continuum emission of O_2 in the atmosphere produces an antenna temperature at centimeter wavelengths, which is also approximately independent of λ ($T \simeq 1$–2 K at sea level). For further details, consult Waters (1976) and Liebe (1985), or for discussions slanted towards CBR observations, Weiss (1980), Costales *et al.* (1986) or Danese and Partridge (1989).

As fig. 4.11(*a*) shows, emission in the broad wings of the 60 GHz band exceeds the non-resonant contribution for all wavelengths less than 1–2 cm, depending on altitude. Given that fact, it is worth noting that observations made as part of a program to determine the CBR spectrum (Smoot *et al.*, 1985) indicate that the O_2 contribution of the 60 GHz band derived from Liebe's (1985) model may actually underestimate O_2 emission by 15% (Danese and Partridge, 1989).

Water Vapor. As the existence of clouds shows, the distribution of H_2O in the atmosphere is highly variable in both space and time, unlike the case for O_2. Since observers can easily avoid making measurements in foggy or cloudy weather, we will ignore any atmospheric emission from liquid droplets or ice crystals, and treat only water vapor. Even water vapor, however, is very non-uniform. As a consequence, H_2O emission is the major source of statistical error or 'noise' in the measurement of T_{atm}.

H_2O is a polar molecule; its microwave spectral lines at 22, 183 and 325 GHz are electric dipole transitions between rotational levels of the ground state (see, e.g. Waters, 1976, or Liebe, 1985). Because these are electric dipole transitions the line strength is large, so the absorption in the H_2O lines is comparable to that in the O_2 lines even though water vapor is a minor constituent of the Earth's atmosphere.

Water vapor also absorbs (and hence reemits) microwave radiation at wavelengths far from its strong lines. Since most CBR measurements are made in the spectral windows between H_2O lines, this continuum absorption is important. Unfortunately, it is not well understood (see Deepack *et al.*, 1980). Several authors have proposed empirical corrections to the line absorption, which fit the measured values of H_2O continuum absorption rather well (e.g., Zammit and Ade, 1981). These models for the continuum absorption have a strong wavelength dependence, λ^{-2}, so the atmospheric emission from H_2O also varies as λ^{-2} far from the strong lines (and where the optical depth is small). The rapid rise in atmospheric emission as the wavelength of observation moves below about 1 cm is due to H_2O continuum reemission (see fig. 4.11b).

Since the amount of H_2O in the atmosphere varies in both space and time, H_2O absorption must be specified in terms of the total amount of H_2O overhead, the *column density*. The units usually employed are 'millimeters of precipitable water vapor'; one imagines that all the water vapor in a vertical column through the atmosphere above the observer has been compressed into liquid form, then measures its depth in millimeters. In good, dry conditions at sea level, the column density of water vapor in the atmosphere is about 10–15 in these units; under comparable conditions at 4000 m altitude, the column density is about 3 mm precipitable. For a number of reasons, the water vapor contribution to T_{atm} is not exactly linearly dependent on the column density of water vapor, as might at first be expected (see references above and Danese and Partridge, 1989, for more detail). The dependence on H_2O column density of the high frequency, short wavelength, emission of the atmosphere is reflected in the difference between the curves of fig. 4.11.

As in the case of Galactic emission, observers cope with the atmosphere by first minimizing it, and then measuring the residual contribution and subtracting it from T_z. To reduce atmospheric emission, ground based observations are made from high altitude sites. Working at high altitude has the direct advantage of reducing the amount of atmosphere overhead; this is particularly true for H_2O since its density falls to 1/e of its sea-level value at very roughly 2500 m altitude. There is an indirect advantage as well: the lower atmospheric pressure results in less pressure broadening of the molecular lines, hence making them narrower as well as weaker. The advantage of working at mountain-top altitudes (about 4000 m) is reflected in fig. 4.11. In addition, since the water vapor emission is variable in both space and time, uncertainties in T_{atm} can be reduced by working at a site that is dry and cold as well as high.

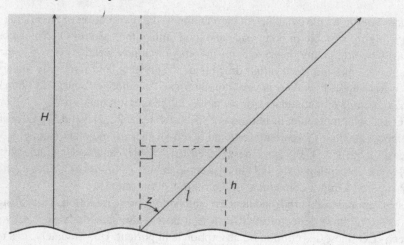

Fig. 4.12   Basis of the technique used to measure T_{atm}; the measured antenna temperature depends on the path length, l, through the atmosphere, which increases as sec z as the antenna is tipped to larger zenith angles, z.

Even at around 4000 m altitude, $T_{atm} \gtrsim T_0$ for $\lambda \lesssim 1.2$ cm, so that the atmospheric contribution to the zenith sky temperature must be measured with care if we are to determine T_0 accurately. T_{atm} is measured by making measurements of the antenna temperature of the sky at different angles from the zenith – a technique radio astronomers refer to as 'horn tipping,' as fig. 4.12 suggests.

As a first approximation, let us assume that the atmosphere may be represented by a stratified, plane-parallel slab. The optical depth along a line of sight at an angle z from the zenith* may be written

$$\tau(z) = \int_0^L k_l(l) \, dl = \int_0^H k_l(h) \ \sec \ z \, dh$$

where k_l is the absorption coefficient in units of reciprocal meters. The optical depth at the zenith, however, is

$$\tau(0) = \int_0^H k_l(h) \, dh.$$

So that, in the plane-parallel approximation,

$$\tau(z) = \tau(0) \sec z. \tag{4.4}$$

If we now add the assumption that $\tau(z) \ll 1$, then, from eqns. (4.3) and (4.4)

$$T_{atm}(z) = T_{atm}(0) \sec z. \tag{4.5}$$

Finally, recall that the CBR itself is quite isotropic, so T_0 may be taken to be independent of direction. Hence we may write

$$T(z) = T_0 + T_{atm}(z) = T_0 + T_{atm}(0) \sec z,$$

$$T(0) = T_0 + T_{atm}(0)$$

---

* The zenith angle z should not be confused with redshift (Chapter 1).

Table 4.1. *Horn-tipping measurements made at* $\lambda = 3$ *cm at about 3800 m altitude. Using eqn. (4.6), these measurements yield* $T_{atm}(0) = 1.19 \pm 0.05$ *K (see Partridge et al., 1984).*

Zenith angle, z	sec z^*	$T(z)$ (K)
0°	1.0	3.80 ± 0.09
32°	1.20	4.03 ± 0.09
43°	1.40	4.27 ± 0.13
51°	1.60	4.51 ± 0.08
55°	1.78	4.79 ± 0.05

* Convolved with beam.

for the antenna temperature observed at zenith angles z and 0, respectively. Hence $T_{atm}(0)$, the antenna temperature of the atmosphere at the zenith, is given by

$$T_{atm}(0) = \frac{T(z) - T(0)}{\sec z - 1}. \qquad (4.6)$$

This expression for $T_{atm}(0)$ is valid only if $\tau(z)$ is small and if all the contributions to $T(z)$ other than atmospheric emission are independent of direction. The latter is true for the CBR, but not for the Galactic contribution; hence corrections for Galactic emission have to be made before eqn. (4.6) is applied (e.g., Partridge *et al.*, 1984). The offset introduced by Galactic emission is not the only complication that we must take into account when using horn tipping to determine T_{atm}. Our derivation of eqn. (4.5) is based on the implicit assumption that the observations are made at an exact value of z; in fact the antennas used typically have beams of 5–12° width, so that they receive radiation from a range of zenith angles. It is therefore necessary to convolve the beam pattern with the secant of the zenith angle to find the actual coefficient of $T_{atm}(0)$ in eqn. (4.5). Finally, the Earth's atmosphere is not in fact a plane-parallel slab – it has curvature, and the curvature makes a correction (small for $z < 60°$) necessary in the sec z dependence of $T_{atm}(z)$.

As eqn. (4.6) shows, a single measurement of $T(z)$ at any zenith angle and another at the zenith would in principle suffice to determine the atmospheric emission at the zenith. However, any error in the measured value $T(z)$ is multiplied by $(\sec z - 1)^{-1}$, so that measurements at large z give more accurate values of $T_{atm}(0)$; on the other hand, measurements at large z increase the chance of side-lobe pickup. Thus, in general, observers have chosen to make observations at a set of values of z, generally on both sides of the zenith, in order to minimize the effect of instrumental asymmetries as well as possible angle-dependent side-lobe pickup, etc. Table 4.1 provides one example of such a data set; from the values in table 4.1, the reader should be able to show that $T_{atm}(0) = 1.19 \pm 0.05$ K.

Because it is the dominant source of error in measurements of the spectrum of the CBR, we have so far emphasized atmospheric *emission*. We should not forget, however, that radiation passing through the atmosphere will also be *absorbed* by it (see eqn. (4.3)). The value of T_0 that we determine at the surface of the Earth must therefore be corrected by a factor e^τ to give the true CBR temperature outside the atmosphere. To make this correction, we must know τ, not just the product $T_{ph}\tau$. An approximate value of τ may be found by assuming that $T_{ph} = 250$ K, so that

$$\tau = T_{atm}(0)/250.$$

For $\lambda \gtrsim 3$ mm, τ found in this way is $\lesssim 0.1$, so that an error of 10–20 K in T_{ph} will introduce less than 1% uncertainty in the final, corrected value of T_0. A more exact value for T_{ph} for observations at a particular wavelength may be found by convolving a vertical temperature profile of the atmosphere with the distribution of O_2 and H_2O (see Costales *et al.*, 1986, or Danese and Partridge, 1989, for instance). At $\lambda = 1$ cm, the effective physical temperature is $T_{ph} = 252$ K; and at $\lambda = 3$ mm, $T_{ph} = 260$ K for an atmospheric model appropriate for summer observations at mid latitudes and at about 4000 m altitude.

For wavelengths < 3 mm, atmospheric emission makes accurate ground-based measurements of the CBR spectrum essentially impossible even at mountain-top altitudes. All measurements* of the CBR spectrum at wavelengths below 3 mm have been made from rockets, balloons or, recently, a satellite. Provided that observations are not begun until the apparatus has left the atmosphere, rocket experiments are entirely free of atmospheric emission. In the case of balloon observations, there is of course some atmosphere above the instrument. At wavelengths < 3 mm (and at high altitudes), oxygen and water vapor are no longer the only molecules contributing to atmospheric opacity. In particular, there are numerous lines of ozone to consider. In fig. 4.13, adopted from Woody and Richards (1979), is shown the calculated emission spectrum expected at a typical balloon altitude of 40 km. Also shown in fig. 4.13 is the observed spectrum of the sky with the CBR component identified; even at 40 km altitude, atmospheric emission swamps the CBR for $\lambda < 0.8$ mm. Rocket and satellite observations thus have provided our only reliable values of T_0 at $\lambda \lesssim 1$ mm.

4.5   Summary of sources of error in direct measurements

Before listing the results of a score of direct measurements of T_0 at different wavelengths, let us summarize all the contributions lumped together in T_F and T_c in eqn. (4.1). From Section 4.3, we have

$$T_c = T_{abs} + T_w + T_{wall} + T^\dagger_{reflect}, \tag{4.7}$$

where

T_{abs} = temperature of the cold-load calibrator surface;
T_w = antenna temperature contributed by Dewar flask window, if any;
T_{wall} = antenna temperature contributed by Dewar flask walls;
$T^\dagger_{reflect}$ = effective antenna temperature of radiation reflected by cold load.

From Section 4.4, we have

$$T_F = T_{gr} + T_{Gal}(\alpha, \delta) + T_{atm}(0) \tag{4.8}$$

for zenith observations, where

T_{gr} = antenna temperature from back-lobe pickup of ground radiation;
T_{Gal} = Galactic emission (depends on direction);
T_{atm} = emission by Earth's atmosphere (for ground based or balloon observations).

---

* A single exception is the measurement in the 1.0–1.4 mm atmospheric window made from Testa Griga in the Alps, at an altitude of 3500 m (Dall'Oglio *et al.*, 1976). Difficulties with the calibration of the bolometric receiver allowed the observers to set an upper limit only on T_0; $T_0 < 2.7$ K.

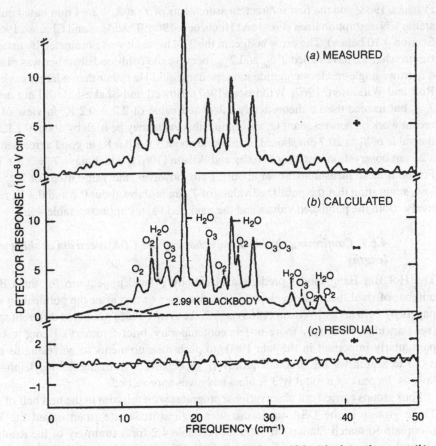

Fig. 4.13   Atmospheric emission at balloon altitudes (~ 40 km) is shown by curve (*b*). While the cm- and mm-wave emission is weak, there are strong O_3 (ozone), O_2 and H_2O at submillimeter wavelengths, starting at ~12 cm^{-1}. From Woody and Richards (1979). In this figure, curve (*a*) shows the raw measurements; curve (*c*) the residual signal after subtraction of the atmosphere and an ~3 K blackbody (the CBR).

All of these quantities in addition to T_{offset} must be measured or calculated before a set of measurements of the zenith sky and a cold-load calibrator can yield a value of T_0.

4.6   Results of direct measurements of T_0 from the ground

Now that we have reviewed the techniques employed for, and the many potential sources of error in, measurements of the CBR spectrum, let us turn to the results of the ground based measurements made over the past 25 years. To avoid unnecessary detail and repetition I have grouped similar observing programs together. I have also chosen to present the results in approximately historical sequence.

Within a year of publication of the detection of the CBR at $\lambda = 7.4$ cm (Penzias and Wilson, 1965), its spectrum had been measured at $\lambda = 3.2$ cm (Roll and Wilkinson, 1966) and at $\lambda = 20.7$ (Howell and Shakeshaft, 1966). As noted in Chapter 2, the values of T_0 obtained, 3.0 ± 0.5 K and 2.8 ± 0.6 K, respectively, were in good agreement with both the 3.5 ± 1.0 K originally reported by Penzias and Wilson (later corrected to 3.3 ± 1.0 K;

Penzias, 1968) and the first indirect measurements of T_0 at $\lambda = 2.64$ mm based on inter-stellar CN absorption lines (Field and Hitchcock, 1966; Thaddeus and Clauser, 1966; see Section 4.10 below). The error budget in the 3.2 cm result was dominated by uncertainties in what we have called T_{offset} and T_{wall}, because the cold-load absorber was placed in a narrow, single-mode waveguide immersed in liquid He rather than a larger cavity (see Roll and Wilkinson, 1967; Wilkinson, 1967). Howell and Shakeshaft did not measure T_{atm}, but instead used a theoretically calculated value of 2.2 ± 0.2 K; in view of more recent work on non-resonant O_2 absorption, this value may be high by about 0.2 K. If so, the value of T_0 at 20.7 cm should be raised to about 3.0 ± 0.6 K, in good agreement with a 21 cm observation made by Penzias and Wilson (1967), which gave $T_0 = 3.2 \pm 1.0$ K. For both these measurements at about 21 cm, however, the value used for T_{Gal} seems low, suggesting that the published values of T_0 are high by about 0.2 and 0.4 K, respectively. Both the published values and the modified values appear in table 4.2.

4.6.1 Confirming the blackbody shape of the CBR spectrum at short wave-lengths

The Hot Big Bang model predicts a blackbody (Planck) spectrum for the CBR. A number of rival theories developed in the first year or two after the publication of the papers by Penzias and Wilson (1965) and Dicke *et al.* (1965) predicted different spectra (see Partridge, 1969, for a more-or-less contemporary, brief summary). Hence it seemed particularly important in the late 1960s to push measurements to wavelengths below 3 cm to search for the gradual 'turnover,' or departure from the λ^{-2} Rayleigh–Jeans law, as the peak of a roughly 3 K blackbody was approached.

Four groups carried out observational programs with this aim in the first half of 1967. Three groups in the USA made use of a high altitude site (then called the White Mountain Research Station) at 3800 m; see table 4.2 for a summary of the results. At about the same time, a group from the Lebedev Physics Institute of Gorki University in the USSR made a low altitude measurement at $\lambda = 8.2$ mm (Puzanov, Salomonovich and Stankevich, 1968).

In part because I was involved in it, and in part because it was the first to introduce most of the techniques now standard in CBR spectral measurements, let me begin with the work of Wilkinson and his colleagues then at Princeton University (Wilkinson, 1967; Stokes *et al.*, 1967). This group employed Dicke-switched, heterodyne receivers (fig. 4.14) to measure T_0 at 3.2, 1.58 and 0.856 cm wavelength. A large diameter cold load was employed to reduce T_{wall}; extensive ground screens were used to reduce T_{gr}. The atmospheric antenna temperature at each wavelength was measured by horn tipping to several zenith angles, using a large movable reflector. This allowed the observers to confirm that T_{atm} obeyed the secant law (eqn. (4.5)).

A novel feature of this experiment is shown in fig. 4.15; the antennas were pointed down, at 45° from the vertical. This allowed the observers to couple the antennas to the cold load without moving the radiometers (thus avoiding position-dependent contributions to T_{offset}). To view the zenith, the antenna beams were reflected by a large, plane mirror; its emission (about 0.1 K; see references for exact values) had to be subtracted from T_z to give T_0.

The basic features of these experiments – use of a large diameter cold load to which the antenna was coupled, use of ground screens, and a measurement of T_{atm} based on

Table 4.2. *The results of all direct measurements of T_0. All but the measurements of Johnson and Wilkinson (1987) and Boynton and Stokes (1974) at $\lambda = 1.2$ and 0.33 cm, respectively, were ground based observations. For the work of the Berkeley group, only the most recent results are given at $\lambda = 8, 3$ and 0.33 cm (De Amici et al., 1990; Kogut et al., 1988; Bersanelli et al., 1989); see table 4.3 for some earlier values.*

Reference	λ (cm)	T_{atm} (K)[a]	T_0 (K)	Section where discussed	Revised[b] T_0 (K)	Comment on revision
Howell and Shakeshaft (1967)	73.5	1.3 (c)	3.7 ± 1.2	4.6.3		
	49.2	1.95 (c)				
Stankevich et al. (1970)	73	n.a.	3.0 ± 0.5	4.6.3		
Sironi et al. (1990)	47	n.a.	3.0 ± 1.2	4.6.3		
	50	1.17 (c)				
Pelyushenko and Stankevich (1969)	30 20.9 } 15		2.5 ± 0.3	4.6.3	~3.7 (?)	Very uncertain correction for possible error in T_{atm}
Levin et al. (1988)	21.3	0.83 (c)	2.11 ± 0.38	4.6.4		
Penzias and Wilson (1967)	21.2	2.3	3.2 ± 1.0	4.6	2.8 ± 1.0 }	Approximate correction to T_{Gal}
Howell and Shakeshaft (1966)	20.7	2.2 (c)	2.8 ± 0.6	4.6	2.8 ± 0.6	
Otoshi and Stelzried (1975)	13	2.3	2.66 ± 0.26	4.6.3	2.76 ± 0.30	See Section 4.6.3
Sironi and Bonelli (1986)	12		2.79 ± 0.15	4.6.4		
De Amici et al. (1990, 1991)	8.0	0.93	2.64 ± 0.07	4.6.4		
Penzias and Wilson (1965)	7.35	3.4	3.5 ± 1.0	4.6	3.3 ± 1.0	Penzias (1968)
Mandolesi et al. (1986)	6.3		2.70 ± 0.07	4.6.4		
Kogut et al. (1990a)	4.0		2.60 ± 0.07	4.6.4		
Roll and Wilkinson (1967)	3.2	3.0	3.0 ± 0.5	4.6		
Stokes et al. (1967)	3.2	1.3	$2.69^{+0.16}_{-0.21}$	4.6.1	2.9 ± 0.2 }	Approximate correction for T_{atm}
Kogut et al. (1988)	3.0	1.2	$2.61^{+0.06}_{-0.06}$	4.6.4	2.62 ± 0.04	
Stokes et al. (1967)	1.58	≃ 4	$2.78^{+0.12}_{-0.17}$	4.6.1		
Welch et al. (1967)	1.58	≃ 4	2.0 ± 0.8	4.6.1	2.45 ± 1.0	Correction to thermodynamic temperature
Johnson and Wilkinson (1987)	1.2	≃ 0	2.783 ± 0.025	4.7		
Ewing et al. (1967)	0.92	≃ 5	3.16 ± 0.26	4.6.1		
De Amici et al. (1985)	0.91	6–7	2.81 ± 0.12	4.6.4		
Wilkinson (1967)	0.86	16–19	$2.56^{+0.17}_{-0.22}$	4.6.1		
Puzanov et al. (1968)	0.82		2.9 ± 0.7	4.6.1	2.9 ± 0.9	Correction of T_0 and T_e to thermodynamic temperature
Kislyakov et al. (1971)	0.36	≃ 15	2.4 ± 0.7	4.6.2		
Boynton et al. (1968)	0.33	11–12	$2.46^{+0.40}_{-0.44}$	4.6.2		
Millea et al. (1971)	0.33	12	2.61 ± 0.25	4.6.2		
Boynton and Stokes (1974)	0.33	1.2	$2.48^{+0.50}_{-0.54}$	4.6.2	3.6 ± 0.5	Approximate correction for T_{atm}
Bersanelli et al. (1989)	0.33	14–16	2.60 ± 0.09	4.6.4		

[a] In this column (c) is used to indicate calculated rather than directly measured values. Note that the measurements were made at many different altitudes, which explains some of the apparent discrepancies in T_{atm}.

[b] These are the suggested revisions discussed in this chapter.

Fig. 4.14   Microwave radiometers used by the Princeton group to measure the CBR spectrum at 3.2, 1.58 and 0.856 cm shown at the White Mountain site in 1967. In the background is a wire-mesh ground screen.

scans at several zenith angles – became standard in the field. Using very similar apparatus, this program was extended to $\lambda = 3.3$ mm a year later by Boynton, Stokes and Wilkinson (1968); see Section 4.6.2.

RADIOMETER

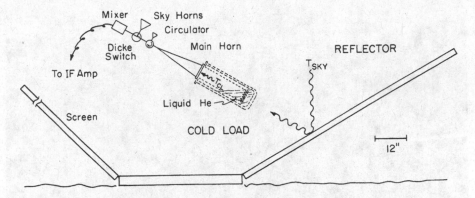

Fig. 4.15   Schematic of the radiometers used by Wilkinson (1967) and by Stokes *et al.*
(1967) to measure the CBR spectrum at 3.2, 1.58 and 0.856 cm wavelength. Note the
use of a reflector; 'T_{sky}' here includes foreground emission.

These observations also produced the first measurements of T_0 accurate to within
about 10%.* They established the existence of a short-wavelength turnover in the CBR
spectrum, expected if its spectrum is thermal (the 8.56 mm point excludes a
Rayleigh–Jeans λ^{-2} power law spectrum by 2–3 times the experimental error).

One puzzle remains in this work – the high value found for T_{atm} at 3.2 cm. Later observations at the same site (Section 4.6.4) and the atmospheric emission models discussed
in Section 4.4.3 above suggest that the value of T_{atm} found by Stokes *et al.* (1967) was
high by a factor of about 20–30%, in which case the value derived for T_0 would be low
by about 0.2 K.

At the same time and at the same site, a group from MIT was making observations at
$\lambda = 0.92$ cm (Ewing, Burke and Staelin, 1967). The technique was basically the same
(including the use of a large diameter He cold load). The MIT group, however, used a
mechanical waveguide switch to interchange sky and cold load. This introduced an
offset, which varied from day to day; the offset was measured to an accuracy of ± 0.10
K each run. The MIT observers also made zenith scans with a horn of 10° beam width
(the zenith observations were made with a smaller horn of 20° beam width). In dry,
cloud-free weather, they found $T_{atm} \approx 4$–5 K, consistent with other 9 mm measurements
and with atmospheric models.

The results given in table II of Ewing *et al.* (1967) appear to be expressed in antenna
temperature; it is not clear how the conversion to thermodynamic temperature for T_0 (or
for the cold load temperature) was made.

The third group in the USA to make observations at the White Mountain Research
Station was based at the University of California, Berkeley (Welch *et al.*, 1967); their
observations were carried out at 1.5 cm. Liquid nitrogen rather than liquid helium was
used for the cold load calibration, opening up the possibility that non-linearities in the

---

* Note that the tables were inadvertently mixed up in Stokes *et al.* (1967) and Wilkinson (1967) – the 3.2 cm
  results appear in table I of the latter paper, the 0.856 cm results in table I of the former.

(a)

(b)

Fig. 4.16   Photograph of the radiometers used in 1982 and 1983 to obtain five precise values of T_0. (a) General layout, showing the short, raised rail line (on left) used to move the instruments into position over the cold load calibrator shown in fig. 4.3. The cold load itself is barely visible beneath the down-pointing 12 cm horn. (b) Detail of the 12 cm radiometer, showing the main horn antenna (here directed vertically) and the reference horn (directed at the North Celestial Pole). The trapezoidal panels are ground screens.

gain of the radiometer could have introduced error into their results. Hence the observers made additional laboratory checks using a liquid helium termination. Although it is not entirely clear from their brief paper, their published result, 2.0 ± 0.8 K, appears to be expressed in antenna temperature; converted to thermodynamic temperature,* it becomes 2.45 ± 1.0 K.

The low altitude measurements of the group from Gorki (Puzanov *et al.*, 1968) exploited the same atmospheric window at $\lambda \simeq 9$ mm used by others. Since their measurements were made from beneath more of the Earth's atmosphere, however, their measured T_{atm} was 16–19 K. These values were determined by tipping the horn to a single zenith angle only, $z = 60°$. A liquid nitrogen cold load was used; the paper does not mention lower temperature calibrations. The value for T_0 (2.9 ± 0.7 K) given by the authors is clearly expressed in antenna temperature; converted to thermodynamic temperature, we find $T_0 = 3.7 \pm 0.9$ K. On the other hand, as Kislyakov *et al.* (1971) note, a correction of 0.8 K in the opposite sense is needed to convert the cold load temperature to antenna temperature. Hence we end up with 2.9 ± 0.9 K for T_0.

4.6.2 Ground-based measurements in the 3 mm atmospheric window

The measurement of Boynton *et al.* (1968) was the first to exploit the 3 mm atmospheric window between the strong O_2 lines. Even at 3400 m altitude in good weather, T_{atm} was found to be 11–12 K, and uncertainty in T_{atm} dominates the error budget of the result, $T_0 = 2.46^{+0.40}_{-0.44}$ K. This measurement lies below a λ^{-2} extrapolation of longer wavelength measurements by 4–5 times the experimental error, but is consistent with a 2.7 K blackbody spectrum.†

The Gorki group also made a spectral measurement in the 3 mm atmospheric window, this time from a high altitude site (Kislyakov *et al.*, 1971). As in their work at 9 mm, this experiment employed a liquid nitrogen calibrator, and the authors implicitly assumed that the gain of their receiver (which employed a parametric amplifier as well as a mixer) was independent of antenna temperature. As noted in Section 4.3.2, gain variations, or saturation, if present, could have introduced systematic error into the results. In this experiment, the cold-load calibrator was placed *above* the antenna, and hence both

---

* As noted in Section 3.2.2, antenna temperature and true thermodynamic temperature T are related by

$$T_A = T[x / (e^x - 1)],$$

where $x = h\nu/kT$. Thus the conversion factor in parentheses itself depends on T. Given a result expressed in antenna temperature, there are two ways to proceed in converting it to thermodynamic temperature. One is to solve the above equation for T; the other is to *assume* a value of T for the CBR derived from *other* measurements and use it to compute x and hence the term in parentheses. The first approach gives $T_0 = 2.45 \pm 1.0$ K for the value of antenna temperature found by Welch *et al.* This is the experimentally self-consistent result. The second approach, with T_0 fixed at 2.75 K, would give $T_0 = 2.39 \pm 0.96$ K. In either case, the result is consistent with longer wavelength measurements.

† A question has arisen (Kislyakov *et al.*, 1971) about the correction for atmospheric self-absorption made by Boynton *et al.* (1968). Kislyakov and his colleagues note that '… the value $T_0 = 2.46 \pm 0.44$ K given in [Boynton *et al.*] is, perhaps, slightly too high since it was obtained by means of a secant law with allowance for atmospheric absorption; this is not completely correct. This may have given rise to a systematic error of about 0.4 K.' First, as shown in Section 4.4.3, no correction to the secant law is required if the exact form of eqn. (4.3) is used, *provided* that the physical temperature of the atmosphere scales with height in the same way in all directions (plane parallel assumption). On the other hand, it is necessary to take the secant-dependent *absorption* of the CBR into account in computing T_{atm}. The thesis of Stokes (1968) makes it clear that both steps were made correctly in the work reported by Boynton *et al.* (1968).

absorber and cryogen had to be contained in a transparent cell rather than a Dewar flask. The observers constructed such a cell of plastic foam. Their paper contains a full description of the means that they employed to determine the quantities corresponding to T_w, T_{wall} and $T^{\dagger}_{reflec}$ in eqn. (4.7). If the radiometer gain was constant, as Kislyakov *et al.* assumed, the measured antenna temperature was 0.95 ± 0.58 K, allowing for 'all possible systematic errors.' Converted to thermodynamic temperature, this becomes 2.4 K with an uncertainty of about 0.7 K.

A few years later Millea *et al.* (1971) made an extended series of 3.3 mm measurements at two roughly 3000 m sites in California. The technique employed was similar to that introduced by Wilkinson (1967). Care was taken to measure all the quantities in eqn. (4.7) and eqn. (4.8), including possible non-linear response in the receiver. The value of T_0 obtained, 2.61 ± 0.25 K, is the most accurate of the 3 mm results, and is in excellent agreement with both longer wavelength measurements and the indirect CN measurements discussed below.

Finally, in 1971 a pioneering experiment to measure T_0 from an airborne platform was made by Boynton and Stokes (1974). They employed a Dicke-switched receiver operating at $\lambda = 3.3$ mm carried aloft to about 15 km altitude; at this altitude, $T_{atm} \lesssim T_0$. A difficulty with this technique is that the primary calibration of the instrument could only be carried out on the ground, before and after the flight. In addition, the measured value of T_{atm}, 1.21 ± 0.37 K, is roughly five times larger than expected from atmospheric emission models (Danese and Partridge, 1989). The cause of this discrepancy is not clear. If we *arbitrarily* change T_{atm} to 0.25 K, adjust T_z accordingly, and make the conversion to thermodynamic temperature, we find $T_0 = 3.6 \pm 0.5$ K instead of the $2.48^{+0.50}_{-0.54}$ K given by Boynton and Stokes.

It is worth noting explicitly that these observations at $\lambda < 1$ cm have since been superseded by others made above the atmosphere (see Sections 4.8 and 4.9 below), and by more recent ground-based measurements (see Section 4.6.4).

4.6.3 *Long wavelength measurements*

We turn now to the opposite end of the CBR spectrum – wavelengths $\gtrsim 30$ cm, where Galactic emission, even at the poles, is as large as or larger than atmospheric emission. The pioneering work at long wavelengths was by Howell and Shakeshaft (1967). Their measurements were made at 49.2 and 73.5 cm using antennas scaled to the wavelength, so that the antenna beams had equal widths of 15°. The measurement technique in most respects was identical to their earlier work (Howell and Shakeshaft, 1966) and to the techniques described in Section 4.6.1 above. A liquid helium cold load was used as the primary calibrator.

Once a few corrections had been applied, the zenith measurements gave the sum of two terms, T_0 and the Galactic contribution, which is wavelength dependent:

$$T(\lambda) = T_0 + T_{Gal}(\lambda) \tag{4.9}$$

Like Howell and Shakeshaft, I have implicitly assumed T_0 to be independent of wavelength.

Their observations were made near the north celestial pole at times when the Galactic plane was not near their beam. To calculate the residual Galactic contribution at each wavelength, Howell and Shakeshaft observed the Galaxy at a number of different locations where the Galactic emission overwhelmed T_0. From these results, they derived the average wavelength dependence of Galactic emission at 49–73 cm wavelength: $T_{Gal} \propto \lambda^{2.8 \pm 0.1}$. Assuming that this same index was appropriate where the Galactic emission was much smaller, they wrote

$$
\left.
\begin{aligned}
T(49) &= T_0 + T_{Gal}(49) \\[2ex]
T(73) &= T_0 + \left(\frac{73}{49}\right)^{2.8} T_{Gal}(49)
\end{aligned}
\right\} \quad (4.10)
$$

These two equations are then solved for T_0 (and $T_{Gal}(49)$ as well). Their value for T_0 in the wavelength range 49–73 cm is 3.7 ± 1.2 K.

At these long wavelengths, the contribution from discrete extragalactic sources cannot be ignored (Section 4.4.2). As it happens, however, the spectral index for typical radio sources is close to the 2.8 index found for the Galaxy. This close agreement, combined with the fact that both beams were the same size on the sky, ensured that any radiation from discrete sources could be lumped in with the Galactic emission when solving the coupled equations (4.10).

Two years later similar observations at 15, 20.9 and 30 cm wavelength were carried out by Pelyushenko and Stankevich (1969) in the Soviet Union. Their instrument was calibrated by a liquid nitrogen load covering part of the main beam; there is no mention of possible non-linearity in the response of the detector. The atmospheric contribution was measured by tipping the horn antennas used to a zenith angle of 45°. There is a suggestion in the paper that the antenna pattern was convolved with sec z; in any case the values for T_{atm} are 50–100% higher than expected from the model described in Section 4.4.3 above. The measurements of T_{atm} may have been affected by side- or back-lobe pickup.

Since observations were made at three frequencies, the results could be used to solve for T_0, the Galactic contribution and the spectral index together. For the temperature spectral index, Pelyushenko and Stankevich find 2.85 ± 0.20, in good agreement with other results, generally at longer wavelength. For T_0 in the wavelength range 15–30 cm, they find 2.5 ± 0.3 K.*

Stankevich also participated in a novel program to measure T_0 using the Moon as a foreground 'screen' to occult background emission from the CBR and the Galaxy (Stankevich et al., 1970). Since the Moon has a much higher temperature than $T_0 + T_{Gal}$, the results depend strongly on the linear response of the receiver. On the other hand, the use of a 'screen' beyond the Earth's atmosphere makes measurement of T_{atm} unnecessary.

The observations were made with the 64 m Parkes telescope in Australia at 47 and 73 cm. Taking 225 K as the Moon's temperature, they find $T_0 = 3.0 \pm 0.5$ K. This value depends only weakly on the spectral index in the 47–73 cm region, and is also insensitive to changes of a few kelvin in the temperature assumed for the Moon.

More recently, Sironi and his colleagues in Milan and Bonn have mounted an ambitious program to remeasure the temperature of the CBR at $\lambda = 50$ cm. The first measurements give $T_0 = 3.0 \pm 1.2$ K; further observations are planned (Sironi et al., 1990). The 50 cm observations were made at 2100 m altitude in an isolated valley of the Alps (to minimize man-made radio frequency interference).

---

* Suppose, in fact, that the Soviet group took too large values for T_{atm}. Then the value for the spectral index derived from their zenith measurements would rise to about 3.5, and T_0 to about 3.7 K.

According to the authors, roughly 2/3 of the error budget arises from systematic effects, such as radiation into the side and back lobes of the horn employed. Ground screens were not employed, perhaps because of the large size of the horn required (about $2 \times 2 \times 3$ m).

Sironi *et al*. did attempt to tip the antenna to measure T_{atm}. They found, however, that angle-dependent changes in T_{gr} and T_{Gal} made it hard to determine the variation of T_{atm} with zenith angle*. As a consequence, Sironi *et al*. adopted a value of $T_{atm} = 1.17 \pm 0.20$ K based on the atmospheric emission model of Danese and Partridge (1989). Also, T_{Gal} was not measured directly, but scaled from the 73 cm map of Haslam *et al*. (1982) using a temperature spectral index of 2.8; the estimated uncertainty in T_{Gal} is ± 0.18 K.

Because the horn was so large, it was clearly impossible to couple it directly to a cold load. Instead, a switch was used to connect the receiver alternately to the antenna and to a liquid helium cold load. Losses in the switch and the cable leading to the cold load (terms analogous to T_{wall} and $T^{\dagger}_{reflec}$ in eqn. (4.2)) contributed to T_c and to the systematic error of the experiment. In addition, since comparison with the cold load was made at a point between the antenna and the receiver, loss and consequent emission in the antenna and in the coaxial cable connecting it to the switch had to be taken into account (unlike the case for calibration carried out at the antenna mouth).†

Finally, although the wavelength employed was 13 cm, so that $T_{Gal} \simeq 0.1 \; T_{atm} \simeq 0.09 \; T_0$, let us include here the measurement of Otoshi and Stelzried (1975). (The observations were actually carried out in 1967 and first presented in an internal report.) A conical horn antenna with low side and back lobes was employed at a sea level site. The antenna was fixed in a vertical position, and hence could not be directly coupled to a cold load calibrator. Instead, a waveguide switch was used to connect the receiver alternately to the antenna and to a waveguide connected to a liquid He cold load. The error budget of this experiment was dominated by uncertainties in the loss and reflections in the waveguide switch and in the waveguide connecting the antenna and the load to the receiver.

In addition, as in the experiment of Sironi *et al*. (1990), two contributions to the measured zenith temperature were calculated rather than measured. T_{Gal} was calculated by scaling a Galactic map made at $\lambda = 73$ cm to $\lambda = 13$ cm using a temperature spectral index of 2.7 (using an index of 2.5 or 2.9 would change T_{Gal} by $\simeq 0.08$ K). In addition, because Otoshi and Stelzried could not tip their horn antenna, they also calculated T_{atm} using a model atmosphere. The value for T_{atm} that they obtained, about 2.2–2.3 K, is only about 0.1 K higher than expected from the more recent model of Danese and Partridge (1989).

The precision of this measurement is high; $T_0 = 2.66 \pm 0.77$ K where the error is a formal 3σ error. It is thus important enough to warrant a careful examination of its accuracy. As noted above, Otoshi and Stelzried may have overestimated T_{atm} by $\simeq 0.1$ K. There is a weak inverse correlation between their calculated values for T_{atm} and the

---

* These uncertainties were surely also present in the observations of Pelyushenko and Stankevich (1969) discussed above.

† These observations were continued at the South Pole at a slightly different wavelength (Sironi, Bonelli and Limon, 1991). At $\lambda = 37$ cm, they find $T_0 = 2.7 \pm 1.6$ K; the error budget is dominated by losses between the antenna and the receiver.

resulting values for T_0, suggesting the possibility that their model for atmospheric absorption overestimated the water vapour contribution to T_{atm}. A case can be made, therefore, for lowering T_{atm} and increasing their value of T_0 by about 0.15 K (well within the $1\sigma = 0.26$ K error of their measurement). If the values of T_{Gal} used by Otoshi and Stelzried for each run are plotted against the resulting values of T_0, a clear correlation is visible; T_0 is high when T_{Gal} is also. This correlation suggests that they underestimated T_{Gal}, especially when $T_{Gal} \gtrsim 0.4$ K. The authors give the right ascension and declination of each run, so their values for T_{Gal} may be compared with other measurements. Such a comparison suggests that the authors underestimated the contribution of the Galaxy at low Galactic latitudes by about 0.2–0.3 K. Making such a change in T_{Gal} would lower T_0 to 2.62 K. An alternative approach is simply to drop the two runs with $T_{Gal} > 0.4$ K. Doing so changes T_0 from 2.66 to 2.61 K. If we include these *post facto* corrections for T_{atm} and T_{Gal}, we find $T_0 = 2.76 \pm 0.30$ K. A 1σ error bar is given for comparison with other measurements.

I have presented the results of these long wavelength measurements of T_0 in detail for several reasons. First, as we shall see in the next chapter, they play an especially important role in setting constraints on a number of physical processes early in the history of the Universe. Second, there has been less activity at this end of the CBR spectrum than at short wavelengths, in part, perhaps, because the accuracy of the observations is limited by an unavoidable external source – the Galaxy – not by the instruments employed. Third, unlike the case for measurements at centimeter and millimeter wavelengths, the newer measurements are not substantially more precise than the results obtained twenty years ago (see table 4.2).

To conclude, the results I have presented above suggest that $T_0 \approx 2.9$ K in the broad wavelength range 20–75 cm, but it is important to recall that this result depends upon the assumption that T_0 is in fact constant over the range of wavelengths studied. The observations now available cannot reveal a change in T_0 *within* the interval $20 \lesssim \lambda \lesssim 75$ cm; but they *can* be compared with measurements at shorter wavelengths, such as the measurement at 13 cm by Otoshi and Stelzried (1975) or the more accurate measurements of T_0 to which we turn next.

4.6.4   The most recent ground-based measurements

The 50 cm measurements by Sironi and his Italian colleagues are a continuation of a large-scale collaborative venture to improve ground-based measurements of T_0 mounted by groups from the University of California, Berkeley; TESRE, a research institute in Bologna; Haverford College; LFCTR, a research institute in Milan; and the University of Padua. The collaboration was established in late 1979, and the first observations were made in 1982, at 3800 m altitude in the White Mountains of eastern California, the same site used for some of the observations described in Section 4.6.1.

The major aim of this program was to search for small-amplitude departures from a blackbody spectrum (see Chapter 5). For that reason, the observations were designed specifically to produce a set of values at several different wavelengths with high internal accuracy, so they could be intercompared even if the absolute level of accuracy was a bit lower.

The first set of observations at five wavelengths $0.33 < \lambda < 12$ cm were carried out in 1982 and 1983 (see Smoot *et al.*, 1985 or Partridge *et al.*, 1986 for summaries). The values of T_0 obtained at 0.33–12 cm were in good agreement with other earlier results, and showed no internal evidence for departures from a Planck curve at 2.75 K (see table 4.3).

To achieve our aim of high internal accuracy, all five instruments used the same, 0.7 m diameter, cold-load calibrator (shown in fig. 4.3); all had similar conical horn antennas with corrugated inner walls to reduce side lobes; the antennas were scaled in such a way that the beam widths were comparable (7.5° at $\lambda = 0.33$ and 0.91 cm and 12.5° at $\lambda = 3.0$, 6.3 and 12 cm); and the measurements were made essentially simultaneously in good weather during the summers of 1982 and 1983. For all five wavelengths, the zenith and cold load temperatures were compared by flipping the antenna 180° to look alternately at the zenith and at the cold load placed directly below.

Careful attention to the design of ground screens reduced T_{gr} to $\lesssim 0.03$ K (see table 4.3 for actual values). In addition, T_{atm} was monitored carefully during the observations. First, while one of the instruments was busy making the comparison between zenith sky and the cold load, the others were used to make horn tipping measurements of T_{atm}. Second, a separate instrument (Partridge et al., 1984), designed specifically to monitor T_{atm} at 3.2 cm, made continual scans, producing a value of T_{atm} accurate to ± 0.03 K every 8 min; thus time variations in atmospheric emission could be followed and corrected for. The large number of simultaneous measurements at different wavelengths by instruments with similar beam widths allowed the group to study O_2 and H_2O contributions to T_{atm} separately and hence to refine models for atmospheric emission (Costales et al., 1986; Danese and Partridge, 1989).

The results of this first set of measurements are summarized by Smoot et al., 1985; in addition, there are more detailed descriptions of the observations at each frequency (see table 4.2 for references).

Certain aspects of this collaborative venture deserve further scrutiny.

First, consider the trend in values for the cold load temperature, T_c, in table 4.3. At short wavelength, T_c falls below the thermodynamic temperature of boiling He because T_c is expressed in antenna temperature. At $\lambda = 12$ cm, on the other hand, the inner diameter (0.7 m) of the Dewar flask containing the liquid helium and the absorber was not large relative to the diameter of the beam defined by the antenna. Hence radiation from the Dewar flask wall was picked up, raising T_c above T_{abs} by about 1.5 K (Sironi et al., 1984). Although Sironi and his colleagues made careful measurements of T_{wall}, uncertainty in this quantity dominated the error budget of the 12 cm observations.* It is also worth noting that T_c was not corrected for the reflection of coherent radiation (see Section 4.3.1) in these results. Since isolators were employed† between the receiver and the antenna in all cases, $T^{\dagger}_{reflec}$ was not large.

Second, consider the offset terms shown in column 4 of table 4.3. These arose when the instruments were inverted from their zenith position to couple them to the cold load below. In its 1982 configuration, the 6.3 cm instrument suffered from a substantial offset asymmetry, 0.96 ± 0.16 K, which completely dominated the error in the results; the instrument was redesigned for the 1983 observations, essentially eliminating the offset (see Mandolesi et al., 1986). More refined measurements in 1983 decreased the value of

---

* T_{offset} was included in these measurements, and hence in the value of T_c, for the 12 cm observations.
† Except for the 6.3 cm radiometer in 1982 only; this deficiency was corrected by 1983.

Table 4.3 Results of the White Mountain collaboration (Section 4.6.4); see Smoot et al. (1985, 1987) or Partridge et al. (1986) for summary and description. Values in columns 3–7 are expressed in millikelvin antenna temperature; the values of T_0 in column 8 are in kelvin, thermodynamic. The values of T_0 are those determined from the first two summers of work only; see Table 4.2 for the updated results.

λ (cm)	Year of observation	T_c (mK)	T_{offset} (mK)	T_{gr} (mK)	T_{atm} (mK)	T_{Gal} (mK)	T_0 (K)
21.3	1986	3800 ± 500	± 90	9 ± 8	850 ± 50	707 ± 150	2.79 ± 0.15
12.0	1982	4999 ± 212	*	12 ± 10	950 ± 50	148 ± 30	
12.0	1983	5518 ± 130	*	6 ± 10	950 ± 50	200 ± 30	
8.2	1986	3735 ± 57	−35 ± 20	30 ± 7	860 ± 122	66 ± 30	2.70 ± 0.07
6.3	1982	3716 ± 31	960 ± 160	30 ± 10	1000 ± 100	40 ± 30	
6.3	1983	3682 ± 10	0 ± 20	20 ± 20	997 ± 70	35 ± 25	2.75 ± 0.08
3.0	1982	3562 ± 13	0 ± 40	0 ± 3	930 ± 160	3 ± 3	
3.0	1983	3562 ± 13	0 ± 30	0 ± 3	1200 ± 130	4 ± 2	
3.0	1984	3562 ± 13	0 ± 30	0 ± 3	1122 ± 120	4 ± 2	
3.0	1986	3568 ± 13	6 ± 17	0 ± 3	1222 ± 65	8 ± 4	2.81 ± 0.12
0.91	1982	3068 ± 9	−100 ± 50	1 ± 1	4850 ± 140	1 ± 1	
0.91	1983		−30 ± 30	1 ± 1	4530 ± 90	1 ± 1	
0.91	1984		−30 ± 30	1 ± 1	4340 ± 90	1 ± 1	
0.33	1982	2083 ± 37	0 ± 50(?)	1 ± 1	12600 ± 570	0 ± 1	2.57 ± 0.14
0.33	1983		10 ± 9	12 ± 15	9870 ± 90	0 ± 1	
0.33	1984		13 ± 5	12 ± 15	11300 ± 130	0 ± 1	
0.33	1986	2087 ± 37	−8 ± 4	12 ± 15	15020 ± 100	0 ± 1	

*Included in T_c (see Sironi et al., 1984).

and error in T_{offset} for the 0.91 and 0.33 cm observations; T_{offset} at 0.33 cm was not directly measured in 1982 (see Witebsky *et al.*, 1986).

Finally, note that Galactic emission was negligible at $\lambda \leq 6.3$ cm, and was measured sufficiently accurately at 12 cm that uncertainty in T_{Gal} made a negligible contribution to the final uncertainty in T_0.

Smoot and his colleagues at Berkeley, and Sironi and his colleagues at Milan, made additional observations at 0.33, 0.91, 3 and 12 cm in the summers of 1984, 1986 and 1987, and the austral summer of 1989. In addition, the Berkeley group added instruments operating at 4, 8 and 21.3 cm wavelength. A summary of most of these observations is given by Smoot *et al.*, (1987); Sironi, Bonelli and Limon (1991); and Kogut *et al.* (1991); additional and more detailed references are given in table 4.2.

As tables 4.2 and 4.3 show, the additional observations change T_0 at the two shortest wavelengths only insignificantly. On the other hand, T_0 measured at 3.0 cm changed from 2.75 ± 0.08 K (Smoot *et al.*, 1985) to 2.61 ± 0.06 K (Kogut *et al.*, 1988). The bulk of the change may be ascribed to a reevaluation of T_{atm}. First, as both Friedman *et al.* (1984) and Kogut *et al.* (1988) note, the value of T_{atm} determined at 3.0 cm in 1982 is almost certainly too low at 0.93 K. On the basis of other measurements at 3.0 and 3.2 cm, Kogut *et al.* revise it upwards to 1.19 K, thus lowering the average value of T_0. In addition, in 1986, T_{atm} was measured at several different zenith angles using the technique and apparatus described in Partridge *et al.* (1984). As in this earlier work, there appears to be a systematic trend in the values of T_{atm} at the zenith derived from measurements at different values of z: as z increases, the apparent value of $T_{atm}(0)$ increases. Partridge *et al.* (1984) ascribe this to side-lobe pickup, which increases as the optical axis moves closer to the horizon. In our data, the effect is to add about 0.02 K to the average value of T_{atm}, and we made a correction for the pickup. Kogut and his colleagues (private communication) have performed a number of tests to look for both side-lobe pickup and asymmetries resulting from a small misalignment of the horn and reflector used. The results do not provide a definite explanation; hence they do not attempt to make such a correction but rather adopt a conservative error estimate of 0.065 K for T_{atm}. This leaves the possibility that the value of T_{atm} they adopt for the 1986 measurements, 1.21 K, may be high by about 0.02–0.03 K. If we *arbitrarily* change T_{atm} by this amount, then recompute the error in T_{atm} and T_0 for their 1986 observations alone, we find that $T_0 = 2.59 \pm 0.06$ K. The weighted average, with this correction, changes to 2.62 ± 0.04 K; i.e., slightly higher but well within the margin of error of the measurements.*

Let us now turn to the new measurements at 4, 8 and 21.3 cm wavelength (Kogut *et al.*, 1990a; De Amici *et al.*, 1988, 1990, 1991; and Levin *et al.*, 1988; respectively). The 4 cm measurements employed a direct receiver and a reconfigured cold load with additional (cold) windows (see Kogut *et al.*, 1990a, for a description). The antenna was carefully shielded from stray radiation from the ground (the ground screens, for instance, reduced the beam pattern at 90° from the optical axis by a factor $\gtrsim 100$). The error budget of this experiment was dominated by uncertainties in the measurement of T_{atm} and T_{offset}. The latter has a value of -52 ± 34 mK, and there is some evidence that it may

---

* Kogut *et al.* (1988) give a weighted average, where the weights are taken to be the inverse of the error in each year's value of T_0; the more precise 1986 measurement dominates the weighted average.

be temperature-dependent, becoming more negative as the antenna temperature drops. If T_{offset} is in fact <-52 mK when the zenith and cold load temperatures are measured, then T_0 should be correspondingly increased above the value of 2.60 ± 0.07 K given by Kogut *et al.* (1990a).*

For the 8 cm measurements made at White Mountain (De Amici *et al.*, 1990), uncertainty in T_{atm} dominated the final error in T_0. The uncertainty in T_{atm}, in turn, is ascribed largely to zenith-angle-dependent radiation into the side lobes of the horn antenna used (as discussed for the 3 cm measurements above). In addition, the value measured in 1986, $T_{atm} = 0.870 \pm 0.108$ K, is about 0.06 K lower than more precise measurements made by the same group one and two years later, and about 0.1 K below the value expected from the atmospheric models of Danese and Partridge (1989). Revising T_{atm} lowers the final value of T_0; on the other hand, the 1988 measurement is anomalously high, increasing the final value of T_0 to 2.64 ± 0.07 K (De Amici *et al.*, 1990). These 8 cm observations were repeated at the South Pole during the austral summer of 1989 (De Amici *et al.*, 1991). In these runs $T_{atm} = 1.109 \pm 0.004$ K, in excellent agreement with the predictions of the model of Danese and Partridge (1989) for that frequency and the South Pole site. The resulting value of T_0 for these observations alone is 2.64 ± 0.07 K; when combined with earlier results, $T_0 = 2.64 \pm 0.06$ K.

Sironi, Bonelli and Limon (1991) also repeated the 12 cm measurements at the South Pole in December 1989. As in the earlier work, a large diameter cold load was employed (as described in Kogut *et al.*, 1990a). Atmospheric emission was measured by horn-tipping. The resulting value $T_{atm} = 1.16 \pm 0.3$ K is about 100–200 mK higher than expected from atmospheric models. As a consequence, the value of T_0 derived by Sironi and his colleagues (2.50 ± 0.34 K) may be low by a corresponding amount. For this reason, I have elected to leave the earlier results (Sironi and Bonelli, 1986) in table 4.2.

The 21.3 cm measurements (Levin *et al.*, 1988) were made with a narrow bandwidth, $\Delta v = 25$ MHz, so that the coherence length (Section 4.3.1) of the instrument $l_c \approx 12$ m was larger than the distance to the absorber in the cold load. Hence reflections dominated the systematic error in this work. The amplitude of the coherent reflection contribution to $T^{\dagger}_{reflec}$ is given as 0.00 ± 0.31 K, but the peak-to-peak variation was about three times larger, and I think the authors are correct to say that '... this measurement [of $T^{\dagger}_{reflec}$] must be viewed somewhat skeptically.' In addition, the value taken for T_{atm} (0.83 ± 0.10 K) was not measured, but extrapolated from measurements at 8 and 3 cm. The final value for T_0 at 21.3 cm is 2.11 ± 0.38 K, significantly lower than the earlier measurements discussed in Section 4.6.3.

4.7   A precise direct measurement of T_0 at $\lambda = 1.2$ cm

The most precise direct measurement of T_0 now available was made by Johnson and Wilkinson (1987) using a radiometer carried aloft to about 26 km altitude by a balloon. At this altitude, and for their wavelength of 1.2 cm, T_{atm} was negligibly small, 2 ± 2 mK. Working at balloon altitude thus enabled the authors to avoid a major source of both systematic and statistical error that is present in ground-based experiments.

The experiment employed a novel type of receiver, not the usual superheterodyne used in the observational programs described earlier – for details, see Johnson (1986) or

* *Note added in proof*: Levin *et al.* (*Ap. J.* **396**, 3 [1992]) now report $T_0 = 2.64 \pm 0.06$ K at $\lambda = 4$ cm.

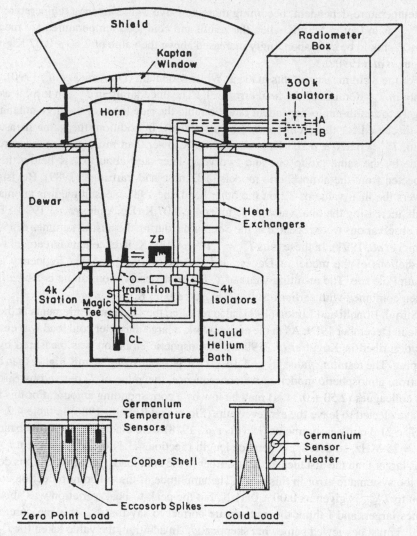

Fig. 4.17   Apparatus used in the balloon-borne spectral measurement of Johnson and Wilkinson (1987). The calibrator or 'zero point load' was moved into place near the throat of the conical horn antenna. 'CL' labels the cold load in the upper drawing.

Kraus (1966, section 7.1). In this receiver, a waveguide element known as a 'Magic T' (Montgomery *et al.*, 1963) was employed in place of a Dicke switch (see fig. 4.17). The cold load itself was an absorbing spike housed in a waveguide; radiation from the walls of the waveguide was reduced to a negligible level by immersing the entire cold load and the Magic T in about 4 K liquid He.

Absolute or 'zero point' calibration of the instrument was carried out by sliding a cryogenically cooled absorbing surface into the throat of the horn (fig. 4.17). Since the calibration was carried out behind the horn, radiation from the walls of the horn was included in T_z. As fig. 4.17 shows, much of the horn was cooled, so that its contribution to T_z was not large; several different means of estimating the emission from the horn give a value of 30–50 mK (Johnson, 1986). The value adopted for horn emission was 49 ± 12 mK.

Cooling the horn required use of a window above the horn to prevent condensation on it. The window also contributes to T_z; this contribution was found to be 35 ± 12 mK.

Finally, the largest contribution to the error budget (±16 mK) came from uncertainty in the reflection coefficients of the horn and the zero point calibrator; since the coherence length of the receiver was large compared to the physical dimensions of the horn throat and calibrator, coherent reflection was involved.

Johnson and Wilkinson (1987) obtain $T_0 = 2.783 \pm 0.025$ K at $\lambda = 1.2$ cm, assuming that errors in horn emission, window emission, reflections, etc. add in quadrature. However, even if we take the *sum* of all the uncertainties, we arrive at a 1σ error of only 0.08 K. This result is in acceptable agreement with ground-level observations made through part or all of the Earth's atmosphere, giving us some confidence that the values for T_{atm} measured from the ground are indeed accurate. If we compare Johnson and Wilkinson's result with some of the most recent and most precise ground-based measurements, however, we see that the latter fall significantly below $T_0 = 2.783$ K.

4.8   Short wavelength measurements from balloons and rockets

Emission from the Earth's atmosphere makes ground-based observations at wavelengths below the atmospheric 'window' at 3 mm very difficult, and only a single ground-based measurement at wavelengths shorter than 3 mm has been attempted. This was the experiment of Dall'Oglio *et al.* (1976) in the 1 mm window, which produced only an upper limit on T_0 of 2.7 K. All other short wavelength direct measurements of the CBR spectrum have been carried out by instruments carried aloft by balloon or rockets.

In addition, all observations at wavelengths < 3 mm have employed bolometric rather than superheterodyne detectors. Hence, in discussing the results, we need to consider the measures used to define the wavelength range of the observations.

4.8.1 Observational difficulties and techniques

If the spectrum of the CBR is indeed close to a Planck curve with $T_0 = 2.75$ K, its brightness will fall rapidly with decreasing wavelength in the Wien region, that is, for λ below about 1 mm. The bolometric detectors now employed for short-wavelength CBR observations have adequate sensitivity to measure a 2.75 K thermal spectrum well into the Wien region. It is rather the measurement of weak signals in the presence of higher temperature backgrounds that is the major hurdle encountered in the observations discussed here. As an illustration, let us consider the ratio of the brightnesses of two blackbodies at $T' = 2.75$ K and $T'' = 5.5$ K at various wavelengths. At $\lambda = 10$ cm, well into the Rayleigh–Jeans region of either spectrum, $B'/B'' = 0.494$, very close to the ratio of the temperatures. At $\lambda = 1$ cm, B'/B'' has fallen to 0.435; and further $B'/B'' = 0.295, 0.068$ and 1.6×10^{-4}, at $\lambda = 3$, 1 and 0.3 mm, respectively. Thus great care is necessary to keep emission from surfaces with $T > 2.75$ K under control.* For instance, the instrument must be carefully shielded from radiation from the roughly 300 K ground. Emission from the antenna walls can also contribute to systematic errors, unless a careful calibration is carried out *outside* the antenna aperture. Hence experimenters have frequently chosen to cool all optical surfaces cryogenically.

---

* As noted in Section 4.2, high temperature backgrounds present particularly severe problems if the filter used to define the short wavelength side of the bandpass has a high frequency leak.

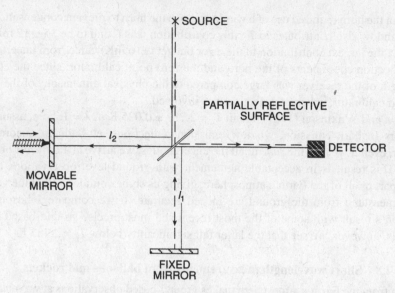

Fig. 4.18   Light from the source follows two paths (solid and dashed lines) through the instrument, then is brought together at the detector.

With bolometric detectors, filters are needed to restrict the range of wavelengths reaching the detector (refer to Section 4.2). In some recent experiments only filters are used to determine the band passes in which spectral measurements are made (see fig. 4.2). In other experiments, a different technique has been used to scan a wide wavelength range, namely Fourier transform spectroscopy. In its simplest form, this technique employs a Michelson interferometer, which has one adjustable arm (fig. 4.18) (see Born and Wolf, 1964, Chapter 7; or for a description of the specific type of interferometer employed in CBR measurements, Martin and Puplett, 1969). As the distance l_2 is uniformly changed, the output of the detector $S(t)$ is just the autocorrelation function of the incoming signal. Thus $S(t)$ is the Fourier transform of the source spectrum $B(v)$. Suppose, for instance, that the source spectrum is dominated by a single spectral line at frequency v. Then the detector output can easily be shown to vary cosinusoidally as l_2 is changed:

$$S(t) = \cos\left(\frac{2\pi|l_2 - l_1|}{\lambda}\right) = \cos\left[\frac{2\pi v}{c} vt\right]$$

where v is the speed at which reflector 2 is moved. This technique allows a wide spectral range $(\Delta\lambda/\lambda \simeq 1)$ to be scanned at a resolution comparable to or better than that achieved by superheterodyne receivers.

An inherent problem in Fourier transform spectroscopy is that any non-linearity in the motion of the reflector or any time variation in the total flux entering the instrument will modify $S(t)$ and hence introduce distortions in its Fourier transform $B(v)$. Such a problem was encountered in one of the observational programs described below (Gush, 1981).

4.8.2   Early results

In part for the reasons discussed above, short wavelength observations of the CBR are very difficult. The early experiments (prior to 1984, say) produced discordant results. For instance, one of the earliest attempts to determine the CBR spectrum at millimeter wavelengths, a rocket observation (Houck and Harwit, 1969), suggested far more flux at wavelengths below 1 mm than expected for a roughly 3 K blackbody. These results have not been borne out by subsequent observations.

These early results are discussed and analyzed in detail in the 1980 review of the CBR by Weiss.* Because that review is both detailed and readily accessible, I will mention here only two of the many balloon and rocket observations made before the mid-1980s.

First, there is the balloon-borne experiment of Woody and Richards (1979, and earlier references therein). The spectrum between 0.25 and 6 mm ($v = 1.7$–40 cm^{-1}) was scanned by a cooled Michelson interferometer. Useful measurements of or upper limits on the CBR were obtained between about 0.4 and 4 mm ($v = 2.5$–24 cm^{-1}). Their work was the first to establish convincingly the decrease in flux expected in the Wien region of a roughly 3 K blackbody. On the other hand, the best-fit value for T_0 in their wavelength interval was 2.96 K, well above the CBR temperature found at longer wavelengths. Furthermore, the measured flux was especially high at $\lambda \simeq 1.6$ mm and then fell more rapidly than the exponential cutoff of a blackbody curve at wavelengths below about 1.1 mm. These spectral features are now accorded very little weight, since these early observations have been superseded by newer results discussed below.

Second, there is the rocket experiment of Gush (1981), which also employed a scanning Michelson interferometer. The data in Gush's 1981 paper were obtained from a flight made in 1978. Although the instrument was separated from the rocket motor and casing as the flight reached apogee, the rocket motor unfortunately remained in the field of view of the instrument. As it scanned the sky, the instrument picked up radiation from the rocket motor, which introduced a spike in the interferogram once each period of revolution. Recall, however, that the measured output signal or interferogram $S(t)$ is the Fourier transform of the spectrum $B(v)$; therefore the periodic spikes in $S(t)$ produced a distortion in the spectrum, which Gush had to model and subtract. Once this had been done, the data were in approximate agreement with a roughly 3 K blackbody. However, the flux near the peak of a roughly 3 K blackbody curve, at $\lambda \simeq 1.0$–1.6 mm, was low relative to a $T_0 = 2.75$ K curve, and, more interestingly, excess flux was observed at shorter wavelengths, $0.4 < \lambda < 1$ mm (fig. 4.19). This qualitative pattern is thus almost the opposite of the millimeter and submillimeter spectrum reported by Woody and Richards (1979)†.

4.8.3   A balloon-borne measurement

In late 1983, Richards and his colleagues from Berkeley launched a redesigned instrument to measure the spectrum of the CBR in the wavelength interval 0.9–4 mm ($v =$

---

* One that he omitted is the balloon-borne measurement at $\lambda = 2$ mm by Grenier *et al.* (1976). Side- and back-lobe pickup was a problem, so the authors could only set a limit of $T_0 \leq 3.4$ K (at 90% confidence) on the thermodynamic temperature of the CBR at 2 mm.

† Gush and his colleagues have now repeated these observations, obtaining a spectrum in excellent agreement with the satellite results to be discussed in Section 4.9, and $T_0 = 2.736 \pm 0.017$ K (Gush *et al.*, 1990).

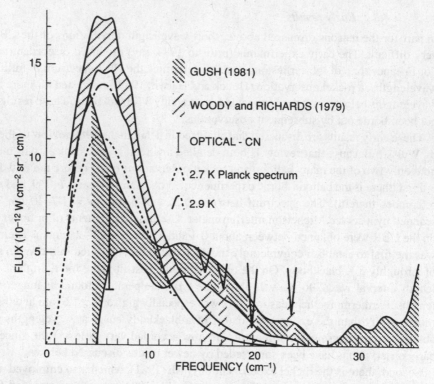

Fig. 4.19   Results of some early balloon and rocket measurements. The submillimeter excess found by Woody and Richards (1979) at ~2–8 cm^{-1} is contrasted with the results of Gush (1981) which fall below a 2.7 K blackbody curve at ~5–10 cm^{-1}.

2.3–11 cm^{-1}). This experiment employed filters to define the wavelength response, not a scanning interferometer (Peterson *et al.*, 1985). It also included an in-flight calibrator maintained at 3.2 K. As for the instrument described in Section 4.7, the calibration was performed between the detector and the flared horn antenna employed (fig. 4.20). It was therefore necessary to determine the emission from the horn, which contributed to T_z. This was done by heating the horn in flight, and measuring the resulting change in flux reaching the detector.* Likewise, the term corresponding to T_{offset} (the switch asymmetry) had to be measured. The latter term was the dominant source of error except in the shortest wavelength band, where the flux from the residual atmosphere above float altitude was about three times the measured CBR signal.

The results of the skillful and important experiment of Peterson *et al.* (1985) are shown in table 4.4 and fig. 4.21 (which also shows the bandpasses of the filters employed). Three general features are apparent in fig. 4.21. First, all five points are consistent with a single value of T_0, 2.78 ± 0.11 K, which in turn agrees with the longer wavelength measurements discussed in Sections 4.6 and 4.7 as well as with the CN measurements to be discussed below. Second, these results fall systematically below the earlier results of Woody and Richards (1979) that are also shown in fig. 4.21. Third, it

---

* For a discussion of problems with this technique and a reanalysis of these results, see Bernstein *et al.* (1990).

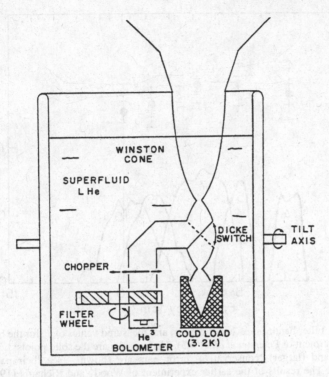

Fig. 4.20   Apparatus flown by Peterson and his colleagues (reproduced with their permission from Peterson *et al.*, 1985).

Table 4.4 *Filter bands and measured values of T_0 for two recent experiments. The measurements at $\lambda = 0.71$ and 0.48 mm (710 and 480 μm) reveal the excess submillimeter flux discussed in Section 4.8.4 and Chapter 5.*

Reference	Center λ (mm)	Bandwidth $\Delta\lambda/\lambda$	T_0 (K)
Peterson *et al.* (1985)	3.5	39%	2.80 ± 0.16
Peterson *et al.* (1985)	2.0	10%	$2.95^{+0.11}_{-0.12}$
Peterson *et al.* (1985)	1.5	31%	2.92 ± 0.10
Peterson *et al.* (1985)	1.14	14%	$2.65^{+0.09}_{-0.10}$
Peterson *et al.* (1985)	1.01	21%	$2.55^{+0.14}_{-0.18}$
Matsumoto *et al.* (1988)	1.16	30%	2.799 ± 0.018
Matsumoto *et al.* (1988)	0.71	21%	2.955 ± 0.017
Matsumoto *et al.* (1988)	0.48	19%	3.175 ± 0.027
Matsumoto *et al.* (1988)	0.26	36%	...

is intriguing that these new observations do suggest the same spectral pattern as found earlier by Woody and Richards, namely a rapid decrease in flux (or equivalently, a lower value of T_0) as the wavelength drops below about 1.2 mm ($\nu \simeq 8$ cm^{-1}). As the experimenters remark, however, the errors in the measurements are not small enough to allow one to claim with certainty that the apparent short wavelength fall-off is a real property of the CBR.

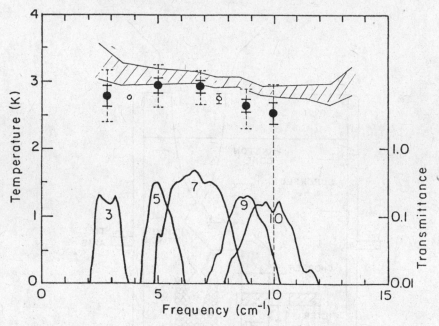

Fig. 4.21   Filter bandpasses (lower curves) and measured values of T_0 for the balloon-borne experiment of Peterson *et al.* (1985). Their results are the solid points; both conventional and (larger) 'conservative' error bars are shown – see their paper for discussion. The results of the earlier experiment of Woody and Richards (1979) are shown as a shaded band. The open circles are some of the CN results discussed in Section 4.10 (adapted from Peterson *et al.*, 1985, with permission).

4.8.4 Recent rocket-borne measurement

The next entry in the field of CBR spectral measurements was an experiment designed by groups from Berkeley and Nagoya, and carried aloft by a sounding rocket (Matsumoto *et al.*, 1988). Six filter bands (table 4.4 and fig. 4.2(*b*)) were employed; dichroic beam splitters were employed to divide the incoming beam. Each separate wavelength band had its own cooled bolometric detector, so observations in all six bands could be made simultaneously. It is a measure of the sensitivity of the detectors that measurements of T_0 to within about 1% accuracy could be obtained in about 5 min of observations.

The observers took a number of steps to avoid or to control sources of systematic error.

1.   At altitudes of several hundred kilometers, T_{atm} is expected to be negligibly small, except possibly in the two shortest wavelength bands where atomic oxygen lines can contribute. To check on T_{atm}, the experimenters continued their observations as the instrument reentered the upper atmosphere. An increase in flux in the two shortest wavelength bands was indeed seen, and the affected data were discarded. It should be noted that these two bands are not involved in the CBR measurements.

2.   The instrument was calibrated before* flight using a variable temperature cold

* Since the instrument was not recovered, no post-flight calibration was possible.

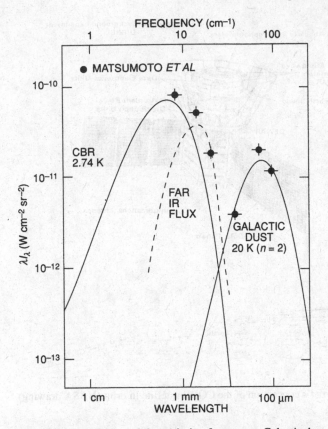

Fig. 4.22   The CBR and the emission from warm Galactic dust (solid lines) compared with the measured fluxes λI_λ obtained by Matsumoto *et al.* (1988). The measurements at 0.71 and 0.48 mm suggest a submillimeter excess or 'far IR' background (dashed lines). The dashed curve has been drawn assuming that the emissivity of the material responsible for the far IR excess flux varies as λ^{-2}, as appears to be true for Galactic dust.

absorber placed over the mouth of the antenna. In flight, the gain of the instrument was monitored with a thermal source located between the antenna and the beam splitters and the detectors.

3.   Emission from the antenna walls was reduced to a negligible level by cooling the entire optical train, including the antenna, to 1.2 K.

4.   As the flight neared apogee, the instrument was ejected by springs from the rocket motor and its accompanying cloud of exhaust gases.

The bandwidth spanned by this instrument is wider than that covered by earlier observations, corresponding essentially to $\lambda \simeq 0.1$–1.2 mm. At the short wavelength end, the brightness of the sky is dominated by emission from interstellar dust at a temperature of about 20 K (see fig. 4.22). As expected, the intensity in the two shortest wavelength bands is strongly position dependent, and is correlated with other indicators of dust density in the Galaxy (Matsumoto, private communication). The longest wavelength band, on the other hand, fell essentially at the peak of a roughly 3 K CBR spectrum, and the measured T_0 at 1.2 mm was 2.799 ± 0.018 K, in reasonable agreement with the CN

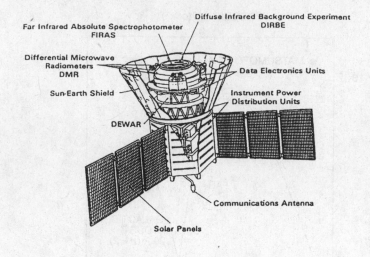

Fig. 4.23   Artist's conception of the COBE satellite in orbit (NASA drawing).

measurements to be discussed below. The major result of the Berkeley–Nagoya experiment, however, is the excess flux observed in the bands centered at 0.71 and 0.48 mm. In these two bands, the observed flux is significantly above that expected for a Planck curve drawn for either 2.75 or 2.799 K. Nor, as fig. 4.22 shows, can radiation from interstellar dust account for this submillimeter excess. As we shall see in the next section, the excess is almost certainly an instrumental effect, despite the pains the experimenters took to check on potential sources of systematic error, such as radio frequency interference, contamination from rocket exhaust, etc. To continue these checks, they plan at least one more flight. Slightly different wavelength bands will be employed, and the instrument will be uncovered only well after it has separated from the rocket.

4.9   Precision measurement by the COBE satellite

Just as this book was nearing completion, the first scientific results from a USA satellite devoted to study of the CBR became available (Mather *et al.*, 1990). The Cosmic Background Explorer (COBE), launched 18 November, 1989, by NASA, is designed to measure both the spectrum and the large-scale isotropy (Section 6.7) of the CBR over a range of wavelengths (see Mather, 1982; Gulkis *et al.*, 1990; Mather *et al.*, 1990; and Boggess *et al.*, 1992). An artist's conception of COBE is shown in fig. 4.23.

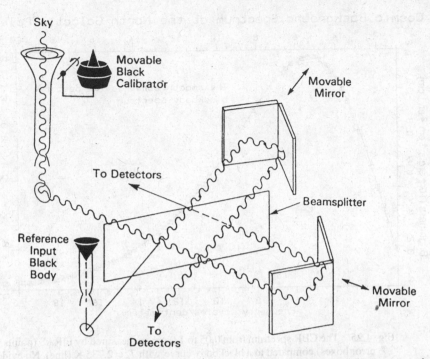

Fig. 4.24   Schematic of the FIRAS instrument flown on COBE; it makes comparative measurements of the flux arriving from the sky and the reference blackbody source. Note that calibration is performed at the horn mouth, and can be carried out in flight. The two arms of the Michelson interferometer are defined by the two dihedral mirrors.

4.9.1   FIRAS

In particular, one of its three instruments is devoted to absolute measurements of the far infrared background, including the wavelength range where excess flux was found by Matsumoto *et al.* (1988). The far infrared background in the broad wavelength range about 10 mm > λ > 0.1 mm can be measured to better than 1% accuracy by a scanning Fourier interferometer (resembling those described in the previous section). This is the FIRAS instrument (the Far Infrared Absolute Spectrophotometer), sketched in fig. 4.24. It is housed in the satellite Dewar flask so that the entire optical train is at cryogenic temperatures. Two features of FIRAS, which allow a crucial improvement in the precision of absolute temperature measurements of the CBR, deserve special mention.

1. FIRAS is a dual-input, differential device. The output power is proportional to the Fourier transform of the *difference* in the spectral power arriving from two inputs. One input is a smoothly flared, cryogenically cooled, horn antenna with $\theta_{1/2} = 7°$ viewing the sky. The other is a similar reference antenna terminated in a temperature-controlled blackbody within the Dewar flask. The temperature of the latter can be adjusted until a null, or nearly null, output interferogram is obtained. Thus uncertainty in the instrument gain is minimized.

2. FIRAS is equipped with a precision calibrator, again temperature-controlled, which can be inserted into the mouth of the sky antenna. This allows a precision external calibration in flight (Mather *et al.*, 1990).

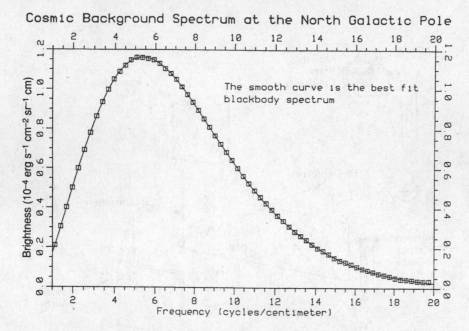

Fig. 4.25   The CBR spectrum from 0.05 to 1.0 cm as measured by FIRAS (points with 2% error boxes) compared to a blackbody curve with $T_0 = 2.735$ K (line). No evidence of spectral distortions in this wavelength range is present. Mather et al. (1990).

4.9.2   *The orbit and orientation of COBE*

COBE was placed in a 900 km orbit, inclined 99° to the equator. The optical axis of FIRAS coincides with the spin axis of the satellite and points away from the Earth (other instruments observe at 30° to the spin axis). The oblateness of the Earth is used to precess the orbital plane of the satellite at 1° per day, so that the spin axis is always inclined to the solar direction by more than 90°. Finally, the inputs to the COBE instruments are protected by a conical 'ground' screen.

As the spin axis precesses, the beam of FIRAS scans all 4π sr of the sky each 6 months.

4.9.3   *Preliminary results from FIRAS at 1 cm ≥ λ ≥ 0.5 mm*

Mather *et al.* (1990) have reported an analysis of 9 *min* of FIRAS data obtained in part of the instrument's wavelength range (0.5–10 mm). In this wavelength range, the fit to a blackbody spectrum is extraordinarily good (fig. 4.25). The value of T_0 deduced for this wavelength range is 2.735 K, with an error of ±0.06 K. A further word on this error and the error boxes in fig. 4.25: they are preliminary, and will surely decrease.* The absolute precision of FIRAS is at least five times better than the preliminary errors given by Mather *et al.* (1990). Some questions about temperature gradients in the reference

---

* *Note added in proof*: the COBE team (Mather *et al.*, *Ap. J.* **420**, 439 [1994]) now reports $T_0 = 2.726 \pm$ 0.010 K at 95% confidence; see Appendix C. FIRAS no longer operates because the cryogens have run out; it worked for more than one year.

and external blackbody calibrators (at the roughly 0.01 K level) must be settled before T_0 can be specified at the roughly 1/2% level or better, for which we may eventually hope.

In the meantime, it is important to note that even if T_0 is specified only with $\pm 2\%$ errors, the data constrain possible non-blackbody *distortions* in the spectrum much more tightly, as fig. 4.25 clearly shows. In particular, there is no evidence in the FIRAS results for the millimeter and submillimeter excess reported by Matsumoto *et al.* (1988) and discussed in Section 4.8 above.

4.9.4 Additional measurements by COBE

FIRAS operates in two spectral bands, 1 cm $\geqslant \lambda \geqslant$ 0.5 mm and 0.5 mm $\geqslant \lambda \geqslant$ 0.1 mm. No data from the shorter wavelength band are yet available. COBE also houses a second photometric instrument, DIRBE (the Diffuse Infrared Background Experiment). It is designed to make spectral measurements from 300 µm (0.3 mm) down to 1 µm. Only *very* preliminary results from DIRBE have been released (Hauser *et al.* 1991). Measurements at a few hundred micrometers suggest that the sky is somewhat darker than IRAS measurements (discussed in Section 4.4.2) suggest. If so, Galactic emission may be less of a problem for submillimeter CBR measurements than previously thought.

Finally, COBE includes three differential microwave radiometers to measure the large-scale isotropy of the CBR; these instruments and the preliminary results from them will be discussed in Chapter 6.

COBE may be the first in a series of satellite experiments devoted to the CBR spectrum; additional missions are under discussion by Russian space scientists (see Section 6.7.1) and at the European Space Agency (see Beckman *et al.*, 1986) as well as at NASA.

4.10   Values of T_0 derived from observations of interstellar molecules

Observations of interstellar molecules, particularly CN (cyanogen), at optical wavelengths provide the most precise and, I will argue, among the most accurate measurements of T_0 at millimeter wavelengths. These are *indirect* measurements; the optical observations reveal directly only the number of molecules in the various energy states of the ground electronic level of a particular molecular species, and from the ratios of these numbers we can then calculate T_0.

4.10.1 Boltzmann's equation

Let us begin by considering a collection of very simple quantum mechanical systems with just two energy states E_A and E_B, with $E_B > E_A$; these systems could be atoms or molecules. Place them in an oven maintained at temperature T and allow them to reach thermodynamic equilibrium. The ratio of the numbers in the two energy states is then given by Boltzmann's equation (derived in most statistical mechanics texts):

$$\frac{n_B}{n_A} = \frac{g_B}{g_A} \exp\left[(E_A - E_B)/kT\right]. \tag{4.11}$$

Here, g_A and g_B are the so-called *statistical weights* of the two energy states, which can readily be calculated from their quantum numbers (in the cases we will consider, $g = 2J + 1$).

Table 4.5 *Results of indirect measurements of T_0 using interstellar molecules. Only the most recent results are shown for CN.*

Reference	Molecule	λ (mm)	Transition used	T_0 (K)
Meyer and Jura (1985)	CN	2.64	R(1) and P(1)	2.70 ± 0.04
Meyer *et al.* (1989)	CN	2.64	R(1) and P(1)	$2.75 \pm 0.03*$
Meyer *et al.* (1989)	CN	2.64	R(1) only	2.77 ± 0.07
Meyer *et al.* (1989)	CN	2.64	P(1) only	2.75 ± 0.08
Meyer and Jura (1985)	CN	1.32	R(2) and P(2)	2.76 ± 0.20
Meyer *et al.* (1989)	CN	1.32	R(2) and P(2)	2.83 ± 0.09
Meyer *et al.* (1989)	CN	1.32	R(2) only	2.82 ± 0.11
Meyer *et al.* (1989)	CN	1.32	P(2) only	2.85 ± 0.16
Crane *et al.* (1986)	CN	2.64	R(1) and P(1)	2.74 ± 0.05
Crane *et al.* (1989)	CN	2.64	R(1) and P(1)	$2.796 ^{+0.014}_{-0.039} *$
Palazzi *et al.* (1990)	CN	1.32	R(2) and P(2)	2.83 ± 0.07
Palazzi *et al.* (1990)	CN	1.32	R(2) only	2.84 ± 0.09
Palazzi *et al.* (1990)	CN	1.32	P(2) only	2.82 ± 0.11
Kaiser and Wright (1990)	CN	2.64	R(1) and P(1)	2.75 ± 0.04
Thaddeus (1972)	CH	0.56		< 5.23
Thaddeus (1972)	CH^+	0.36		< 7.35
Kogut *et al.* (1990b)	H_2CO	2.1		3.2 ± 0.9

* Revisions of results of preceding line; the revised versions of T_0 are about 0.05 K larger because of smaller corrections for e^- excitation, as discussed in Section 4.10.

It is easy to see from eqn. (4.11) that energy state B will be appreciably populated only if $E_B - E_A \lesssim kT$. More to the point, we see that if g_A and g_B are known, T can be found directly from a measurement of the population ratio n_B/n_A.

Now consider a collection of the same two-state systems suspended in interstellar space and bathed in the (approximately isotropic) thermal radiation of the CBR, with $T_0 \simeq 2.75$ K. If energy state B lies about $3 k$ ergs or less above the lower energy state A, state B will be populated. Furthermore, provided that the population ratio n_B/n_A is determined entirely by the radiation field, a measurement of n_B/n_A gives T_0. There are at least three diatomic molecules* common in interstellar clouds that have low-lying rotational energy states which can be populated by the thermal radiation of the CBR; these are CN, CH and CH^+ (table 4.5). We will concentrate on CN (cyanogen), since measurements of the population ratios in the other two are very difficult, and thus far have provided only upper limits on T_0.

4.10.2 Energy level diagram of CN

In fig. 4.26 is shown the energy level diagram for CN. I have labeled the various energy states using the conventional spectroscopic notation, including quantum number J, but the labels are not important for our purposes. What is important is the energy difference $E_1 - E_0$ – it corresponds to a wavelength $\lambda = hc/(E_1 - E_0)$ of 2.64 mm. CN has an additional rotational state with $(E_2 - E_1) = 2 (E_1 - E_0)$; the corresponding wavelength of the transition is 1.32 mm. Thus measurements of the population ratios n_1/n_0 and n_2/n_1 permit us to determine T_0 at 2.64 and 1.32 mm, respectively. The ratios of statistical weights appearing in eqn. (4.11) are $g_1/g_0 = 3$ and $g_2/g_1 = 5/3$, respectively.

* See Herzberg (1945) for a treatment of the spectroscopy of diatomic molecules.

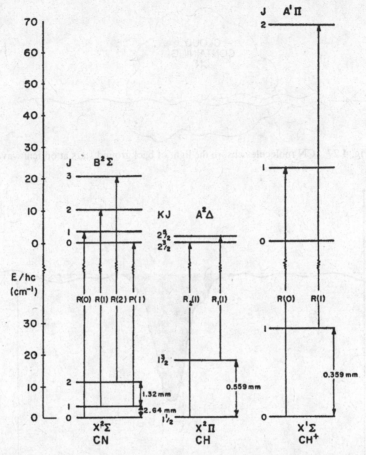

Fig. 4.26   Energy level diagram of the CN molecule, and others used to measure T_0 (from Thaddeus, 1972). The optical absorption lines are labeled R(O), etc.

These low-lying energy states are connected to a much higher energy excited level by electronic transitions; the wavelength corresponding to these electronic transitions is about 3900 Å (fig. 4.26). Thus the population ratios of the various rotational states of the lowest energy level of CN may be measured by observing the *absorption* lines of CN at about 3900 Å. The electronic transitions allowed by quantum mechanical selection rules are identified in fig. 4.26. Note that the absorption lines R(1) and P(1) provide two independent means of determining the population in energy level 1.

4.10.3   *The optical observations*

CN molecules are found in abundance in interstellar clouds, and in some cases these clouds lie in the line of sight between us and a luminous background star (fig. 4.27). The optical CN lines of interest then appear as narrow absorption features imposed on the stellar spectrum. In fact, these absorption lines were detected 50 years ago by Adams (1941), McKellar (1941) and others in the spectra of several stars, and the existence of absorption lines originating from energy states 1 and 2 (with $J = 1$ and 2) was noted as

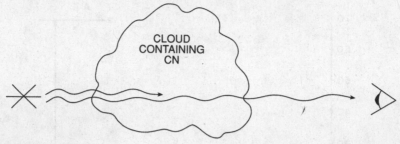

Fig. 4.27   CN molecules absorb the light of background stars at optical wavelengths.

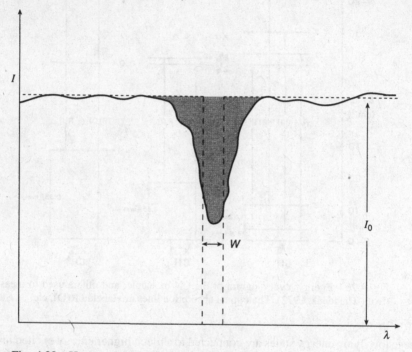

Fig. 4.28   How equivalent width w is defined. The shaded area is set equal to $I_0 w$, where I_0 is the continuum intensity in the neighborhood of the line.

a puzzle. McKellar (1941) even estimated a value of a few kelvin for the excitation temperature required. Not until 25 years later, immediately following Penzias' and Wilson's discovery of the CBR, did it become clear what was exciting the low-lying energy states of CN (Field and Hitchcock, 1966; Shklovskii, 1966; Thaddeus and Clauser, 1966, all independently made the connection). These and other observers reexamined old measurements and made new observations to refine values of n_1/n_0; these early results are summarized in the review of Thaddeus (1972).

Spectroscopists use *equivalent width* w (operationally defined in fig. 4.28) to measure the strength of spectral lines. If the lines are unsaturated, the population of the state producing the absorption will be directly proportional to the measured equivalent width and inversely proportional to the oscillator strength of the transition: $n \propto w/f$. Meyer and Jura (1985) give the appropriate values of f. Hence, for unsaturated lines, n_1/n_0 would follow from the ratio of the measured w values for the corresponding absorption lines, say R(1) and R(0). In some interstellar molecular clouds, however, the abundance of CN is large enough to saturate the R(0) line, and the simple proportion breaks down.* As a consequence, a small correction in the value of T_0 calculated from the observed equivalent widths is required. Even for dense clouds, the required corrections are only –0.1 K or so (note the negative sign), and can be found to an accuracy of a few hundredths of a kelvin (see, for instance, Meyer and Jura, 1985; and more recently Meyer and Roth, 1990).

In addition, even large, dense interstellar clouds produce very narrow CN absorption lines (w is 0.01–0.03 Å for the R(0) line and is still smaller for lines originating on energy states 1 and 2). This line width is comparable with or smaller than the spectral resolution of instruments used to measure the spectrum (the 'instrumental width'), so the observed line widths must be deconvolved to give true estimates of the equivalent widths of the CN lines (see Crane *et al.*, 1986).

4.10.4 Correction for local (non-equilibrium) excitation

Thus far, we have assumed that the populations of the three energy states in the ground level of CN are determined entirely by the thermal radiation of the CBR, so that eqn. (4.11) holds exactly. We must now consider the possibility that other processes might alter the population ratios. Collisions between electrons and the CN molecules can do so (other, less significant, mechanisms are discussed by Thaddeus, 1972). In particular, e^- collisions can raise the population in states 1 and 2, which would lead us to overestimate T_0 if we used eqn. (4.11) with no correction for such collisional excitation. Unfortunately, the magnitude of the required correction depends on the electron density in the interstellar cloud, n_e, and this quantity can vary from cloud to cloud (hence e^- collisions are referred to as *local* excitation). For three clouds often used to determine T_0 from the CN lines, Meyer and Jura (1985) show that $n_e \simeq (1-2) \times 10^{-4}$ cm^{-3}, resulting in corrections of –0.04 to –0.09 K to T_0. The uncertainty in these corrections is again a few hundreths of a kelvin.

Two points may be made about the corrections for local excitation. The first is that both this correction and the correction required for saturation lower T_0. Thus the uncorrected values of T_0 calculated directly from eqn. (4.11) are firm upper limits on the CBR temperature. The second point is that the local excitation correction cannot be measured directly, but must be calculated from estimates of n_e. How confident can we then be in the resulting values of T_0? Personally, I find the close agreement between the final, corrected values of T_0 derived from measurements of several different clouds (with different values of n_e) comforting. In addition, there is a way to ensure that e^- collisions do not play a major role in populating states 1 and 2 – use a radio telescope to look for *emission* at 2.64 mm (and 1.32 mm) wavelength (Penzias *et al.*, 1972). If the population of state 1 exceeds its thermal equilibrium value because of electron excitation in a particular cloud, that cloud will produce an emission line at $\lambda = 2.64$ mm. Penzias and his colleagues used a heterodyne receiver tuned to $\lambda = 2.64$ mm to search for such a line in one high density interstellar cloud. Their results (adjusted to $T_0 = 2.75$ K) imply that the correction for e^- excitation in this dense cloud is about 0.17 K or less in amplitude. Recently

* See a description of the *curve of growth* in an astrophysics text such as Aller (1953).

published work by Crane, Hegyi, Kutner and Mandolesi (1989) has improved the sensitivity of such searches by a factor of about three. In one less dense interstellar cloud frequently used for CN observations, the cloud lying in front of ζ Oph, they find *no* evidence for e^- excitation, and place an upper limit of about 0.03 K on its amplitude. If these observations are correct, it would appear that the electron-excitation corrections given earlier in this section may be too large in amplitude; if so, values of T_0 derived from earlier (pre-1989) CN measurements should be raised by about 0.05 K.

4.10.5 Results

The results of three recent observational programs to determine T_0 from measurements of the optical absorption lines of CN are shown in table 4.5 (Meyer and Jura, 1985; Meyer *et al.*, 1989; Crane *et al.*, 1986, 1989; Palazzi *et al.*, 1990; Kaiser and Wright, 1990, respectively). The tabulated values are those given by the observers; I have not attempted to take into account the possible overestimate of e^- excitation that has just been discussed.

The values of T_0 at $\lambda = 2.64$ mm are more precise than all but the very best direct measurements. Are they accurate – i.e., free of systematic error – as well? The agreement between the various groups' results is encouraging, as is the agreement from one cloud to another. The absence of a 2.64 mm radio emission line in some clouds implies that any error arising from e^- excitation must be small. Finally, it is worth emphasizing that the indirect measurements, employing molecular 'thermometers' lying light years from us, produce values of T_0 in good agreement with the results of both COBE and direct, ground-based radiometric observations.* Since the direct and indirect measurements employ radically different observational techniques, the agreement in the values of T_0 gives us confidence that the systematic errors in *both* techniques are well understood.

Since the population ratio n_2/n_1 is very small, the R(2) and P(2) lines are very weak. In addition the 3873.37 Å R(2) line happens to fall near a weak (and apparently variable – Crane *et al.*, 1986) absorption line produced by some component of the Earth's atmosphere. For both reasons, values of T_0 at 1.32 mm are less precise than those at 2.64 mm. On the other hand, no saturation corrections are required, and the corrections for e^- excitation are smaller. It is intriguing that the values of T_0 at 1.32 mm are typically 50–100 mK (or 1σ) above both the values found at 2.64 mm and the COBE measurements at $\lambda \simeq 1$ mm.

4.10.6 Other molecules

Two other diatomic molecules, CH and CH$^+$, have low-lying excited states of the ground level which can be probed by optical absorption lines. The energy differences $E_B - E_A$ correspond to wavelengths of 0.56 and 0.36 mm in CH and CH$^+$, respectively. It is easy to see from eqn. (4.11) that the excited state will be very lightly populated in either case if $T_0 \simeq 2.75$ K, and hence observations will be very difficult. On the other hand, observations of CH and CH$^+$ absorption lines could in principle allow us to check the COBE short-wavelength observations. The currently available measurements (Thaddeus,

---

* Palazzi, Mandolesi and Crane (1992) have very recently pointed out that CN measurements typically give T_0 approximately 80 ± 30 mK above the COBE results; they found no convincing explanation for this discrepancy.

1972) provide only upper limits on T_0, however, and these are so rough that they cannot discriminate between a blackbody spectrum and a spectrum with a far infrared excess as suggested by the results of Matsumoto *et al.* (1988). Nevertheless, I hope that new attempts to measure the CH and CH$^+$ absorption lines will soon be made, using the sophisticated optical techniques that have allowed astronomers to improve the accuracy of the 1.32 mm CN measurement by nearly a factor of ten since the pioneering work of Hegyi *et al.* (1972).

Kogut *et al.* (1990b) have recently obtained $T_0 = 3.2 \pm 0.9$ K at $\lambda = 2.1$ mm using radio wavelength observations of the absorption lines of orthoformaldehyde (H$_2$CO). Very substantial corrections for collisional excitation were required – see Kogut *et al.* (1990b) for details.

4.10.7 Measurements of T_0 at high redshift

Soon after the first indirect measurements of T_0 became available, it was realized that, in principle, such measurements could be carried out in cosmologically distant clouds. Thus – again in principle – we could measure the CBR temperature at earlier epochs when it would have been higher (Bahcall and Woolf, 1968; Bahcall *et al.*, 1973). From Section 1.3, we have

$$T(t) = T_0[z(t) + 1] \approx \left(\frac{t_0}{t}\right)^{2/3} T_0$$

for the Einstein–de Sitter model.

Meyer and his colleagues (1988) have recently attempted to make just such a measurement, by searching for an absorption line originating on a low-lying (fine structure) energy state of carbon atoms. In neutral carbon, $E_B - E_A$ corresponds to $\lambda = 0.61$ mm. The optical absorption line is at a rest frame wavelength of 1657 Å, and hence can be observed only in systems at redshifts > 1; Meyer *et al.* employed a cloud at $z = 1.776$ lying between us and a quasar. Their measurements can only set upper limits on the temperature of the CBR at that redshift, in part because of uncertainty in the collisional excitation rate. The result is ≤ 16 K. Thus $T_0 < 16/(1+1.776) = 5.76$ K now. Once again, we see that the observations are not yet precise enough to pin down the CBR temperature. It would obviously be valuable to extend such measurements to larger redshifts and to additional atoms or molecules. Improved measurements could constrain T_0 and would also confirm in a very direct fashion the cosmological origin of the CBR.

4.11 Limits on T_0 from measurements of the dipole moment in the CBR

For completeness, we note here that measurements of the dipole moment in the angular distribution of the CBR (especially at short wavelengths) may be used to constrain its spectrum. This point is elaborated further in Section 6.6.3. Here we note only that the short wavelength measurements of the dipole moment carried out by Halpern *et al.* (1988) may be used to find values of T_0 at the same wavelengths *provided* that the CBR spectrum is assumed to be blackbody. In that case, we find $T_0 = 2.86 \pm 0.26$ K in the wavelength range 1.3 mm < λ < 8 mm, in good agreement with results given earlier, and $T_0 = 3.01 \pm 0.31$ K in the range 0.55 mm < λ < 2 mm. On the other hand, as Halpern and his colleagues point out, their observations are not in good agreement with a spectrum like the one shown in fig. 4.22, which has a sharply rising temperature as ν increases; see Halpern *et al.* (1988) and Section 6.6.3 for a further discussion.

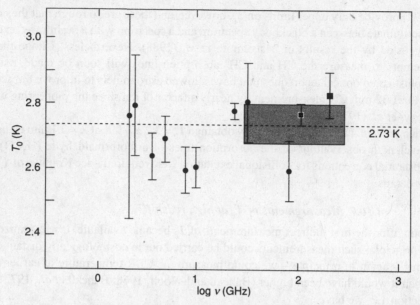

Fig. 4.29   The most accurate determinations of the CBR spectrum (from tables 4.2, 4.4 and 4.5). The shaded area represents the COBE measurement (Mather *et al.*, 1990); recall that the large error bars shown are only preliminary. CN measurements are shown as squares; at $\lambda = 2.64$ mm, I have averaged the best values from table 4.5.

4.12   Summary

Are the many measurements that we have of the spectrum of the CBR consistent with a single value of T_0, that is, with a blackbody spectrum? Entries from tables 4.2, 4.4 and 4.5 are combined in fig. 4.29 and provide an apparent answer: over a wavelength range of nearly 1000, the observed spectrum is indistinguishable from a Planck curve. To answer that crucial question in more detail, however, it is useful to divide the observational material into two parts by wavelength. First let us consider all the measurements at $\lambda > 2$ mm, regardless of the experimental techniques used, but omitting for the moment the FIRAS results from the COBE satellite and the newest measurements of Gush *et al.* (1990). Although it is a risky proposition to combine data taken in so many different ways by different groups, each using different means to calculate experimental error, let us form both unweighted and weighted averages of all the measurements at wavelengths > 2 mm *except* the COBE and rocket values. The results are given in the top line of table 4.6. In forming the weighted average, I have not used the usual weighting factor proportional to $1/\sigma^2$ (where σ^2 is the experimental variance in a particular measurement). Weights proportional to $1/\sigma^2$ are appropriate when statistical errors dominate. In most of these measurements, however, the error budget is dominated by *systematic* rather than statistical error. Thus I have arbitrarily chosen to weight each value by $1/\sigma$, taking the values of σ from tables 4.2, 4.4 and 4.5. I have also computed the weighted and unweighted averages of the *revised* measurements taken from table 4.2; these results appear on the second line of table 4.6. All four averages are in satisfactory agreement, and we may conclude that $T_0 = 2.75 \pm 0.03$ K at $\lambda > 2$ mm. This result is in excellent agreement with the satellite and rocket measurements in the 2–10 mm range.

Table 4.6 *Average values of T_0 for various sets of measurements drawn from tables 4.2, 4.4 and 4.5. The tabulated errors are standard deviations of the means of the values of T_0 used.*

Measurements used	Unweighted average T_0 (K)	T_0 weighted by 1/error (K)
All unrevised values with $\lambda > 2$ mm*	2.751 ± 0.062	2.728 ± 0.026
Revised values, $\lambda > 2$ mm*	2.829 ± 0.065	2.748 ± 0.026
Subset of 12 most precise values, 0.2 cm $< \lambda \le 13$ cm*	2.707 ± 0.024	2.732 ± 0.014
Unrevised values 1 mm $\le \lambda \le 2$ mm*	2.790 ± 0.054	2.791 ± 0.013
COBE FIRAS value 10 mm $\ge \lambda \ge 0.5$ mm	2.726 ± 0.010	
Gush *et al.* (1990)	2.736 ± 0.017	

*COBE and other rocket results excluded.

I have also selected the twelve measurements with lowest error in the wavelength range 0.2–13 cm (again omitting the COBE numbers): here the average value for T_0 is somewhat lower, $T_0 = 2.70$–2.73 K, because the measurements by the Berkeley group at 8, 4, 3 and 0.33 cm play a larger role in the average (for references, see table 4.2). The χ^2 for eleven degrees of freedom exceeds about 25* for the unweighted average, making it clear that these twelve 'best' measurements are not internally consistent. The inconsistency is between the relatively low values of T_0 determined at 8, 4, 3 and 0.33 cm by the Berkeley group on the one hand, and the relatively high values found from the CN measurements at 2.64 mm and by Johnson and Wilkinson (1987) at 1.2 cm on the other. We do not know which set of measurements lies closer to the true value of T_0, although the COBE and CN results marginally favor the latter (see fig. 4.29). It is apparent, however, that there is no systematic *trend* of T_0 with λ in the wavelength range 0.2–13 cm (or, though the longer wavelength measurements are less accurate, in the wider wavelength range 0.2–75 cm).

Next, let us look at measurements of T_0 in the narrow wavelength range 1–2 mm, including the 1.32 mm CN results. From tables 4.4 and 4.5 we find $T_0 = 2.79$ (line 4 of table 4.6), marginally higher than the value at $\lambda > 2$mm. That upward trend as λ decreases is continued by the two submillimeter measurements of Matsumoto *et al.* (1988) with $T_0 = 2.955 \pm 0.017$ and 3.175 ± 0.027 K at $\lambda = 0.71$ and 0.48 mm, respectively. On the other hand, the precision satellite and rocket measurements (Mather *et al.*, 1990; Gush *et al.*, 1990) show no wavelength-dependent trend in temperature, and are consistent with $T_0 = 2.73 \pm 0.03$ K over the entire wavelength range 10–0.5 mm, which completely brackets the peak of a 2.5–3 K blackbody. I am inclined to accord these more recent results more weight, and take $T_0 = 2.73$ K as the appropriate value for 10 mm $\ge \lambda$ ≥ 0.5 mm. That value, in turn, is in excellent agreement with $T_0 = 2.732 \pm 0.014$ K derived from a weighted average of the best measurements in the wavelength interval 13 cm $\ge \lambda \ge 0.2$ cm.

* The value of χ^2 depends sensitively on the magnitude of the error assumed for the 2.64 CN measurement of Crane *et al.* (1989).

Thus the measurements currently available suggest no significant departures from a Planck curve at any wavelength. As expected from the Big Bang model, the CBR does indeed have a blackbody spectrum, and we know its temperature to a precision of 1% or better. Henceforth, I will take 2.73 K as the value of T_0.

A few very recent measurements are summarized in Appendix C, 'Recent results'.

References

Adams, W. S. 1941, *Ap. J.*, **93**, 11.

Aller, L. H. 1953, *Astrophysics*, Ronald Press, New York.

Bahcall, J. N., and Wolf, R. A. 1968, *Ap. J.*, **152**, 701.

Bahcall, J. N., Joss, P. C., and Lynds, R. 1973, *Ap. J. (Lett.)*, **182**, L95.

Beckman, J. E., Chanin, G., Torre, J.-P., Melchiorri, F., Lizon-Tati, J., Olthof, H., and Wyn-Roberts, D. 1986, in *Gamow Cosmology*, eds. F. Melchiorri and R. Ruffini, North-Holland, Amsterdam.

Bernstein, G. M., Fischer, M. L., Richards, P. L., Peterson, J. B., and Timusk, T. 1990, *Ap. J.*, **362**, 107.

Bersanelli, M., Witebsky, C., Bensadoun, M., De Amici, G., Kogut, A., Levin, S. M., and Smoot, G. F. 1989, *Ap. J.*, **339**, 632.

Bielli, P., Pagana, E., and Sironi, G. 1983, *Proc. ICAP*, **1**, 509.

Boggess, N. W. *et al.* 1992, *Ap. J.*, **397**, 420.

Born, M, and Wolf, E. 1964, *Principles of Optics*, 2nd ed., Pergamon Press, Oxford.

Boynton, P. E., and Stokes, R. A. 1974, *Nature*, **247**, 528.

Boynton, P. E., Stokes, R. A., and Wilkinson, D. T. 1968, *Phys. Rev. Lett.*, **21**, 462.

Costales, J. B., Smoot, G. F., Witebsky, C., De Amici, G., and Friedman, S. D. 1986, *Radio Sci.* **L1**, 47.

Crane, P., Hegyi, D. J., Mandolesi, N., and Danks, A. C. 1986, *Ap. J.*, **309**, 822.

Crane, P., Hegyi, D. J., Kutner, M. L., and Mandolesi, N. 1989, *Ap. J.*, **346**, 136.

Dall'Oglio, G., Fonti, S., Melchiorri, B., Melchiorri, F., Natale, V., Lombardini, P., Trivero, P., and Sivertsen, S. 1976, *Phys. Rev. D*, **13**, 1187.

Danese, L., and Partridge, R. B. 1989, *Ap. J.*, **342**, 604.

De Amici, G. *et al.* 1988, *Ap. J.*, **329**, 556.

De Amici, G. *et al.* 1991, *Ap. J.*, **381**, 341.

De Amici, G., Smoot, G. F., Friedman, S. D., and Witebsky, C. 1985, *Ap. J.*, **298**, 710.

De Amici, G., Bensadoun, M., Bersanelli, M., Kogut, A., Levin, S., Smoot, G. F., and Witebsky, C. 1990, *Ap. J.*, **359**, 219.

Deepack, A., Wilkerson, T. T., and Ruhnke, L. H. 1980, *Atmospheric Water Vapor*, Academic Press, New York.

Dicke, R. H., Peebles, P. J. E., Roll, P. G., and Wilkinson, D. T. 1965, *Ap. J.*, **142**, 414.

Ewing, M. S., Burke, B. F., and Staelin, D. H. 1967, *Phys. Rev. Lett.*, **19**, 1251.

Field, G. B., and Hitchcock, J. L. 1966, *Phys. Rev. Lett.* **16**, 817.

Friedman, S. D., Smoot, G. F., DeAmici, G., and Witebsky, C. 1984, *Phys. Rev. D*, **29**, 2677.

Gorenstein, M. V., Muller, R. A., Smoot, G. F., and Tyson, J. A. 1978, *Rev. Sci. Instr.*, **49**, 440.

Grenier, P., Roucher, J., and Talureau, B. 1976, *Astron. Astrophys.*, **53**, 249.

Gulkis, S., Lubin, P. M., Meyer, S. S., and Silverberg, R. F. 1990, *Scientific American*, **262**, 122.

Gush, H. P. 1981, *Phys. Rev. Lett.*, **47**, 745.

Gush, H. P., Halpern, M., and Wishnow, E. H. 1990, *Phys. Rev. Lett.*, **65**, 537.

Halpern, M., Benford, R., Meyer, S., Muehlner, D., and Weiss, R. 1988, *Ap. J.*, **332**, 596.

Haslam, C. G. T., Salter, C. J., Stoffel, H., and Wilson, W. E. 1982, *Astron. Astrophys. Suppl.*, **47**, 1.

Hauser, M. G. *et al.* 1991, in *After the First Three Minutes*, eds. S. S. Holt, C. L. Bennett, and V. Trimble, American Institute of Physics, New York.

Hegyi, D., Traub, W., and Carleton, N. 1972, *Phys. Rev. Lett.*, **28**, 1541.

Herzberg, G. 1945, *Molecular Spectra and Molecular Structure*, Prentice Hall, New York.

Houck, J. R., and Harwit, M. 1969, *Science*, **164**, 1271.

Howell, T. F., and Shakeshaft, J. R. 1966, *Nature*, **210**, 1318.

Howell, T. F., and Shakeshaft, J. R. 1967, *Nature*, **216**, 753.

Johnson, D. G. 1986, Ph.D. Thesis, Princeton University.

Johnson, D. G., and Wilkinson, D. T. 1987, *Ap. J.* (*Lett.*), **313**, L1.

Kaiser, M. E., and Wright, E. L. 1990, *Ap. J.* (*Lett.*), **356**, L1.

Kislyakov, A. G., Chernyshev, V. I., Lebskii, Yu. V., Mal'tsev, V. A., and Serov, N. V. 1971, *Soviet Astron. J.*, **15**, 29.

Kogut, A., Bersanelli, M., De Amici, G., Friedman, S. D., Griffith, M., Grossan, B., Levin, S., Smoot, G. F., and Witebsky, C. 1988, *Ap. J.*, **325**, 1.

Kogut, A., Bensadoun, M., De Amici, G., Levin, S., Smoot, G. F., and Witebski, G. 1990a, *Ap. J.*, **355**, 102.

Kogut, A., Petuchowski, S. J., Bennett, C. L., and Smoot, G. F. 1990b, *Ap. J.* (*Lett.*), **348**, L45.

Kogut, A. *et al.* 1991, in *After the First Three Minutes*, eds. S. S. Holt, C. L. Bennett and V. Trimble, American Institute of Physics, New York.

Kraus, J. D. 1966, *Radio Astronomy*, McGraw-Hill, New York.

Lawson, K. D., Mayer, C. J., Osborne, J. L., and Parkinson, M. L. 1987, *Monthly Not. Roy. Astron. Soc.*, **225**, 307.

Levin, S. M., Witebsky, C., Bensadoun, M., Bersanelli, M., De Amici, G., Kogut, A., and Smoot, G. F. 1988, *Ap. J.*, **334**, 14.

Liebe, H. J. 1985, *Radio Sci.*, **20**, 1069.

Lubin, P. M., and Villela, T. 1985, in *The Cosmic Background Radiation and Fundamental Physics* ed. F. Melchiorri, Editrice Compositori, Bologna.

Mandolesi, N. *et al.* 1986, *Ap. J.*, **310**, 561.

Martin, D. R., and Puplett, E. 1969, *Infrared Physics*, **10**, 105.

Mather, J. C. 1982, *Optical Engineering*, **21**, 769.

Mather, J. C., *et al.* 1990, *Ap. J.* (*Lett.*), **354**, L37.

Matsumoto, T., Hayakawa, S., Matsuo, H., Murakami, H., Sato, S., Lange, A. E., and Richards, P. L. 1988, *Ap. J.*, **329**, 567.

McKellar, A. 1941, *Publ. Dominion Astrophys. Observatory*, **1**, 251.

Meyer, D. M., and Jura, M. 1985, *Ap. J.*, **297**, 119.

Meyer, D. M., and Roth, K. C. 1990, *Ap. J.*, **349**, 91.

Meyer, D. M., Roth, K. C., and Hawkins, I. 1989, *Ap. J.* (*Lett.*), **343**, L1.

Meyer, D. M., Black, J. H., Chaffee, F. H., Foltz, C. B., and York, D. G. 1988, *Ap. J.* (*Lett.*), **308**, L37.

Millea, M. F., McColl, M., Pedersen, R. J., and Vernon, F. L. 1971, *Phys. Rev. Lett.*, **26**, 919.

Montgomery, C. G., Dicke, R. H., and Purcell, E. M. 1963, *Principles of Microwave Circuits*, Boston Technical Lithographers, Boston, Massachusetts (reprinted 1965 by Dover).

Otoshi, T. Y., and Stelzried, C. T. 1975, *I.E.E.E. Trans. Instrumentation Measurements*, **24**, 174.

Pacholczyk, A. G. 1970, *Radio Astrophysics*, W. H. Freeman, San Francisco.

Palazzi, E., Mandolesi, N., and Crane, P. 1992, *Ap. J.*, **398**, 53.

Palazzi, E., Mandolesi, N., Crane, P., Kutner, M. L., Blades, J. C., and Hegyi, D. J. 1990, *Ap. J.*, **357**, 14.

Partridge, R. B. 1969, *American Scientist*, **57**, 3.

Partridge, R. B., Cannon, J., Foster, R., Johnson, C., Rubinstein, E., Rudolph, A., Danese, L., and DeZotti, G. 1984, *Phys. Rev.* D, **29**, 2683.

Partridge, R. B. *et al.* 1986, in *The Cosmic Background Radiation and Fundamental Physics*, ed. F. Melchiorri, Italian Physical Society, Bologna.

Pelyushenko, S. A., and Stankevich, K. S. 1969, *Soviet Astron. J.*, **13**, 223.

Penzias, A. A. 1968, *I.E.E.E. Trans. Microwave Theory Tech.* **16**, 608.

Penzias, A. A., and Wilson, R. W. 1965, *Ap. J.*, **142**, 419.

Penzias, A. A., and Wilson, R. W. 1967, *Science*, **156**, 1100.

Penzias, A. A., Jefferts, K., and Wilson, R. W. 1972, *Phys. Rev. Lett.*, **28**, 772.

Peterson, J. B., Richards, P. L., and Timusk, T. 1985, *Phys. Rev. Lett.*, **55**, 332.

Puzanov, V. I., Salomonovich, A. E., and Stankevich, K. S. 1968, *Soviet Astron. J.*, **11**, 905.

Roll, P. G., and Wilkinson, D. T. 1966, *Phys. Rev. Lett.*, **16**, 405.

Roll, P. G., and Wilkinson, D. T. 1967, *Annals of Physics*, **44**, 289.

Shklovskii, I. S. 1966, *Astron. Circular* (USSR Academy of Sciences), **364**, 1.

Sironi, G., and Bonelli, G. 1986, *Ap. J.*, **311**, 418.

Sironi, G., Bonelli, G., and Limon, M. 1991, *Ap. J.*, **378**, 550.

Sironi, G., Inzani, P., and Ferrari, A. 1984, *Phys. Rev.* D, **29**, 2686.

Sironi, G., Limon, M., Marcellino, G., Bonelli, G., Bersanelli, M., Conti, G., and Reif, K. 1990, *Ap. J.*, **357**, 301.

Smoot, G. F., De Amici, G., Friedman, S. D., Witebsky, C., Mandolesi, N., Partridge, R. B., Sironi, G., Danese, L., and De Zotti, G. 1983, *Phys. Rev. Lett.*, **51**, 1099.

Smoot, G. F., De Amici, G., Friedman, S., Witebsky, C., Sironi, G., Bonelli, G., Mandolesi, N., Cortiglioni, S., Morigi, G., Partridge, R. B., Danese, L., and De Zotti, G. 1985, *Ap. J. (Lett.)*, **291**, L23.

Smoot, G., Bensadoun, M., Bersanelli, M., De Amici, G., Kogut, A., Levin, S., and Witebsky, C. 1987, *Ap. J. (Lett.)*, **317**, L45.

Stankevich, K. S., Wielebinski, R., and Wilson, W. E. 1970, *Australian J. Phys.*, **23**, 529.

Stokes, R. A. 1968, Ph.D. Thesis, Princeton University.

Stokes, R. A., Partridge, R. B., and Wilkinson, D. T. 1967, *Phys. Rev. Lett.*, **19**, 1199.

Thaddeus, P. 1972, *Ann. Rev. Astron. Astrophys.*, **10**, 305.

Thaddeus, P., and Clauser, J. F. 1966, *Phys. Rev. Lett.*, **16**, 819.

Van Vleck, J. H. 1947, *Phys. Rev.*, **71**, 413.

Verschuur, G. L. and Kellermann, K. I., eds., 1974, *Galactic and Extra-galactic Radio Astronomy*, Springer-Verlag, New York.

Waters, J. W. 1976, in *Methods of Experimental Physics: Astrophysics. Part B. Radio-telescopes*, ed. M. L. Meeks, Academic Press, New York.

Weiss, R. 1980, *Ann. Rev. Astron. Astrophys.*, **18**, 489.

Welch, W. J., Keachie, S., Thornton, D. D., and Wrixon, G. 1967, *Phys. Rev. Lett.*, **18**, 1068.

Wilkinson, D. T. 1967, *Phys. Rev. Lett.*, **19**, 1195.

Witebsky, C. 1978, COBE Report No. 5013, unpublished.

Witebsky, C., Smoot, G. F., De Amici, G., and Friedman, S. D. 1986, *Ap. J.*, **310**, 145.

Woody, D. P., and Richards, P. L. 1979, *Phys. Rev. Lett.*, **42**, 925; also 1981, *Ap. J.*, **248**, 18.

Zammit, C. C., and Ade, P. A. 1981, *Nature*, **293**, 550.

5

What we learn from observations of the CBR spectrum

The many measurements of the spectrum of the CBR discussed in the preceding chapter are consistent with a Planck spectrum with $T_0 = 2.73 \pm 0.02$ K over a wavelength range $0.1 \text{ cm} \lesssim \lambda \lesssim 75$ cm. Only the submillimeter observations reviewed in Section 4.8.4 provided any evidence for a significant deviation from a thermal spectrum, and these appear to be erroneous. What conclusions may we draw from the essentially thermal spectrum of the CBR? This chapter provides some answers to that question. It is presented as an introduction to, not an exhaustive treatment of, the processes that determine the CBR spectrum. There are a number of reviews, which treat these topics in more detail, such as Danese and De Zotti (1977), Sunyaev and Zel'dovich (1980) and Bond (1988).

We begin by noting the conditions under which we would expect an exactly thermal spectrum to have been produced early in the Hot Big Bang, then consider a number of physical processes that could have distorted an initially thermal spectrum. We also consider the possibility that one or more additional 'cosmic' backgrounds may be present, adding to the CBR at wavelengths below about 1 mm. Finally, we investigate the constraints that the spectral measurements of Chapter 4 place on these processes.

5.1 Why is the CBR spectrum so closely thermal?

In earlier chapters, I have frequently stated that the CBR should have a thermal spectrum if it is indeed a relic of the Hot Big Bang, and indeed that the CBR spectrum is the best evidence we have for the Hot Big Bang model. It is now appropriate to look at that argument in more detail.

For the observed CBR spectrum to be thermal at the present epoch, two conditions must be met:

1. in some earlier epoch, the matter and radiation components of the Universe were in thermal equilibrium, so that a Planck spectrum was established;
2. expansion of the Universe between that epoch and the present did not alter the spectrum.

We shall show that both conditions can be met by the Hot Big Bang model.

5.1.1 The thermal equilibrium of the early Universe

Neutral atoms interact only minimally with microwave radiation; hence the Universe is now transparent at the wavelengths used for CBR studies. In much earlier epochs, however, corresponding to a redshift of about 1000 or more, matter in the Universe was ionized, and free electrons in the plasma scattered the radiation. The process involved was Thomson scattering. This process changes the direction of a photon without changing its frequency, at least to first order; thus pure Thomson scattering by itself cannot establish a thermal spectrum. We need to look still further back in the history of the Universe to find conditions appropriate for formation of a thermal spectrum of the CBR.

What physical conditions were necessary to establish a thermal radiation spectrum in the expanding Universe? We will later see that it is useful to ask this question in a slightly different form: under what conditions would a radiation field with an initially *non-Planckian* spectrum relax to a blackbody form? Two conditions were required for creation of a thermal spectrum early in the history of the Universe. One was a mechanism (or set of mechanisms) to create photons and/or redistribute their energy. The second requirement was that the reaction rates for these mechanisms were large compared to the expansion rate, so that the mechanisms had time to act. Both conditions were met early in the history of the Universe.

It will help to make a couple of simplifying assumptions at the start. The first is to begin our consideration of thermalization after the last massive particle–antiparticle pairs annihilated. Unless one or more of the three families of neutrinos has mass, the last such annihilation event (involving known leptons) was the $e^+ + e^- \rightarrow 2\gamma$ transition at $z \simeq 10^9$ when the Universe was about 1 min old. As we will soon see, no matter what spectrum the radiation had at $z \simeq 10^9$, thermal equilibrium was restored in the next two months of the history of the Universe or earlier; in fact, however, the close coupling between matter and radiation via $e^+ + e^- \leftrightarrows 2\gamma$ and Coulomb scattering surely kept the radiation close to thermal equilibrium up to the epoch corresponding to $z \simeq 10^9$. Thus there is some support for the second assumption, namely that the radiation spectrum at $z \simeq 10^9$ was not grossly distorted. It may be shown that even arbitrarily large distortions would have relaxed to thermal equilibrium (see, e.g., Burigana *et al.*, 1991), so we lose no generality in making this simplifying assumption.

Now let us follow the subsequent history of the radiation field and its interactions with the residual electrons and baryons. At $z \lesssim 10^9$, essentially three physical processes could create photons or alter their energy: thermal bremsstrahlung (or free–free emission); the ordinary Compton effect, in which electrons scatter photons, changing their energy; and radiative Compton (or 'double Compton') scattering, in which a second photon is produced in the electron–photon collision (see Lightman, 1981; or for specific reference to the present issue, Danese and De Zotti, 1982).* The first of these processes involves the deflection of e^- by protons (or other positively charged particles) and hence depends on the baryon density. The Compton effect depends on the energy (and therefore the temperature T_e) of the electrons, as well as their number density. Thus, qualitatively, all these mechanisms were more effective at larger redshifts when the density and temperature were higher.

Because the Thomson scattering cross-section σ_T is large (6.65×10^{-25} cm^2), the reaction rate for Compton scattering was large before recombination. As Kompaneets (1957), Weymann (1965), Zel'dovich and Sunyaev (1969), and Peebles (1971) show,

---

* Thomson scattering, mentioned above, is the classical, non-relativistic asymptote of Compton scattering, for $h\nu/kT \ll 1$.

Compton scattering alone could produce *kinetic* equilibrium between e^- and the photons that they scattered on a time scale given by

$$t_c = \frac{m_e c}{n_e \sigma_T kT_e} \approx 1.1 \times 10^{28} \left(\frac{2.7}{T_0}\right)\left(\frac{T}{T_e}\right)\left(\Omega_b h^2\right)^{-1}(z+1)^{-4}\ \text{s}, \qquad (5.1)$$

where m_e is the electron mass and n_e the number density of e^-; Ω_b is the baryon density; and T is the radiation temperature, which for small perturbations of the spectrum is of order $T_0(z+1)$. This time scale was short relative to the expansion time scale (so that *kinetic* equilibrium could be produced) for

$$z_c \geq 2.2 \times 10^4 \left(\frac{T}{T_e}\right)^{1/2}\left(\Omega_b h^2\right)^{-1/2} \qquad (5.2)$$

Since T/T_e is not known *a priori* unless the exact nature of the spectral distortion is given, we can only evaluate (5.2) approximately. What we *can* say, given the assumption that the initial spectrum at $z \simeq 10^9$ was not too distorted, is that kinetic equilibrium could easily be established in the Universe for any redshift $z \gtrsim 10^4$–10^6, say. Note that there was no need to create additional photons to establish kinetic equilibrium (Compton scattering, like Thomson scattering, preserves photon number).

We next make use of a standard result from kinetic theory, namely that the spectrum of a radiation field with a fixed number of photons and in kinetic equilibrium with matter is a Bose–Einstein spectrum (see, e.g., Reif, 1965). Thus the photon occupation number (number per mode) is

$$\eta = \frac{1}{\exp\left(\dfrac{h\nu}{kT_e}+\mu\right)-1}. \qquad (5.3)$$

The quantity μ is the *chemical potential*, which is zero for an exact blackbody spectrum.

Hence we are left with the following conclusion: at any redshift greater than z_c given in eqn. (5.2), an arbitrary spectrum would have relaxed to the Bose–Einstein form given in (5.3) and shown in fig. 5.1.

Next, I will argue that at sufficiently early times (or large z) μ tended to zero, leaving us with a Planck function. In other words, at sufficiently large z, not only *kinetic* equilibrium but true *thermal* equilibrium could be produced. To establish thermal equilibrium, photons must have been *created* as well as moved about in frequency. The two mechanisms on which we can call are thermal bremsstrahlung and radiative Compton scattering; both can change the photon occupation number η and hence alter μ. The relaxation of a Bose–Einstein to a Planck spectrum via thermal bremsstrahlung was first studied by Kompaneets (1957). Later, Danese and De Zotti (1982) added the effect of radiative Compton (or double Compton) scattering. This was in fact the dominant mechanism in the cosmological context unless $\Omega_b \gtrsim 0.1\ h^{-2}$. In the notation used by Burigana *et al.* (1991), we have for the evolution of μ

$$\frac{d\mu}{dt} = -\frac{1.7}{\phi(\mu)M(\mu)} \int_0^\infty \left[\left(\frac{\partial\eta}{\partial t}\right)_B + \left(\frac{\partial\eta}{\partial t}\right)_{RC}\right] x_e^2\ dx_e, \qquad (5.4)$$

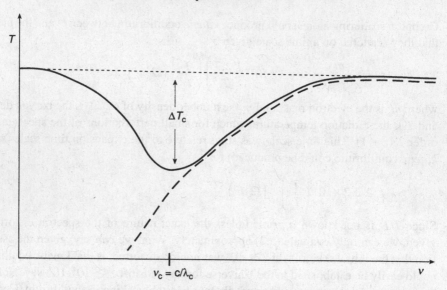

Fig. 5.1   A Bose–Einstein spectrum, characterized by $\mu > 0$ (heavy dashed line). As noted in Section 5.2.2, bremsstrahlung 'fills in' the Bose–Einstein spectrum at long wavelengths, bringing T back into equilibrium, as shown by the solid line. The magnitude of the temperature dip at $\lambda \sim \lambda_c$ depends directly on μ (eqn. (5.10)).

where $x_e \equiv h\nu/kT_e$ and

$$\phi(\mu) = \frac{1}{2.404} \int_0^\infty \frac{x_e^2 \, dx_e}{\exp\,(x_e + \mu) - 1},$$

$$M(\mu) = 3\frac{d \ln f(\mu)}{d\mu} - \frac{4 \, d \ln \phi(\mu)}{d\mu},$$

with

$$f(\mu) = \frac{1}{6.494} \int_0^\infty \frac{x_e^3 \, dx_e}{\exp\,(x_e + \mu) - 1}.$$

In (5.4), $(\partial\eta/\partial t)_B$ is the production rate of new photons by the bremsstrahlung process and $(\partial\eta/\partial t)_{RC}$ the corresponding rate for the radiative Compton process.

Both $(\partial\eta/\partial t)_B$ and $(\partial\eta/\partial t)_{RC}$ can be evaluated for plasma with cosmic abundances of H and ^{4}He. Numerically, in units of reciprocal seconds,

$$\left(\frac{\partial\eta}{\partial t}\right)_B \approx 4.2 \times 10^{-24} \left(\frac{T}{T_e}\right)^{7/2} (z+1)^{5/2}(\Omega_b h^2)^2 \, \frac{g(x_e)}{x_e^3 \, e^{x_e}}[1 - \eta(e^{x_e} - 1)], \tag{5.5}$$

where $g(x_e)$ is a Gaunt factor of order unity (see Burigana *et al.*, 1991, for detailed evaluation); and

$$\left(\frac{\partial\eta}{\partial t}\right)_{RC} \approx 3.3 \times 10^{-39} \left(\frac{T}{T_e}\right)^{-2} f(\mu)(\Omega_b h^2)(z+1)^5 \, \frac{1}{x_e^3 \, e^{3x_e/2}}[1 - \eta(e^{x_e} - 1)]. \tag{5.6}$$

Several authors have investigated solutions to (5.4)–(5.6) analytically and numerically (see, for instance, Chan and Jones, 1975a, for early work; Danese and De Zotti, 1982, and references therein; Bond, 1988). Here, I will quote the results of the most recent numerical integration (Burigana *et al.*, 1991), which show that $\mu \to 0$ for redshifts above a limiting value

$$z_{th} \approx 1.5 \times 10^6 \, \mu_0^{0.11} (\Omega_b h^2)^{-0.39}, \tag{5.7a}$$

where μ_0 specifies the initial chemical potential, assumed not to be very large ($\mu_0 < 1$). Burigana *et al.* evaluate the same set of equations for arbitrarily large distortions ($\mu_0 > 1$), and find

$$z_{th} \approx 1.8 \times 10^6 (\Omega_b h^2)^{-0.36}. \tag{5.7b}$$

If the baryon density is high, $\Omega_b h^2 > 0.3$, thermal bremsstrahlung dominates, and we find instead (for $\Omega_b = 1$):

$$z_{th} \approx 2.5 \times 10^6 \, \mu_0^{0.17} \text{ or } 2.75 \times 10^6, \tag{5.7c}$$

for μ_0 small and large, respectively. In other words, no matter what form the radiation spectrum might have had at $z > z_{th}$, by a redshift of $z_{th} \simeq 10^6–10^7$, it acquired a *thermal*, blackbody spectrum. This range of redshift corresponds to an epoch of up to about eight weeks after the Big Bang.

Let me note here explicitly that energy added to the matter or radiation *after* the epoch corresponding to z_{th} cannot necessarily be thermalized. Mechanisms that can operate at $t > t_{th}$ to produce lasting distortions in the spectrum will be treated in Section 5.2 below.

5.1.2 Maintaining the thermal spectrum in the expanding Universe

We now must show that the Planck spectrum generated by thermal bremsstrahlung and other processes at $z \gtrsim 10^6$ was *maintained* in the subsequent expansion of the Universe.

First, let us consider the situation when matter and radiation in the expanding Universe were tightly coupled *and* the radiation energy density dominated that of matter. Then the tight coupling and the large heat capacity of the radiation field kept the temperature of the matter equal to the CBR temperature. Thermal equilibrium was maintained, and the CBR spectrum remained blackbody (or Bose–Einstein if $\mu \neq 0$).

At some point, however, the energy density of radiation dropped below the energy density of matter. Since $\rho_\gamma / \rho_{matter}$ was proportional to $(z + 1)$, and $\rho_\gamma = a T_0^4 / c^2$ now (eqn. (1.16)), it may easily be shown that the redshift at which ρ_γ dropped below ρ_{matter} was

$$(z+1) = \rho_0 c^2 \, (a T_0^4)^{-1} = \left(\frac{3 \, H_0^2}{8 \pi \, G} \, \Omega \right) c^2 (a T_0^4)^{-1} \approx 4 \times 10^4 \, (\Omega h^2),$$

for $T_0 = 2.73$ K (as seen in Section 1.6.6). Somewhat later[*], at $z \simeq 1000$ (Section 1.6.5), the primeval plasma recombined and the CBR photons suddenly decoupled from the matter. If the decoupling had been instantaneous, all CBR photons would reach us from a surface of last scattering at a single well-defined temperature $T_s = T_0(z_s + 1)$. Recombination, however was not instantaneous, so the surface of last scattering had a finite thickness (using redshift as a measure of distance, the thickness was $\Delta z_s \simeq 100$), as first shown by Zel'dovich, Kurt and Sunyaev (1968) and by Peebles (1968); see also

---

[*] If the product $\Omega h^2 \lesssim 0.03$, decoupling will occur first.

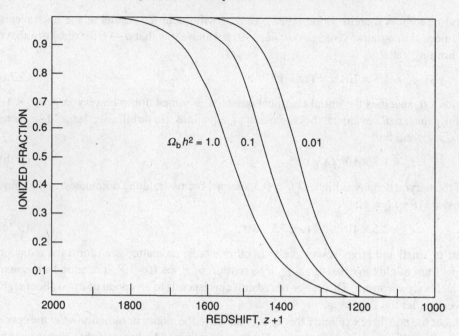

Fig. 5.2   The fraction of ionized hydrogen as a function of redshift for several values of the baryon density (adapted from Kolb and Turner, 1990). Note the relatively rapid drop in the ionization (and hence relatively sudden decoupling) for any value of Ω_b.

Jones and Wyse (1985). In fig. 5.2 are shown typical recombination curves; note that redshift of recombination depends slightly on proton and electron density and hence on Ω_b (see Peebles, 1971 or Kolb and Turner, 1990). We now need to ask whether an initially blackbody spectrum could be maintained through recombination and the consequent decoupling of matter and radiation. Yes, almost exactly.* The argument is made nicely by Weinberg (1972), who writes down the present energy density of the background in terms of the probability P that a photon at a particular present frequency v will survive from an earlier time t to the present:

$$u_0(v)\,dv = \frac{8\pi h v^3\,dv}{c^3} \int_0^{t_0} \left[\exp\left(\frac{hv(z(t)+1)}{kT(t)} \right) - 1 \right]^{-1} \times \frac{d}{dt}\,P(t_0, t, v)\,dt \quad (5.8)$$

(Weinberg's eqn. (15.5.3)). Here, T is the temperature of matter at time t. The quantity $P \to 0$ at sufficiently early time and approached unity for $t \gg t_s$, the epoch of last scattering. As Weinberg remarks, the right-hand side of (5.8) is a 'weighted average of Planck blackbody distributions.' Next, he points out that the large value of n_γ/n_b (Section 1.6.2) ensured that the heat capacity of the radiation was large enough to keep matter and radiation cooling at the same rate. Thus $T \propto (z + 1)$ for both until well after decoupling was complete. The quantity in square brackets in eqn. (5.8) then becomes independent of time and may be taken outside the integral. Thus $u_0(v)$ had a Planckian form whether or not decoupling was instantaneous.

* The very small distortions of the spectrum are considered in Section 5.2.6; they are completely undetectable with current technology.

After the CBR photons last scattered, at a redshift $z_s \simeq 1000$, they propagated freely in the expanding Universe. Tolman in 1934 first studied the free propagation of thermal radiation in an expanding Universe, showing that expansion did not alter the blackbody spectrum. The photon occupation number remained fixed at

$$\eta = \frac{1}{\exp{(h\nu / kT)} - 1} \, .$$

Once again, both ν and T varied as $(z + 1)$, so the Planck spectrum was maintained, but T decreased to its present value T_0.

Note that these arguments do not depend on any particular value of z_s or even of Δz_s. Thus the spectrum will remain essentially blackbody* even if the CBR was rescattered at some redshift $\ll 1000$ by reionized matter in the Universe.

5.1.3 Summary

Let us briefly review the arguments of Sections 5.1.1 and 5.1.2. Providing that the spectrum was not distorted by addition of energy *after* an epoch corresponding to $z = z_{th}$ (roughly $z_{th} = 2 \times 10^6$ or $T \simeq 5 \times 10^6$ K), the Hot Big Bang produced an exactly thermal spectrum. If no heat was added later, that spectrum was preserved until the present even though the Universe expanded in volume by a factor of about 10^{19}.

It is worth noting explicitly that the good match between the observations and a Planck curve thus provides direct evidence that the Universe was both about a factor 2×10^6 hotter than it is at present and about 10^{19} times denser. Spectral measurements of the CBR thus allow us to probe the physical conditions of the Universe back to an epoch a few weeks or months after the Big Bang.

5.2   The origin of (small) spectral distortions

In Section 5.1.1, we laid the groundwork for discussion of some of the physical processes that can in principle distort the blackbody spectrum of the CBR. Since no distortions have yet been observed, these processes clearly played only a minor role. Indeed, it may be asked why we need to consider them at all. The answer is that we will learn how to turn upper limits on various classes of spectral distortions into interesting limits on a variety of energy generating mechanisms operating in the wide redshift range from about 10^6 to 1.

Among the processes that could result in spectral distortions are the following.

1.   Energy released as primordial anisotropy, shear or vorticity was damped (e.g., Rasband, 1971). The lost kinetic energy appeared as heat, which could in turn distort the CBR spectrum.
2.   Energy released as smaller scale ($l \ll ct$) vorticity, turbulence, pressure waves or gravity waves were damped (e.g., Ozernoi and Chernin, 1968; Chan and Jones, 1975b; and Daly, 1991).
3.   Gravitational energy released as galaxies or other large-scale structures formed (Sunyaev and Zel'dovich, 1972a).

---

* Again, this may not have been exactly so; see Section 5.2.3.

4.   The (radiative) decay of long-lived exotic particles (Silk and Stebbins, 1983; Fukugita, 1988; Field and Walker, 1989) or the annihilation of particle–antiparticle pairs (with $m < m_e$) after the epoch corresponding to $z \simeq 10^9$ (Stecker and Puget, 1972; Jones and Steigman, 1978).

5.   Evaporation of primordial Black Holes.

6.   Decay of superconducting cosmic strings (Section 1.7.4; see Ostriker and Thompson, 1987).

7.   Scattering of the CBR photons by electrons with a kinetic temperature higher than the CBR temperature – the inverse Compton effect, first explored in this context by Sunyaev and Zel'dovich (1972b).

In principle, such processes could have produced distortions of arbitrary magnitude. The observations, however, tell us that we need to consider only small amplitude distortions, and we will restrict our attention to these.

5.2.1 Energy release at $z > z_{th}$

As we have seen above, energy released at redshifts $> z_{th} \simeq (1\text{–}2) \times 10^6$ could increase the temperature of the CBR, but could not distort the spectrum – no matter what happened at earlier epochs, the radiation at $z \simeq z_{th}$ has a thermal spectrum. Consequently, the shape of the observed CBR spectrum can tell us nothing about heating processes at times earlier than $t_{th} \approx 2$ months or less. For instance, the annihilation of particle–antiparticle pairs with $m \gtrsim 0.01 m_e$ would merely have increased T without altering the spectrum. Likewise, viscous damping of large-scale anisotropic or vortex motions (see e.g., Doroshkevich, Zel'dovich and Novikov, 1968) is likely to have occurred at times much earlier than t_{th}, thus leaving no fingerprint in the CBR spectrum.

On the other hand, we do have some constraints on such heating mechanisms in the redshift range $2 \times 10^8 \gtrsim z \gtrsim 2 \times 10^6$. These constraints are set by the observed abundances of light nuclei (see Section 1.6.4). The results in that section, shown in fig. 1.4 and fig. 1.10, were calculated by extrapolating the present temperature back to the epoch of nucleosynthesis at $z = (2\text{–}4) \times 10^8$ using $T(t) = T_0[z(t) + 1]$. That relation held, of course, only if no heat was added between t and t_0. Suppose now that heat *had* in fact been added in that time interval t to t_0; then the actual temperature at the synthesis epoch would have been $T(t) = \varepsilon T_0[z(t) + 1]$ with $\varepsilon < 1$. Since $n_\gamma \propto T^3(t)$, and nucleosynthesis results depend on n_b/n_γ, the result is to modify the limits the observations of ^{2}H and ^{7}Li set on Ω_b as follows:

$$0.01 h^{-2} \varepsilon^3 \leq \Omega_b \leq 0.015 \, h^{-2} \varepsilon^3.$$

If ε were less than about 0.4, even the upper limit on Ω_b set by nucleosynthesis would be less than the density of baryonic matter observed in galaxies alone, $\Omega_b \gtrsim 0.01$ (Trimble, 1987). Thus (unless h happens to be < 0.4) the temperature of the Universe cannot have been raised by more than a factor of about 2.5 in the redshift interval of roughly 2×10^8 to 2×10^6.

5.2.2 Energy release in the interval $z_{th} > z > z_c$

Here we may use the results of Section 5.1.1 directly: Compton scattering ensured kinetic equilibrium and hence a Bose–Einstein spectrum until z_c, but neither radiative

Compton scattering nor bremsstrahlung could produce a completely thermal spectrum at redshifts $< z_{th}$.

We need now to take account of a feature of bremsstrahlung skipped over earlier in this chapter: its frequency dependence. (The frequency dependence of bremsstrahlung emission has already been noted in Section 3.7.1.) While the temperatures in the redshift interval $z_{th} - z_c$ were about 100 times higher than those encountered in interstellar HII regions, it was still approximately true that the bremsstrahlung reaction rate was proportional to $\nu^{-2.1}$. Hence bremsstrahlung was more effective at producing thermal equilibrium at long wavelengths than at short ones.* Thus a spectrum of the characteristic form shown by the solid line in fig. 5.1 was produced by the injection of heat in the redshift interval $z_{th} > z > z_c$. At $\lambda < \lambda_c$ the spectrum is Bose–Einstein; at $\lambda > \lambda_c$, the spectrum is thermal.

The value of λ_c and the resulting shape of the spectrum have been investigated in detail by a number of authors (e.g., Chan and Jones, 1975a, b; Illarionov and Sunyaev, 1975; Danese and De Zotti, 1977, 1982). Spectra like those represented in fig. 5.1 can be parameterized by three quantities. The first of these is the chemical potential μ, which describes the spectrum at $\lambda < \lambda_c$. The second is λ_c itself, and the third is ΔT_c, the temperature 'step' at about λ_c, which we will see is simply proportional to μ. For relatively small deposits of heat compared to the energy density of the CBR, that is for $\Delta E \ll E \equiv aT^4$, we have the following approximate relations, first derived by Illarionov and Sunyaev (1975):

$$\mu = 1.4\frac{\Delta E}{E} = 1.85 \times 10^{14}\frac{\Delta E}{T^4}, \tag{5.9}$$

with ΔE in erg cm^{-3}. Since the CBR temperature decreased with time, the present amplitude of μ depends on when the extra heat energy ΔE was added to the radiation field. Conversely, for a fixed limit on μ, ΔE can be larger if the heat were added earlier when the redshift and hence temperature were larger; see fig. 5.3. The amplitude of the temperature 'step' at λ_c for small perturbations ($\mu \ll 1$ or $\Delta E/E \ll 1$) is given approximately by Burigana *et al.* (1991) from numerical calculations:

$$\Delta T_c / T = 3.2\frac{\Delta E}{E}(\Omega_b h^2)^{-2/3} = 2.3\mu(\Omega_b h^2)^{-2/3}. \tag{5.10}$$

Finally, again approximately,

$$\lambda_c \approx 2.2(\Omega_b h^2)^{-2/3} \text{ cm}. \tag{5.11}$$

Since $\Omega_b h^2 \simeq 0.02$, we expect to see the temperature step at about 30 cm.

5.2.3 Energy release or transfer at $z < z_c$

At redshifts lower than the critical value given by eqn. (5.2), redistribution of photon energy by Compton scattering became impossible. Thus kinetic equilibrium could not be maintained if energy were added to the radiation field at $z < z_c$ (or an epoch later than roughly 100 years). It follows that the perturbed spectrum will not be Bose–Einstein, but will instead reflect the wavelength dependence of whatever physical process was responsible for addition of energy to the CBR. Thus, in contrast to the situation at $z > z_c$

---

* Only at $z > z_{th}$ was the reaction rate high enough to produce a thermal spectrum at *all* wavelengths.

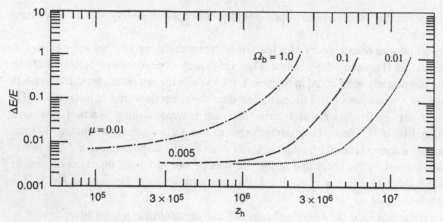

Fig. 5.3   The relation between μ and $\Delta E/E$ is not exactly proportional: if the heating occurs early at a large redshift z_h, μ can be smaller for a given value of $\Delta E/E$ (from Burigana, *et al.*, 1991).

covered in Sections 5.2.1 and 5.2.2 above, we will need to consider one by one the various processes that could have added energy to the CBR, since each could have produced a different and characteristic imprint on the perturbed spectrum.

Among the processes that could alter the CBR spectrum at $z < z_c$ are emission by warm dust (a process known to occur in our Galaxy; see Section 3.8.2); decay of elementary particles into one or more photons; line emission at the epoch of recombination, $z \simeq 10^3$; and the inverse Compton or Sunyaev–Zel'dovich effect. It may be argued whether some of these processes resulted in a *distortion* of the CBR spectrum or merely an *addition* to it. Historically, some of these processes were first introduced as sources of separate 'cosmic backgrounds'; in some cases even as alternatives to the Hot Big Bang itself. For that reason, I have elected to treat all processes that *added* photons to the CBR separately in Section 5.3; and to treat here only processes that *scattered* or *redistributed* photons in frequency without increasing their number.

5.2.4 Inverse Compton scattering (the Sunyaev–Zel'dovich effect)

The first of these processes is inverse Compton scattering (reviewed by Rybicki and Lightman, 1979 and Sunyaev and Zel'dovich, 1980), in which photons are scattered by electrons, as in ordinary Compton scattering. In the case of *inverse* Compton scattering, however, the kinetic temperature of the electrons is assumed to be large relative to the CBR temperature, unlike the ordinary case where γ-rays are scattered from electrons in a metal. Scattering from 'hotter' electrons boosts the energy of scattered photons without changing their number; the effect on an initially blackbody spectrum is shown schematically in fig. 5.4.

Note that this process requires the electron temperature to be greater than the CBR temperature T; hence the inverse Compton effect could operate only if heat were supplied to the electrons.* A variety of mechanisms, including star formation or release of gravitational potential energy as galaxies or clusters of galaxies formed, were possible sources of heat; we will examine some of these later in this chapter.

---

* In the absence of heating, the temperature of the matter, and hence T_e, was no larger than the CBR temperature, and will drop below it well after recombination (at $z \simeq 150$; Sunyaev and Zel'dovich, 1980).

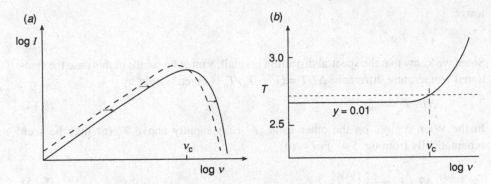

Fig. 5.4 Distortions resulting from inverse Compton scattering of the CBR. (*a*) Since the process conserves photon number but increases the photon energy, an initially blackbody curve (dashed) merely 'slides' to higher frequency, as shown. (*b*) The resulting curve for temperature $T(\nu)$, if $y = 0.01$. See text for numerical relationships between ΔT and the *y*-parameter. In both cases, the undistorted spectrum is shown dashed.

Now let us look at the inverse Compton effect in a bit more detail, and relate the properties of the hot electrons to the distortion produced in the CBR spectrum. We will make several simplifying assumptions; see Zel'dovich and Sunyaev (1969), Danese and De Zotti (1977) and Wright (1979) for more detailed treatments. We assume that the hot electrons were non-relativistic and were in equilibrium at kinetic temperature T_e (which could be a function of redshift), and that the CBR spectrum itself was initially blackbody and characterized by a temperature $T = T_0 (z + 1)$. We will also assume here that there was no bulk motion of the hot electrons relative to the CBR. Finally, we neglect the effect on the spectrum of other processes, such as bremsstrahlung.

With these assumptions, the inverse Compton process was characterized by a single parameter, the rate at which photons gained energy by scattering from the hot electrons:

$$\alpha(z) = \sigma_T n_e(z) c \,\frac{kT_e(z)}{m_e c^2} \tag{5.12}$$

(just the inverse of the time scale given in eqn. (5.1)). The total gain of energy now visible is the integral of $\alpha(z)$ from the redshift of heating of the electrons (z_h) to the present. This integral is the so–called '*y*-parameter' first introduced by Zel'dovich and Sunyaev (1969) and not to be confused with ^{4}He abundance:

$$y = -\int_t^{t_0} \alpha \, dt = \int_0^{z_h} \alpha \frac{dt}{dz} \, dz. \tag{5.13}$$

From eqn. (5.12) and eqn. (5.13), we see that *y* is proportional to the electron pressure (proportional to $n_e T_e$).

We next need to relate *y* to the shape of the spectrum. We will make use of the work of Zel'dovich and Sunyaev (1969), Illarionov and Sunyaev (1975) and Danese and De Zotti (1977), to which the reader is referred for further details. These detailed solutions establish what we see schematically in fig. 5.4: the spectrum 'slides' along in frequency. In particular, at wavelengths much greater than the peak wavelength of the spectrum*, the intensity remains proportional to ν^2, but the brightness temperature is

* That is, in the Rayleigh–Jeans region of the unperturbed spectrum.

lower:

$$T' = T_0 e^{-2y}.$$

Since we know that the spectral distortion is small, y must be small; in this case the fractional temperature difference $\Delta T/T = (T' - T_0)/T_0$ is given simply by

$$\Delta T / T_0 = -2y. \tag{5.14}$$

In the Wien region, on the other hand, T' rises rapidly above T_0 (as may be seen schematically from fig. 5.4). For $y \ll 1$,

$$\Delta T / T_0 \approx +y \left(\frac{h\nu}{kT} - 2 \right). \tag{5.15}$$

At one particular frequency, $T' = T_0$, and the sign of the distortion changes. In the limit of small y, that characteristic frequency is given by

$$\nu_c = \frac{3.83 \, kT_0}{h} \tag{5.16}$$

or $\nu_c = 220$ GHz (corresponding to 1.4 mm wavelength) for $T_0 = 2.73$ K. It is easy to see qualitatively from eqn. (5.14) and eqn. (5.15) or from fig. 5.4 that ν_c will increase slightly as y increases (by about 10% as y rises to 0.05). It has been noted – for instance by Fabbri *et al.* (1978) and Rephaeli (1980) – that the change in sign of the distortion at ν_c provides both a characteristic signature of inverse Compton distortion and a means of measuring T_0, the thermodynamic temperature of the undistorted spectrum.

The derivation of eqns. (5.14) and (5.15) was based on the assumption that the hot electrons scattering the radiation were non-relativistic. That assumption begins to lose its validity if T_e increased beyond 3×10^8 K (or $kT_e/m_e c^2 \gtrsim 0.05$), as shown for one specific model of the hot electron gas by Wright (1979). Fabbri (1981) has investigated numerically some of the consequences of relaxing this assumption. For instance, eqns. (5.13) and (5.15) may be modified as follows for mildly relativistic electrons:

$$\Delta T / T_0 = -2y \left(1 - 1.5 \frac{kT_e}{m_e c^2} \right),$$

$$\nu_c = 3.83 \left(1 + 1.1 \frac{kT_e}{m_e c^2} \right) \frac{kT_0}{h}. \tag{5.17}$$

The qualitative nature of the spectral distortion, however, remains much the same. It is worth remarking that the plasma that we know exists in clusters of galaxies has $T_e < 3 \times 10^8$ K; the temperature of a generally distributed, more tenuous, intergalactic plasma is not known, but energetic considerations make it unlikely that it exceeds about 10^9 K. Thus relativistic corrections do not play a large role in calculations of the perturbed spectrum. It is worth adding, however, that if intergalactic hydrogen is spread throughout the Universe, it must be highly ionized. Otherwise (Gunn and Peterson, 1965) it would produce strong absorption in the light of high redshift quasars. If the ionization were thermal, $T_e \gg 10^5$ K would be required.

The Sunyaev–Zel'dovich effect in clusters of galaxies. One known site of hot electrons is the intergalactic plasma in clusters of galaxies, responsible for the observed X-ray emission of clusters (e.g., Lea *et al.*, 1973; Mushotzky *et al.*, 1978). Temperatures of order $10^7 - 10^8$ K may be inferred from the X-ray spectra. If we can estimate the value of the electron number density $n_e(r)$ as a function of distance from the cluster center, we can integrate eqn. (5.13) to find y. For rich clusters, y is typically about 10^{-4} along a line of sight through the cluster center. Thus, in the direction of rich clusters, we expect the CBR in the Rayleigh–Jeans region to be 'cooler' by several tenths of a millikelvin. This 'dip' in the CBR temperature has indeed been detected in several clusters of galaxies by making differential measurements (see Section 7.10). These observations raise the intriguing possibility that inverse Compton distortions produced by *inhomogeneous* heating or distribution of plasma could have produced detectable anisotropies in the CBR even if the overall value of y were too small to detect from the absolute measurements of the spectrum, a point to which we return in Chapter 8.

5.2.5 Rayleigh scattering

In this chapter and elsewhere, we have remarked that scattering of photons by neutral atoms was too small to have any effect on propagation of the CBR or on its spectrum. As La (1989) has recently shown, however, that is not quite true; Rayleigh scattering by neutral hydrogen atoms can become appreciable at $\lambda < 1$ mm (recall that the Rayleigh scattering cross-section is proportional to λ^{-4}). Since this form of scattering does not alter photon energy, it had no effect on the CBR spectrum, but it could have altered photon direction and hence smoothed out anisotropies (again, only at $\lambda < 1$ mm).

5.2.6 Line emission at recombination

Up to this point, we have ignored line emission in treating recombination of the primeval plasma at $z \simeq 1000$. In effect, we have made the assumption that e^- and p^+ combined to produce atomic hydrogen in its ground state directly. Recombination was actually considerably more complex, and proper account must be taken of the transitions to and from excited states of hydrogen. (We can neglect other atomic species, since their number densities are $< 0.1 n_H$). If recombination occurred to a level of principal quantum number n, transitions to a lower state m, with $m = 1$ to $n - 1$, were possible, and line emission resulted. This recombination radiation affected the rate of recombination (see Peebles, 1968, and Zel'dovich *et al.*, 1968); it also left an imprint on the CBR spectrum. The Lyman lines (for which $m = 1$) added to the CBR spectrum at $\lambda < 200$ μm. A somewhat larger distortion resulted from two-photon emission from the 2S ($n = 2$) to the 1S ($m = 1$) level. These photons formed a continuum with (rest frame) wavelengths in the range 1216 Å to infinity; hence the distortion they produced extends across the CBR spectrum (but is swamped by a 3 K thermal spectrum for all $\lambda \gtrsim 300$ μm). Finally, as Dubrovich (1975) pointed out, transitions between higher excited states ($n = 10$–15, with $m = n - 1$) could have produced lines that would be redshifted into the centimeter wavelength range at present.

It is easy to see that the resulting line emission was weak; that is, these lines will only minimally perturb the observed CBR spectrum. Each transition in a hydrogen atom produced one photon, but the total number of CBR photons exceeds n_H by $n_\gamma/n_H \approx n_\gamma/n_b \simeq 10^9$.

Hence the thermal CBR spectrum overwhelms recombination line photons except at very short wavelengths, far in the Wien region of a roughly 3 K spectrum, where the

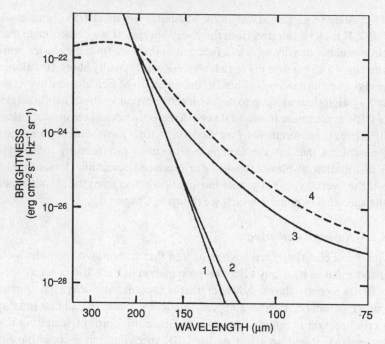

Fig. 5.5   Distortions in the (far) Wien region of the CBR spectrum arising from line and two-photon continuum emission during recombination. Curve 1 is the undistorted Planck curve (in this case for $T_0 = 3$ K) and curves 2 and 3 show the additions from recombination lines. Curve 4 (dashed) is the two-photon (2S → 1S) emission alone (from Peebles, 1968).

CBR brightness is falling rapidly (see fig. 5.5). Dubrovich (1975) estimates $\Delta T/T \simeq 10^{-6}$–$10^{-4}$ for lines in the 1–10 cm region of the spectrum, well below our ability to detect them. As Lyubarski and Sunyaev (1983) later showed, however, much larger line intensities may result if the CBR spectrum was not purely blackbody at recombination. In particular, if the y–parameter (eqn. (5.13)) was non-zero, there was a relative increase in the number of short wavelength photons (fig. 5.4), and more substantial distortions were produced. Even for y as large as 0.05 (already ruled out by the observations), however, $\Delta T/T$ in lines is only about 10^{-3}. Furthermore, the line-emission perturbations will appear at submillimeter wavelengths, where accurate spectral measurements are difficult, and Galactic and intergalactic foregrounds dominate.

5.3   Later addition of photons to the cosmic background

We now turn to processes that operated after recombination, and that could have direct-ly *added* photons to the CBR. We will focus on processes that produced photons in the wavelength range about 0.05–50 cm in which the CBR appears, and thus neglect the X-ray background, possible optical and near-infrared backgrounds from primeval galaxies, etc.

Two processes have been suggested and developed in some detail: (1) far-infrared emission by dust (or graphite or iron needles) heated by starlight; and (2) radiative decay of long-lived elementary particles.

5.3.1 Emission by dust

The suggestion that radiation by warm dust, either in galaxies or intergalactic, might have produced some or all of the observed microwave background has been with us since the announcement of the 3 K radiation 25 years ago (see Section 2.6.3). Initially, dust emission was invoked (e.g., Hoyle and Wickramasinghe, 1967; Layzer, 1968) to explain the entire microwave background. That view is now accepted by only a few (see, e.g., Hawkins and Wright, 1988; Hoyle, 1989).

On the other hand, distortions of the underlying CBR spectrum seem much more likely, and a large number of groups and individuals have worked on this problem. Among these are Rees (1978), Rowan-Robinson et al., (1979), Aiello et al. (1980), Puget and Heyvaerts (1980), Carr (1981), Bond et al. (1986, 1991), McDowell (1986), Hayakawa et al. (1987), Adams et al. (1989), and Draine and Shapiro (1989). As a historical aside, it is interesting to note that enthusiasm for dust emission models has run in cycles, with peaks in about 1967, 1980 and 1989. The first of these peaks followed the discovery of the CBR. The second followed the balloon measurements of Woody and Richards (1979), which appeared to show significant excess flux near the peak of the CBR spectrum (Section 4.8.2). The third burst of interest was triggered by the Berkeley–Nagoya rocket experiment (Matsumoto et al., 1988) discussed in Section 4.8.4, which again gave evidence of a short wavelength excess flux.

The basic mechanism involved is simple: dust was heated by ultraviolet or optical emission from an early generation of stars, quasi-stellar objects, or Very Massive Objects (Carr et al., 1984). It then reradiated, and this radiation (with the appropriate redshift applied) added to or possibly produced the observed microwave background. For this reradiated flux to contribute in the wavelength range 0.05–50 cm, the dust temperature T_d and the effective emission redshift z_d needed to fall in approximately the following interval: $0.06 \leq T_d/(z_d + 1) \leq 60$ K.

The major problem* faced by dust models is approximating a blackbody spectrum over a wide range of wavelengths. Consider, for instance, a model in which T_d was fixed, but radiation from dust occurred in a range of epochs t_1 to t_2. Then the presently observed energy density will be the integral of blackbody spectra from t_1 to t_2, as in eqn. (5.8). We know that the sum of blackbody curves is not itself a blackbody. To achieve a spectrum approximating a blackbody, $P(t)$ would have to have been close to a delta function; that is, the Universe had to become essentially opaque at a sharply defined epoch or redshift. The alternative is to allow T_d to vary smoothly with redshift so that the ratio $T_d/(z + 1)$ happened to stay constant (just the situation described in Section 5.1.2 above for the radiation temperature).

Thus far, we have implicitly assumed that the spectrum of the radiation from dust at a temperature T_d was in fact blackbody. That too is very unlikely. We know that the warm dust in our Galaxy does not radiate like a blackbody. Emission at long wavelengths falls well below a Planck curve. Indeed, the emissivity of interstellar dust is quite well modeled by a power law in wavelength, with index $\gamma \simeq -1$ or -2 (Section 4.4.2). The physical explanation for this wavelength-dependent emissivity is straightforward:

---

* I ignore here two other potential difficulties with such models: the energy supply needed to heat the dust, especially if z_d was large (see Lacey and Field, 1988), and the need for large amounts of dust (Draine and Shapiro, 1989), comparable to the entire density of heavy elements in the Universe.

the interstellar dust particles are very small, with characteristic sizes much less than the wavelengths at which the microwave background is observed. Hence the particles are inefficient radiators of long wavelength radiation. To me, that simple fact is the fundamental flaw in any model attempting to use reradiation by warm dust to explain *all* of the observed microwave background (see Rowan-Robinson *et al.*, 1979, or Wright, 1982, for detailed discussions).

There is an ingenious way around this problem, however; make one or more of the dimensions of the dust particles larger. Spheres or disks with radii 10–100 μm would do, but the total mass required would be prohibitive. So Layzer and Hively (1973), Narlikar *et al.* (1976), Rana (1981) and Wright (1982) suggested *needles*. Wright, for instance, suggested graphite needles of characteristic size $0.1\mu m \times 0.1\mu m \times 20$ μm. The emissivity of needles, especially if they are made of a conductor rather than an insulator, is much less wavelength dependent than the emission of roughly spherical particles of the same cross-section or same mass. Thus needles can do a better job of matching the observed microwave spectrum. The requirements on such models remain severe, however, and I believe that more detailed calculations are required to show that they can explain the good agreement between values of T_0 measured at about 10 cm and values measured at a few millimeters.

The wavelength dependence of the emission of dust is a problem if one attempts to explain the *entire* microwave background. On the other hand that same property is a virtue if one attempts to explain *distortions* in the spectrum. Hence, as noted above, interest in dust models surged after each report of an apparent non-Planckian distortion in the microwave spectrum. For instance, Rowan-Robinson *et al.* (1979) and Aiello *et al.* (1980) made use of the specific wavelength dependence of emission from silicate grains to explain the distortion reported by Woody and Richards (1979). Recall from Section 4.8.2 that their reported spectrum cut off more sharply at $\lambda \leq 1$ mm than a Planck curve at a temperature of about 3 K. Rowan-Robinson, Aiello and their colleagues invoked the spectral properties of silicate grains to explain that cutoff. Note, however, that their explanation works only if the dust emission reaches us from a narrow range of redshifts (to avoid smearing out the spectrum).

Once doubt was cast on the Woody–Richards results, interest in silicate grains waned. Roughly a decade later, the report of a submillimeter excess by Matsumoto *et al.* (1988) quickly revived dust models. As many authors, including Hayakawa *et al.* (1987), Hawkins and Wright (1988), Adams *et al.* (1989), and Bond *et al.* (1991), pointed out, a submillimeter excess could naturally be explained by emission from warm dust at redshifts of order 10–100. The wavelength dependent emissivity of small dust particles would again be a virtue, ensuring that the excess flux appeared only at short wavelengths. By appropriate adjustment of T_d and z_d, the reported excess flux at $\lambda \leq 1$ mm could be explained. A wide range of models of the physical properties of the dust and of mechanisms to heat it, together with a review of the literature, is presented by Bond *et al.* (1991), to which the reader is referred for further details. Most of these models produce rather sharply peaked distortions, with maxima at 300–1000 μm (0.3–1 mm) and full half widths of several hundred micrometers. The spectral distortions from one set of models are shown in fig. 5.6, adopted from Bond *et al.* (1991).

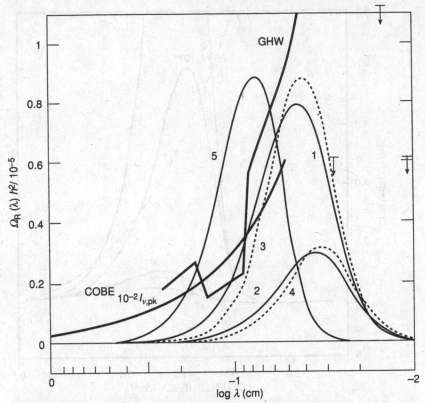

Fig. 5.6 Five models for emission by dust, shown as light lines or dashes (adapted with permission from Bond *et al.* 1991). The vertical axis is in units of energy density as a fraction of $\rho_c c^2$. Also shown (heavy lines) are the upper limits on spectral distortions derived from COBE (Mather *et al.*, 1990) and Gush *et al.* (1990). In addition, *very preliminary* upper limits from the DIRBE experiment on COBE are indicated by arrows. Models 2 and 4 have $\Omega_d = 10^{-6}$ – see Bond *et al.* (1991) for details.

5.3.2 Particle decay

The decay of elementary particles could have added energy to the CBR radiation field and hence distorted its spectrum. In addition, if the decay were radiative and occurred at low enough redshift, photons would have been added directly to the underlying spectrum. We will focus primarily on the latter process here. The decaying particles involved could have been massive neutrinos, photinos or many other 'exotic' particles; for generality, we will simply refer to them as X-particles.

The results of Section 5.1.1 may be used to show that energy released by particle decay at $z > z_{th}$ given by eqn. (5.7) merely altered the temperature of the CBR and did not change its spectrum. Likewise, energy or photons released in the interval $z_{th} > z > z_c$ produced a Bose–Einstein spectrum. If the half-life $t_{1/2}$ of the particles was less than the epoch corresponding to $z_c(t \simeq 10$ years, depending on $\Omega_b)$, their decay thus left no specific trace in the spectrum.

Larger values of $t_{1/2}$, corresponding to decays at $z < z_c$, could have produced an inverse Compton distortion or led to direct addition of photons to the underlying thermal CBR

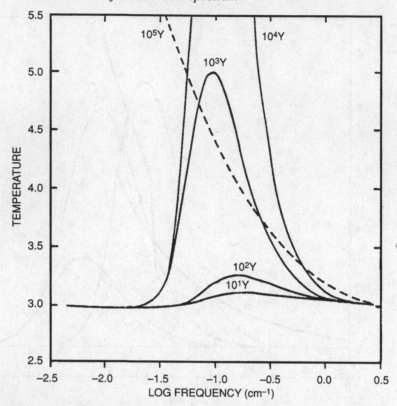

Fig. 5.7   Spectral distortions produced by the radiative decay of particles of mass $m_X = 1$ eV (from Silk and Stebbins, 1983). The distortion is rather sharply peaked at $\lambda \sim 10$ cm. For much more massive particles, the distortion has the opposite sign (Kawasaki and Sato, 1986; see Section 5.4.2). Lifetimes are given in years.

radiation field. In the particular case of radiative two-particle decay, X $\rightarrow$ photon + daughter, the photon emerged with a fixed frequency. The form of the resulting spectrum, including the redshift, can then readily be calculated, and is shown schematically in fig. 5.7. If the daughter particle was much less massive than the decaying X-particle, the emerging photon had a frequency given by $h\nu = m_X c^2/2$. The bulk of the decay photons will then be detected at

$$\nu_0 \approx \nu[z(t_{1/2})+1]^{-1} \approx \frac{m_X c^2}{2h}\left(\frac{t_{1/2}}{t_0}\right)^{2/3} \approx 0.02(m_X / 1\,\text{eV})(t_{1/2} / 1\,\text{year})^{2/3}\,\text{GHz},$$

(5.18)

for $t_0 = 2/3H_0^{-1}$ and $H_0 = 100$ km s^{-1} per megaparsec. Thus only a certain range of combinations of m_X and $t_{1/2}$ would supply photons to the microwave background. Note also that decay of particles with $m_X \gtrsim 27$ eV would have produced ionizing radiation. Any free electrons thus formed would have inverse-Compton scattered the CBR.

It is worth adding that there is another class of decaying particle models, those with $t_{1/2} \gtrsim 10^{24}$ s or $\gtrsim 3 \times 10^6 t_0$ (e.g., Sciama, 1988). Since only a small number of decay photons were produced at $t < t_0$, the corresponding spectral distortions of the CBR are small. These models are better constrained by optical and ultraviolet observations (e.g., Kimble *et al.*, 1981).

5.4 Observational limits

Now that we have laid out several different processes that could in principle have distorted a purely blackbody spectrum of the CBR, let us see what constraints the newest and most accurate spectral measurements place on such processes. As we shall see, the observational evidence suggests that the thermal history of the Universe was remarkably simple and calm. Spectral measurements of the CBR permit us to study that history over a wide range of redshifts, roughly a few to a few million. None of the many energy-generating processes suggested in Sections 5.2 and 5.3 appear to have left a significant trace in the CBR spectrum.

The single most useful measurement is the result from the FIRAS instrument of the COBE satellite (Section 4.9). As noted in Chapter 4, we do not yet have the final limits on spectral distortions in the 0.01–1 cm range, which FIRAS will eventually produce; hence the numerical constraints in this section should be regarded as preliminary.

5.4.1 Limit on the y-parameter and its consequences

The FIRAS instrument is ideally suited to detect or set upper limits on the short wavelength distortions produced by inverse-Compton scattering. As fig. 4.25 shows, no distortion exceeding about 1% of a 2.735 K blackbody flux is consistent with the data. Mather *et al.* (1990) have used the observed spectrum to set an upper limit of about 10^{-3} on the parameter y, which characterizes inverse-Compton distortions. Mather (private communication) reported in early 1991 that the limit on y had been improved to

$$y \leq 4 \times 10^{-4}, \tag{5.19}$$

and I expect a further improvement of a factor of 5 once the FIRAS observations are carefully analyzed and corrected for Galactic emission.*

An upper limit of 4×10^{-4} already sets stringent limits on the number density and temperature of diffuse intergalactic electrons. For instance, if the plasma in clusters of galaxies had been hotter or denser than now observed, the hot electrons in clusters alone could have produced values of $y \gtrsim \text{few} \times 10^{-4}$.

The same limit on y also has interesting consequences for models of the X-ray background. That background, at about 1 keV to 1 MeV, is well observed, but poorly understood (see review by Setti, 1990). Some, perhaps most, of the background is certainly contributed by unresolved discrete sources (see, e.g., Schmidt and Green, 1986; Giacconi and Zamorani, 1987). Some, however, may be produced by bremsstrahlung from a hot, $T_e = 10^6$–10^9 K, intergalactic plasma (Field and Perrenod, 1977; Guilbert and Fabian, 1986, for instance). That plasma will produce an inverse Compton distortion of the CBR spectrum. The X-ray luminosity per volume of hot plasma is proportional to n_e^2 $T_e^{1/2}$ (e.g., Blumenthal and Tucker, 1974), and y in turn is proportional to $n_e T_e$, as we have seen. Hence, unless the intergalactic plasma is strongly clumped (allowing $\overline{n_e^2} \gg (\bar{n}_e)^2$), there is a direct connection between X-ray luminosity and the y-parameter for a plasma of known temperature, T_e. If clumping is neglected, the models of Field and Perrenod

* *Note added in proof*: Mather *et al.* (Ap. J. **420**, 439 [1994]) give $y \leq 2.5 \times 10^{-5}$.

suggest $T_e \simeq 4 \times 10^8$ K and $y \gtrsim 10^{-2}$, clearly violating the observational constraints. Put another way, inequality (5.19) implies that no more than a few percent of the observed X-ray background may be ascribed to homogeneously distributed intergalactic plasma at $T_e \simeq 4 \times 10^8$ K.

On the other hand, our present upper limit on y is not yet stringent enough to rule out the existence of intergalactic plasma altogether. The Gunn–Peterson test mentioned in Section 5.2.4 above requires only that $T_e > 10^5$ K. Suppose we fix $T_e = 10^6$ K and assume that essentially all the baryonic matter in the Universe is in fact intergalactic plasma. Then $\rho_{IGM} = \Omega_b \rho_c$ or

$$n_e = \frac{3 H_0^2 \Omega_b}{8 \pi G m_p} \approx 5 \times 10^{-7} \text{cm}^{-3},$$

with $\Omega_b = 0.05$. To estimate y crudely, we may substitute these values and $t = 0$ into eqn. (5.13). The result is $y \simeq 5 \times 10^{-7}$. In fact, we should properly have taken account of the cosmological evolution of n_e and T_e; had we done so, we would have found $y \simeq 10^{-5}$, integrating out to $z_h = 10$. This value of y still lies below the best limits we can hope for from COBE.

Equally small values of y are predicted from the energy released at $z \approx 3$–300 by formation of galaxies by gravitational collapse or emission from star-forming galaxies (Section 1.7). The explosive scenario for galaxy formation (Ikeuchi, 1981; Ostriker and Cowie, 1981; see also Section 1.7.2) allows for more energy release and hence potentially larger values of y. As we will see in Chapter 8, however, that scenario is much more stringently tested by limits on the small-scale anisotropy of the CBR than by the spectral observations considered here.

5.4.2 Limits on dust emission and particle decay models

The results obtained from the FIRAS instrument aboard COBE and those from the rocket flight of Gush *et al.* (1990) have also strongly constrained possible distortions arising from decay of particles early in the Universe or from dust emission.

In 1983, Silk and Stebbins examined the spectral distortions resulting from particle decay in some detail. They also used other astrophysical constraints to limit the range of m_X and $t_{1/2}$ (see also Raffelt, 1989). The findings may be summarized as follows: for particles with $m_X < m_e$, and with the expected cosmic abundance, $t_{1/2}$ must be $\lesssim 30$ years. That limit was set assuming that spectral distortions did not exceed about 10%; as we shall see, the limits on $t_{1/2}$ are now better. One other window remains open, at $(m_X/1 \text{ eV})$ $(t_{1/2}/1 \text{ year})^{2/3} \approx 10^5$ where the decay photons would be swamped not by the CBR but by interstellar dust emission.

Several years later, this result was refined by Kawasaki and Sato (1986), who included double Compton scattering (Section 5.1.1) and a more realistic value of the baryon density in their calculations. For masses of the X-particle below 100 keV, they find the following approximate upper limit on the half life of the X-particle:

$$t_{1/2} < 0.05(1 + m_X^{-2})^{-1/2} \text{ years}$$

with m_X in keV. As these authors note, the main reason for the more stringent limit is their choice of $\Omega_b \lesssim 0.1$.

There is one additional feature of interest in this paper. Kawasaki and Sato show that the decay of low mass particles (say 50 eV) produced a positive 'bump' in the CBR like the ones shown in fig. 5.7. On the other hand, the decay of more massive particles (say 10 keV) produced a 'dip' instead. Clearly, there must then have been a particular value of m_x for which the distortion is minimized; from their paper, it appears to be about 1 keV.

More recently Wang and Field (1989) reexamined the possibility that decay of about 10^4 eV particles with $t_{1/2} = 300$–1000 years could explain the submillimeter excess and simultaneously the roughly 2 μm background also reported by Matsumoto and his colleagues. A narrow range of particle properties was allowed, but the model clearly violates the more recent COBE limits on spectral distortions.

Likewise, most models involving reemission by warm dust fail to agree with the COBE spectral measurements. The models presented in fig. 5.6 (from Bond *et al.*, 1991) include the few that are consistent with the observed CBR spectrum. In general, as Bond *et al.* show, such models violate the observations unless the cosmic density of dust is very low ($\Omega_d \lesssim 10^{-6}$) and the redshift of maximum reemission was very low. Is a limit like $\Omega_d \lesssim 10^{-6}$ interesting cosmologically? From Section 1.6.4 (or see Section 5.5.1 below), we know that the density of *all* baryonic matter corresponds to $0.01 \lesssim \Omega_b \lesssim 0.1$, so $\Omega_d/\Omega_b \lesssim 10^{-4}$. In stars where the abundance of heavy elements – those able to form dust – has been measured, we find that heavy elements* make up $10^{-4} - 3 \times 10^{-2}$ of the total mass. Thus a value of $\Omega_d/\Omega_b \lesssim 10^{-4}$ is small, but acceptable.

Most models involving more reasonable values of Ω_d produce spectral distortions larger than seen by COBE, as shown in detail by Bond, Carr and Hogan (1991). Indeed, as these authors state, 'The current COBE limits [$y < 10^{-3}$] already rule out [many] models ... and most will be eliminated (or confirmed) with their next order-of-magnitude "improvement."'

It goes without saying that the dust models postulated by Rowan-Robinson *et al.* (1979), Hayakawa *et al.* (1987), and others to explain various reported short wavelength 'excesses' are ruled out by the new FIRAS results from COBE.

Finally, let me note that the constraints imposed by COBE on such dust models do not depend strongly on the distribution of the dust – whether it is, for instance, intergalactic or instead concentrated in galaxies. On the other hand, angular variations in the dust emission do depend on how the dust is distributed. As Bond *et al.* (1991) note, upper limits on the anisotropy of the microwave sky, especially at small angular scales and high frequency, can sharply constrain models in which the reemitting matter is located in dusty galaxies (Section 8.7.4).

5.4.3 *Limits on* μ

The characteristic signature of a Bose–Einstein spectrum is a dip in observed temperature at wavelength just below $\lambda_c \simeq 30$ cm (Section 5.2.2). Only under extreme conditions ($\Omega_b = 1$, $h > 1$) can λ_c be moved as short as about 1 cm, the longest wavelength accessible to COBE. Nevertheless, the measurements of T_0 by COBE and by Gush *et al.* (1990) did allow us to sharpen our limit on Bose–Einstein distortions by establishing a precise value of T_0 with which to compare the longer wavelength measurements discussed in Sections 4.6 and 4.7.

---

* 'Heavy' here is astronomer's jargon for any element of atomic mass > 4.

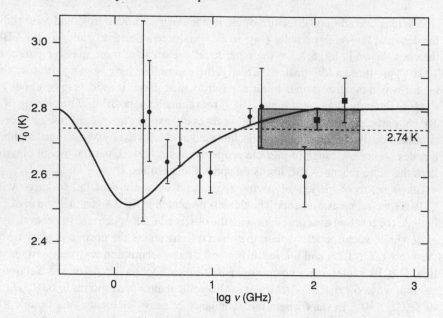

Fig. 5.8   The measured spectrum (points plus shaded area) of the CBR compared to a Bose–Einstein spectrum with $\mu = 5 \times 10^{-3}$ and $T_0 = 2.8$ K. Clearly, larger values of μ would not be consistent with the observations. Equally clearly, better measurements at $\nu = 0.3$–3.0 GHz ($\lambda = 10$–100 cm) are needed.

The current best limit on μ, derived from this comparison, is (Mather, private communication)*

$$\mu \lesssim 5 \times 10^{-3}. \tag{5.20}$$

A Bose–Einstein spectrum with $\mu = 5 \times 10^{-3}$ is shown in fig. 5.8, along with the COBE results and some of the high quality long wavelength measurements discussed in Sections 4.6 and 4.7. That figure makes it clear that more stringent limits on μ – or its detection – will require improved measurements of T_0 at $\lambda \simeq 10$–30 cm. For all the reasons discussed in Section 4.6, such measurements will not be easy; Galactic foreground emission will have to be subtracted carefully. Nevertheless, in both the USA and Italy, attempts are being made to refine our knowledge of the CBR spectrum at wavelengths $\gtrsim 10$ cm. We may hope for eventual sensitivity to values of μ as low as about 10^{-3}.

Even the current limit (5.20), however, tells us that less than 0.4% of the background thermal energy density could have been added by any process operating in the redshift interval $z_{th} > z > z_c$.

5.4.4 Summary

The observational limits on both y and μ (or on spectral distortions in general) may be combined to set constraints on *any* form of energy emission into the CBR at *any* time later than the epoch corresponding to z_c (a few weeks or months after the Big Bang). Burigana and his colleagues (1991) have done such calculations. Sample results from

---

* Now improved to $|\mu| \leq 3.3 \times 10^{-4}$ (see footnote, p. 179, and Appendix C).

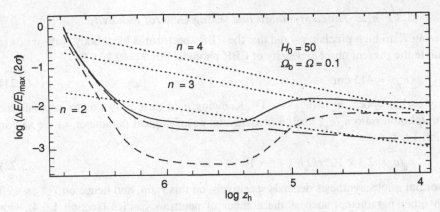

Fig. 5.9   Overall limits (2σ or 95% confidence) on energy input to the CBR as a function of the redshift z_h at which the energy injection occurred (from Burigana *et al.*, 1991). The solid line shows the limits based on the preliminary upper limits on y and μ given by Mather *et al.* (1990). If inequalities (5.19) and (5.20) were used, the limits would be a factor 2 lower. The long dashed line shows the eventual sensitivity a further analysis of the COBE data may provide. The short dashed lines shows what a factor 10 improvement in sensitivity of spectral measurements at $\lambda \sim 10$ cm could do.

their work are shown in fig. 5.9. The 2σ limit shown (solid line) is based on $y < 10^{-3}$ and $\mu < 10^{-2}$, the preliminary values given by Mather *et al.* (1990); as we have noted, both upper limits are now about a factor of two lower. Two features of the limiting curve deserve attention. First, the constraints rapidly grow weaker as $z_h \rightarrow z_{th} \simeq$ few $\times 10^6$, for reasons explained in Section 5.1. Second, the most stringent constraints are at $z_h = 10^5$–10^6; these constraints arise from the limit on the parameter μ.

Also shown in fig. 5.9 are a series of dotted lines representing the energy released by damping of adiabatic density perturbations before recombination (see Section 1.6.5). Curves for different power spectra (n) are shown; clearly, the present observational constraints favor $n < 2$, at least for this $\Omega_b = \Omega = 0.1$ low density model. Constraints on a high density model ($\Omega_b = 0.1$; $\Omega = 1.0$, say) are less stringent.* Nevertheless, it is important to stress that density perturbations must have been present at some level in order to produce galaxies and the other large-scale structure that we see today (Section 1.7). These perturbations were certainly damped at $z \gtrsim 10^3$, adding energy to the CBR; hence *some* distortion in the CBR spectrum must eventually show up. All we can say now is that no reliable evidence for such a distortion is yet available.

5.5   What we learn from the value of T_0 directly

Up to this point in Chapter 5, we have been examining possible distortions of the CBR spectrum and what they – or rather their absence – can tell us about the early Universe. In addition to the tight limits on spectral distortions reviewed in Section 5.4, however, we also know the absolute value of the temperature of the CBR to an accuracy of about 1%. That is the only fundamental cosmological parameter known to anything like that precision; compare the factor-of-two uncertainty in H_0 (Section 1.2.2).

---

* The calibration is explained in Chapter 8. For further details, see Barrow and Coles (1991) or Daly (1991).

5.5.1 n_b/n_γ, nucleosynthesis and testing General Relativity

Knowing T_0 to high precision – and that the CBR spectrum is blackbody – allows us to calculate the present number density of CBR photons. For $T_0 = 2.735$ K,

$$(n_\gamma)_0 = 415 \text{ cm}^{-3}. \tag{5.21}$$

In any earlier epoch, $n_\gamma = (n_\gamma)_0 (z + 1)^3$. Knowing n_γ in turn allows us to calculate the baryon/photon ratio n_b/n_γ, itself a fundamental cosmological parameter, as we saw in Section 1.6.2;

$$n_b/n_\gamma = 2.3 \times 10^{-8}(\Omega_b h^2) \approx 5 \times 10^{-10} h^2. \tag{5.22}$$

Primordial nucleosynthesis depends sensitively on this ratio, and hence on T_0^3, as well as on other parameters such as the number of neutrino species (section 1.6.4). Our precise knowledge of T_0 ensures that uncertainty in n_b/n_γ contributes only about 3% to the error budget in nucleosynthesis calculations.

The excellent agreement between predicted abundances of light elements and astronomical observations thus allows us to test other factors in the nucleosynthesis calculations. As we have already seen in Chapter 1, the astronomical observations allow us to place a firm upper limit on the number of light lepton families $\mathcal{N}_l \leq 4$ (a result now confirmed by particle physicists, who find $\mathcal{N}_l = 3.1 \pm 0.2$ (Denegri, Sadoulet and Spiro, 1990)). With both n_b/n_γ and $\mathcal{N}_l$ fixed, the observations then become a test of the validity of General Relativity early in the history of the Universe. The results displayed in fig. 1.4 and fig. 1.10 are based on General Relativity. Had other theories of gravity been used, such as the scalar–tensor theory of Brans and Dicke (1961), the predicted abundances would have agreed less well with the observations. In particular, variants of standard gravity theory that either speed up or slow down expansion at $t \simeq 100$ s tend to overproduce or underproduce ^{4}He by an unacceptably large factor.

5.5.2 Effect of the CBR on ultrahigh energy cosmic rays

Knowledge of n_γ also allows us to calculate the mean free path of high energy cosmic rays through the CBR radiation field. Shortly after the discovery of the CBR, a number of physicists and astronomers pointed out that CBR photons would interact with cosmic ray particles in potentially observable ways. Interactions with high energy e^- were investigated by Felten (1965) and Hoyle (1965); with p^+ by Greisen (1966), Zatsepin and Kuz'min (1966) and Stecker (1968); and with γ-rays by Gould and Schréder (1966) and Jelley (1966); respectively.

These interactions depend on the energy of the cosmic rays; in all cases, the cross-sections increase with energy, so that the mean free path decreases with energy. For instance, for p^+ cosmic rays, the mean free path through a 2.73 K blackbody radiation field is limited to about 10 Mpc at a proton energy of 10^{11} GeV, because of photo-pion production on the CBR. Thus, if 10^{11} GeV cosmic rays originated at cosmic distances, we should see a cutoff in the energetic p^+ flux at about 10^{11} GeV, within the range of present detection techniques (e.g., Baltrusaitis *et al.*, 1985).

Suppose for a moment that the number density of cosmic microwave photons were much higher than about 400 cm^{-3}, as would have been true, for instance, if some of the early rocket measurements referred to in Section 4.8.2 had proven correct. Then the mean free path would have been proportionally shorter or the energy cutoff at lower energy. Just this argument was used by Apparao (1968) and Encrenaz and Partridge (1969) to argue that the early rocket measurements were probably erroneous*.

Measurements of the high energy end of the e^-, p^+, ν and γ-ray cosmic ray spectra will be improved by large area detectors now operating (e.g., the 'Fly's Eye' described in Baltrusaitis *et al.*, 1985) or planned. The presence and possibly the nature of the high energy cutoff resulting from interactions with CBR photons should allow us to learn more about particle interactions at ultrahigh energies, about the source ('local' or cosmic?) of high energy cosmic rays, or both (see Hill and Schramm, 1985, or Berezinky and Grigor'eva, 1988, for recent discussions).

References

Adams, F. C., Freese, K., Levin, J., and McDowell, J. C. 1989, *Ap. J.*, **344**, 24.

Aiello, S., Cecchini, S., Mandolesi, N., and Melchiorri, F. 1980, *Lettere al Nuovo Cimento*, **27**, 472.

Apparao, M. V. K. 1968, *Nature*, **219**, 709.

Baltrusaitis, R. M. *et al.* 1985, *Phys. Rev. Lett.*, **54**, 1875.

Barrow, J. D., and Coles, P. 1991, *Monthly Not. Roy. Astron. Soc.*, **248**, 52.

Beresinskii, V. S., Bulanov, S. V., Dogiel, V. A., Ginzburg, V. L., and Ptuskin, V. S. 1990, *Astrophysics of Cosmic Rays*, North Holland, Dordrecht.

Berezinsky, V. S., and Grigor'eva, S. I. 1988, *Astron. Astrophys.*, **199**, 1.

Blumenthal, G. R., and Tucker, W. H. 1974, in *X-Ray Astronomy*, eds. R. Giacconi and H. Gursky, D. Reidel, Dordrecht.

Bond, J. R. 1988, in *The Early Universe*, eds W. G. Unruh and G. W. Semenoff, D. Reidel, Dordrecht.

Bond, J. R., Carr, B. J., and Hogan, C. 1986, *Ap. J.*, **306**, 428.

Bond, J. R., Carr, B. J., and Hogan, C. 1991, *Ap. J.*, **367**, 420.

Brans, C., and Dicke, R. H. 1961, *Phys. Rev.*, **124**, 125; see also Dicke, R. H. 1968, *Ap. J.*, **152**, 1.

Burigana, C., Danese, L., and De Zotti, G. 1991, *Astron. Astrophys.*, **246**, 49; *Ap. J.*, **379**, 1.

Carr, B. J. 1981, *Monthly Not. Roy. Astron. Soc.*, **195**, 669.

Carr, B. J., Bond, J. R., and Arnett, W. D. 1984, *Ap. J.*, **277**, 445.

Chan, K. L., and Jones, B. J. T. 1975a, *Ap. J.*, **195**, 1.

Chan, K. L., and Jones, B. J. T. 1975b, *Ap. J.*, **200**, 461.

Daly, R. A. 1991, *Ap. J.*, **371**, 14.

Danese, L., and De Zotti, G. 1977, *Revista del Nuovo Cimento*, **1**, 277.

Danese, L., and De Zotti, G. 1982, *Astron. Astrophys.*, **107**, 39.

Denegri, D., Sadoulet, B., and Spiro, M. 1990, *Rev. Mod. Phys.*, **62**, 1.

Doroshkevich, A. G., Zel'dovich, Ya. B., and Novikov, I. D. 1968, *Soviet Phys. J.E.T.P.*, **26**, 408.

Draine, B. T., and Shapiro, P. R. 1989, *Ap. J. (Lett.)*, **344**, L45.

Dubrovich, V. K. 1975, *Soviet Astron. J. Lett.*, **1**, 196.

Encrenaz, P., and Partridge, R. B. 1969, *Astrophys. Lett.*, **3**, 161.

Fabbri, R. 1981, *Astrophys. Space Sci.*, **77**, 529.

Fabbri, R., Melchiorri, F., and Natale, V. 1978, *Astrophys. Space Sci.*, **59**, 223.

Felten, J. E. 1965, *Phys. Rev. Lett.*, **15**, 1003.

Field, G. B., and Perrenod, S. 1977, *Ap. J.*, **215**, 717.

Field, G. B., and Walker, T. P. 1989, *Phys. Rev. Lett.*, **63**, 117.

---

* There are, of course, other 'background' photons, which could interact with cosmic rays – starlight, for instance, or infrared emission from dusty galaxies throughout the Universe. In all such cases, however, the number density of photons is several orders of magnitude less than n_γ, and the interaction rate correspondingly smaller. The thermal background of low energy neutrinos also interacts with ultrahigh energy cosmic rays, especially neutrinos (see e.g., Beresinskii *et al.*, 1990).

Fukugita, M. 1988, *Phys. Rev. Lett.*, **61**, 1046.
Giacconi, R., and Zamorani, G. 1987, *Ap. J.*, **313**, 20.
Gould, R. J., and Schréder, G. P. 1966, *Phys. Rev. Lett.*, **16**, 252.
Greisen, K. 1966, *Phys. Rev. Lett.*, **16**, 748.
Guilbert, P. W., and Fabian, A. C. 1986, *Monthly Not. Roy. Astron. Soc.*, **220**, 439.
Gunn, J. E., and Peterson, B. E. 1965, *Ap. J.*, **142**, 1633.
Gush, H. P., Halpern, M., and Wishnow, E. H. 1990, *Phys. Rev. Lett.*, **65**, 537.
Hawkins, I., and Wright, E. L. 1988, *Ap. J.*, **324**, 46.
Hayakawa, S., Matsumoto, T., Matsuo, H., Murakami, H., Sato, S., Lange, A. E., and Richards,
     P. L. 1987, *Publ. Astron. Soc. Japan*, **39**, 941.
Hill, C. T., and Schramm, D. N. 1985, *Phys. Rev.* D, **31**, 564.
Hoyle, F. 1965, *Phys. Rev. Lett.*, **15**, 131.
Hoyle, F. 1989, in *Highlights in Gravitation and Cosmology*, eds. B. R. Iyer *et al.*, Cambridge
     University Press, Cambridge.
Hoyle, F., and Wickramasinghe, N. C. 1967, *Nature*, **214**, 969.
Ikeuchi, S. 1981, *Publ. Astron. Soc. Japan*, **33**, 211.
Illarionov, A. F., and Sunyaev, R. A. 1975, *Soviet Astron. J.*, **18**, 691.
Jelley, J. V. 1966, *Phys. Rev. Lett.*, **16**, 479.
Jones, B. J. T., and Steigman, G. 1978, *Monthly Not. Roy. Astron. Soc.*, **183**, 585.
Jones, B. J. T., and Wyse, R. F. G. 1985, *Astron. Astrophys.*, **149**, 144.
Kawasaki, M., and Sato, K. 1986, *Phys. Lett.* B, **169**, 280.
Kimble, R., Bowyer, S., and Jakobsen, P. 1981, *Phys. Rev. Lett.*, **46**, 80.
Kolb, E. W., and Turner, M. S. 1990, *The Early Universe*, Addison-Wesley, Redwood City,
     California.
Kompaneets, A. S. 1957, *Soviet Phys. J.E.T.P.*, **4**, 730 (1956, *Zh. Eks. Teor. Fys.*, **31**, 876).
La, D. 1989, *Ap. J.*, **341**, 575.
Lacey, C., and Field, G. B. 1988, *Ap. J.* (*Lett.*), **330**, L1.
Layzer, D. 1968, *Astrophys. Lett.*, **1**, 99.
Layzer, D., and Hively, R. 1973, *Ap. J.*, **179**, 361.
Lea, S. M., Silk, J., Kellogg, E., and Murray, S. 1973, *Ap. J.* (*Lett.*), **184**, L105.
Lightman, A. P. 1981, *Ap. J.*, **244**, 392.
Lyubarski, Y. E., and Sunyaev, R. A. 1983, *Astron. Astrophys.*, **123**, 171.
Mather, J. C., *et al.* 1990, *Ap. J.* (*Lett.*), **354**, L37.
Matsumoto, T., Hayakawa, S., Matsuo, H., Murikami, H., Sato, S., Lange, A. E., and Richards,
     P. L. 1988, *Ap. J.*, **329**, 567.
McDowell, J. C. 1986, *Monthly Not. Roy. Astron. Soc.*, **223**, 763.
Mushotzky, R. F., Serlemitsos, P. J., Smith, B. W., Boldt, E. A., and Holt, S. S. 1978, *Ap. J.*,
     225, 21.
Narlikar, J. V., Edmunds, M. G., and Wickramasinghe, N. C. 1976, in *Far Infrared Astronomy*,
     ed. M. Rowan-Robinson, Pergamon Press, Oxford.
Ostriker, J. P., and Cowie, L. L. 1981, *Ap. J.* (*Lett.*), **243**, L127.
Ostriker, J. P., and Thompson, C. 1987, *Ap. J.*, **323**, L97.
Ozernoi, L. M., and Chernin, A. D. 1968, *Soviet Astron. J.*, **11**, 907.
Peebles, P. J. E. 1968, *Ap. J.*, **153**, 1.
Peebles, P. J. E. 1971, *Physical Cosmology*, Princeton University Press, Princeton, New Jersey.
Puget, J. L., and Heyvaerts, J. 1980, *Astron. Astrophys.*, **83**, L10.
Raffelt, G. 1989, *Ap. J.*, **336**, 61.
Rana, N. C. 1981, *Monthly Not. Roy. Astron. Soc.*, **197**, 1125.
Rasband, S. N. 1971, *Ap. J.*, **170**, 1.
Rees, M. J. 1978, *Nature*, **275**, 35.
Reif, F. 1965, *Fundamentals of Statistical and Thermal Physics*, McGraw-Hill, New York.
Rephaeli, Y. 1980, *Ap. J.*, **241**, 858.
Rowan-Robinson, M., Negroponte, J., and Silk, J. 1979, *Nature*, **281**, 635.
Rybicki, G. B., and Lightman, A. P. 1979, *Radiative Processes in Astrophysics* (Chap. 7), John
     Wiley, New York.
Schmidt, M., and Green, R. F. 1986, *Ap. J.*, **305**, 68.
Sciama, D. W. 1988, *Monthly Not. Roy. Astron. Soc.*, **230**, 13.

Setti, G. 1990, in *I.A.U. Symposium 139*, eds. S. Bowyer and C. Leinert, Kluwer, Dordrecht.

Silk, J., and Stebbins, A. 1983, *Ap. J.*, **269**, 1.

Stecker, F. W. 1968, *Phys. Rev. Lett.*, **21**, 1016.

Stecker, F. W., and Puget, J. L. 1972, *Ap. J.*, **178**, 57.

Sunyaev, R. A., and Zel'dovich, Ya. B. 1972a, *Astron. Astrophys.*, **20**, 189.

Sunyaev, R. A., and Zel'dovich, Ya. B. 1972b, *Comments Astrophys. Space Phys.* **4**, 173.

Sunyaev, R. A., and Zel'dovich, Ya. B. 1980, *Ann. Rev. Astron. Astrophys.*, **18**, 537.

Tolman, R. C. 1934, *Relativity, Thermodynamics and Cosmology*, Clarendon Press, Oxford.

Trimble, V. 1987, *Ann. Rev. Astron. Astrophys.*, **23**, 425.

Wang, B., and Field, G. B. 1989, *Ap. J. (Lett.)*, **345**, L9.

Weinberg, S. 1972, *Gravitation and Cosmology*, John Wiley, New York.

Weymann, R. 1965, *Phys. Fluids*, **8**, 2112.

Woody, D. P., and Richards, P. L. 1979, *Phys. Rev. Lett.*, **42**, 925.

Wright, E. L. 1979, *Ap. J.*, **232**, 348.

Wright, E. L. 1982, *Ap. J.*, **255**, 401.

Zatsepin, G. T., and Kuz'min, V. A. 1966, *Soviet Phys. J.E.T.P. Lett.*, **4**, 78.

Zel'dovich, Ya. B., Kurt, V. G., and Sunyaev, R. A. 1968, *Soviet Phys. J.E.T.P.*, **28**, 146.

Zel'dovich, Ya. B., and Sunyaev, R. A. 1969, *Astrophys. Space Sci.*, **4**, 301.

6

Searches for anisotropy in the CBR on large angular scales

A characteristic feature of the CBR, noted at the time of its discovery by Penzias and Wilson (1965), is its approximate isotropy (see Appendix A). Approximately equal intensity in all directions is expected if the radiation is a relic of the Hot Big Bang. On the other hand, there are a variety of mechanisms that can induce small amplitude variations in intensity, or anisotropies, into an initially uniform CBR; some of these were outlined in Chapter 2 and will be discussed in detail in Chapter 8.

Careful measurements of the angular distribution of the CBR have therefore been pursued both to confirm the cosmic, Hot Big Bang, origin of the CBR and to search for small amplitude anisotropies imprinted in it. In this chapter we deal with observations of the angular distribution of the CBR on the largest angular scales, $\theta \gtrsim 10°$, and in particular with the dipole and quadrupole moments of the CBR. The value of about 10° for the boundary between 'large' scale anisotropies (discussed in this chapter) and smaller scale anisotropies (Chapter 7) is obviously rather artificial. When we turn in Chapter 8 to the implications of the measurements of and upper limits on CBR anisotropies, we will be drawing on the results of both Chapters 6 and 7. On the other hand, many of the observational techniques used in the search for large-scale anisotropies differ in fundamental ways from those that can be used on smaller angular scales. The dominant sources of systematic error also differ. On these grounds, more experimental than theoretical, I have elected to divide the descriptions of the observations into two chapters.

6.1 Multipole expansion of CBR anisotropy

Any observed anisotropy in the CBR can in principle be fitted by a sum of spherical harmonics. The two lowest harmonics are the dipole and quadrupole moments; these two moments are the most interesting large-scale anisotropies in the CBR for cosmology. With some groups (e.g., Lubin *et al.*, 1985), it has become conventional to express the three independent parameters of the dipole moment as T_x, T_y and T_z, and the five independent parameters of the quadrupole moment as $Q_1 - Q_5$. In terms of the two orthogonal celestial coordinates, right ascension α and declination δ, the angular distribution of the CBR may be expressed by

$$T(\alpha,\ \delta) = T_0 + T_x\ \cos\delta\ \cos\alpha + T_y\ \cos\delta\ \sin\alpha + T_z\ \sin\delta$$

$$+ Q_1\left(\frac{3}{2}\sin^2\delta - \frac{1}{2}\right) + Q_2\ \sin2\delta\ \cos\alpha + Q_3\ \sin2\delta\ \sin\alpha \qquad (6.1a)$$

$$+ Q_4\ \cos^2\delta\ \cos2\alpha + Q_5\ \cos^2\delta\ \sin2\alpha.$$

It is sometimes more useful to describe the distribution of the CBR in Galactic co-ordinates, with the zero point of Galactic longitude (l) and latitude (b) coincident with the center of our Galaxy. Then the surface $b = 0°$ is coincident with the plane of our disk-shaped Galaxy. In these coordinates,

$$T(l, b) = A_{00} + A_{11} \cos b \, \cos l + B_{11} \cos b \, \sin l$$

$$+A_{10} \sin b + A_{20} \left(\frac{3}{2} \sin^2 b - \frac{1}{2} \right) + \frac{3}{2} A_{21} \sin 2b \, \cos l$$

$$+\frac{3}{2} B_{21} \sin 2b \, \sin l + 3 A_{22} \cos^2 b \, \cos 2l$$ (6.1.b)

$$+3 B_{22} \cos^2 b \, \sin 2l.$$

We shall often want to talk about the amplitude of the dipole or quadrupole moment without worrying about its alignment. We thus introduce the quantity T_1 defined by

$$T_1 = (T_x^2 + T_y^2 + T_z^2)^{1/2} = (A_{11}^2 + B_{11}^2 + A_{10}^2)^{1/2}$$ (6.2a)

as the amplitude of the dipole. The corresponding quantity T_2 for the amplitude of the quadrupole moment is not simply the square root of ΣQ_i^2 because the basis vectors (the functions $((3/2) \sin^2 \delta - 1/2)$, etc.) are orthogonal but not *orthonormal* (see Scaramella and Vittorio, 1990, for instance). The r.m.s. amplitude of the quadrupole moment is instead given by

$$T_2 = \left[\left(\frac{4}{15} \right) \left(\frac{3}{4} \, Q_1^2 + Q_2^2 + Q_3^2 + Q_4^2 + Q_5^2 \right) \right]^{1/2}.$$ (6.2b)

To include higher order moments, we may expand $T(\alpha, \delta)$ or $T(l, b)$ using spherical harmonics, e.g.,

$$T(\alpha, \delta) = \sum_{n, m} C_{nm} Y_n^m (\alpha, \delta),$$ (6.3)

where $n = 1$ specifies the dipole moment, $n = 3$ the octopole moment, and so on. The values of the coefficients C_{nm} may be found from the integrals

$$C_{nm} = \int T'(\alpha, \delta) Y_n^m (\alpha, \delta) \, d\Omega,$$ (6.4)

where $T'(\alpha, \delta)$ is the measured temperature distribution. It should be apparent that eval-uating C_{nm} for values of $n \geq 2$ will be difficult if $T'(\alpha, \delta)$ is measured over only part of the sky. As in the case of the dipole amplitude T_1, we may define the r.m.s. amplitudes of higher order moments as

$$T_n = \left[\frac{1}{4\pi} \sum_{m=-n}^{n} C_{nm}^2 \right]^{1/2}.$$ (6.5)

In the particular case of the quadrupole moment, eqn. (6.5) produces the same numer-ical value for T_2 as given by eqn. (6.2b) above. I will use this definition of T_2 through-out. It is important to note, however, that different definitions of the quadrupole moment appear in many theoretical and experimental papers (as noted by Scaramella and Vittorio, 1990). Some groups, for instance, work directly with the sum of the squares of the coefficients of the second spherical harmonic, $\Sigma_m C_{2m}^2$. Others report not the r.m.s.

amplitude of T_2 but the amplitude of the individual components, which are typically $\sqrt{5}$ smaller.

In this book (and particularly in table 6.2 later) I have tried to put all the reported upper limits on a common footing. The values given are of T_2 as defined by eqn. (6.2b) or eqn. (6.5).

6.2   Differential measurements

In principle, the angular distribution of the CBR could be determined from a series of absolute measurements such as those described in Chapter 4, made in different directions. Even the best spectral measurements, however, have an accuracy of only about 1%. We can map the CBR far more accurately by making *differential* measurements, that is by finding the difference between the CBR intensities measured in two different directions. Comparative measurements of this kind, of course, provide no information about the mean temperature of the CBR, T_0, in eqn. (6.1a). On the other hand, many of the systematic errors inherent in absolute measurements are sharply reduced. As we shall see, differential measurements of the angular distribution of the CBR can have errors as small as one part in 10^5 of T_0.

Searches for large-scale anisotropies in the CBR have generally employed a dual-beam technique, in which the input to a receiver is switched rapidly between two different directions in the sky (Section 3.4.4). The angle between the beams is typically 60°–90°.* In searches for large angular scale anisotropies, the angular resolution of the individual beams is not important; generally beam widths of order 5°–10° are employed.

6.2.1   Beam-switching by reflection

An early example of this technique is provided by the first attempt to measure the large-scale distribution of the CBR (Partridge and Wilkinson, 1967; see fig. 6.1). A pyramidal horn antenna was used to define the beam. The beam was directed due south towards a position 8° below the celestial equator (that is, – 8° declination). Periodically, a reflecting sheet was raised to deflect the beam back towards the north celestial pole, a fixed point on the celestial sphere. The zenith angles of these two beams were equal, so both beams passed through the same thickness of the Earth's atmosphere. During the course of a day, measurements could thus be made of the intensity of the CBR along a circle at – 8° declination relative to its intensity at a fixed point in the sky. In this experiment, any anisotropy in the CBR would appear as time-dependent variations in the intensity differences.

In addition to the obvious problem that solar heating or other diurnal effects could mimic a true anisotropy in the CBR, this technique suffered from the disadvantage that it produced measurements at one declination only, and thus could not measure all three components of the dipole moment. Modern observing programs allow for rotation of the equipment about the local vertical, permitting much wider sky coverage. An example is the balloon-borne instrument operating at 3 mm wavelength constructed by Lubin and Villela (1985) and shown in fig. 6.2. As in the earlier instrument, a single horn antenna was employed, and a reflector was used to define the reference beam.

---

* An exception is the experiment of Fabbri *et al.* (1980), which used a 6° angle.

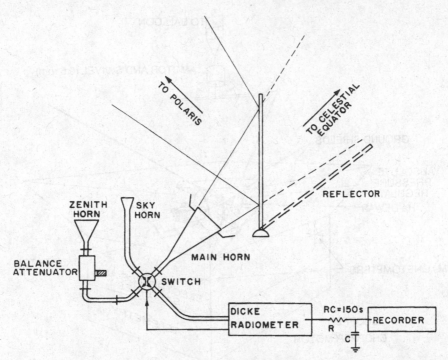

Fig. 6.1   Schematic of the apparatus used by Partridge and Wilkinson (1967) to search for anisotropies in the CBR. With the movable reflector in its lowered position (dashed), the instrument scanned a strip near the celestial equator. When the reflector was raised, the instrument viewed the north celestial pole ('Polaris'). Note that the reference horn was directed towards the zenith, *not* a fixed point on the sky.

6.2.2   *Symmetrical antennas*

An alternative approach involves use of two identical antennas, separated by a fixed angle. This technique has been used by the Princeton and MIT groups (Fixsen *et al.*, 1983; Halpern *et al.*, 1988), as well as by a recent NASA satellite (Section 6.7) and is shown schematically in fig. 6.3. The two antennas are arranged symmetrically about the vertical, so that the zenith angles of the two beams defined by the antennas are equal. The input to the receiver can be rapidly switched from one antenna to another using an electrically driven chopper or switch (a so-called Dicke switch, as described in Section 3.4.4). A variant of this technique was employed by the Soviet satellite experiment to be discussed below: that instrument employed a small reference horn with its optical axis parallel to the spin axis of the spacecraft and a larger, higher resolution, horn antenna with an axis at 90° to the spin axis (Strukov and Skulachev, 1984).

6.2.3   *Producing a map from differential measurements*

As noted above, any of these techniques produces measurements of temperature *differences* if the CBR is anisotropic. Hence the observational data need to be unfolded before a map of the CBR can be produced or comparisons made with mathematical models such as eqn. (6.1) or eqn. (6.4). This process is not easy because the temperature of any single element of the sky can appear in many measured temperature differences. In order to find the actual distribution of temperature across the sky from a set of measured

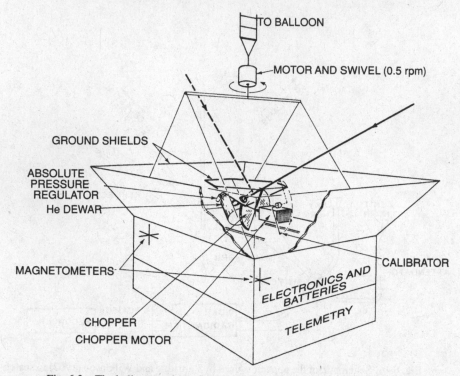

Fig. 6.2   The balloon package flown by Lubin and Villela (1985). The chopper consists of two reflective vanes, so the overall operation resembles that of the instrument shown in fig. 6.1; here, however, the whole instrument rotated every 2 min around the local vertical. Note the use of ground shields.

temperature differences, a matrix inversion may be used and the parameters of a model distribution such as eqn. (6.1) modified until a least-squares fit is obtained. It is difficult to obtain unambiguous results with this technique if either the coverage of the sky is incomplete or the data contain systematic offsets of the sort introduced by some of the effects discussed in Section 6.4 below.

Another approach that has been used (for instance in analysis of the data obtained by the Soviet satellite experiment described in Section 6.7.1) is to construct a numerical model of the experimental parameters such as receiver noise, angle between the beams and sky coverage, and then to apply this model to an ideal, noise-free signal containing one or more multiple moments. The resulting *model* temperature differences are compared with the *observed* temperature differences, and the amplitudes of the coefficients C_{nm} of the ideal, model signal adjusted until a best fit is obtained. Since there is noise in both model differences and observed differences, it is necessary to make many such comparisons to obtain accurate estimates of the coefficients C_{nm}. Note that this process takes account of incomplete or uneven coverage of the sky.

6.3   Receivers

Both heterodyne and bolometric receivers* have been used in the search for large-scale structure in the CBR. Heterodyne receivers are intrinsically narrow band, with

---

* The properties of these two types of receivers, and the advantages and disadvantages of each, are laid out in Chapter 3; see also Weiss (1980).

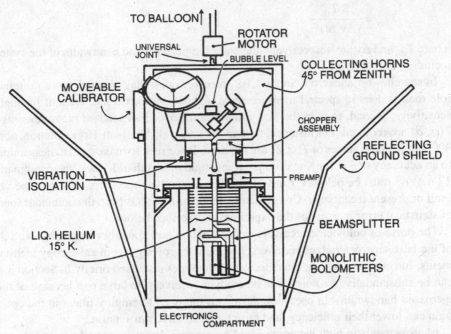

TO BALLOON↑

ROTATOR MOTOR

UNIVERSAL JOINT

BUBBLE LEVEL

COLLECTING HORNS 45° FROM ZENITH

MOVEABLE CALIBRATOR

CHOPPER ASSEMBLY

REFLECTING GROUND SHIELD

VIBRATION ISOLATION

PREAMP

BEAMSPLITTER

LIQ. HELIUM 15° K.

MONOLITHIC BOLOMETERS

ELECTRONICS COMPARTMENT

Fig. 6.3   The experimental arrangement used by Halpern *et al.* (1988). Note the two symmetrical horn antennas, each directed 45° from the zenith. The entire instrument was rotated in flight about the vertical axis.

$\Delta v/v \lesssim 0.1$, and are preferred for observations at wavelengths $\gtrsim 3$ mm. Wide-band bolometric detectors are used at wavelengths $\lesssim 3$ mm. Since the bolometric detectors are intrinsically very broad band, filters are needed in the optical train to define the effective bandwidth (see, for instance, Halpern, 1983; and Section 4.2). Filters are also needed to block emission from the Galaxy, which overwhelms the CBR radiation at both long wavelengths, $\lambda \gtrsim 30$ cm, and short wavelengths, $\lambda \lesssim 0.5$ mm. Since the Galaxy is a strongly anisotropic source, any long wavelength or short wavelength leakage through the filters could introduce systematic errors into measurements of large-scale anisotropy in the CBR.

6.4   Sources of error and means to limit them

Galactic emission is just one of the potential sources of error in such experiments. In this section, several sources of errors peculiar to anisotropy measurements are described, and so are some of the methods that observers have employed to minimize them. For convenience, we will consider statistical and systematic errors separately, though in some cases both can arise from the same source.

6.4.1   *Statistical error – instrumental and atmospheric noise*

All receivers are intrinsically noisy. Expressed in antenna temperature, the r.m.s. noise for differential measurement of duration Δt s is (eqn. (3.24)):

$$T_{rms} = \frac{2\,T_{sys}}{(\Delta v\,\Delta t)^{1/2}},$$

where T_{sys} and Δv are, respectively, the noise temperature and bandwidth of the system, including the receiver.

In searches for anisotropy in the CBR, receiver noise plays a much more prominent role than it does in spectral measurements, for two reasons. First, we aim for higher sensitivity; second, to map the sky, a large number of independent measurements of $T(\alpha, \delta)$ are required. For instance, to map the entire sky at about $10°$ resolution, about 400 independent values of $T(\alpha, \delta)$ are required. If we seek to make such a measurement to an accuracy of ± 0.1 mK at each point in a time of less than 1 week, the combination $T_{sys}/(\Delta v)^{1/2}$ must be below 1.9 mK Hz$^{-1/2}$. For heterodyne receivers, this is at the very limit of present technology. Consequently, receiver noise has been the dominant source of statistical error in many of the experiments described below.

The noise of bolometric receivers at $\lambda \lesssim 3$ mm is now roughly comparable with that of the best superheterodyne receivers; hence their growing use in anisotropy measurements. Intrinsically, bolometric detectors of the types discussed briefly in Section 3.4.5 can be substantially less noisy than heterodyne receivers (in large part because of their increased bandwidth). In operation, however, the need to employ filters in the optical train can lower their efficiency, and raise their effective r.m.s. noise.

Improvements in both heterodyne and bolometric detectors over the next few years should make it possible to obtain values of $T_{rec}/(\Delta v)^{1/2} \lesssim 1$ mK Hz$^{-1/2}$.

Even with a perfect receiver, observers would have to confront the noise produced by emission from the Earth's atmosphere. Were the atmosphere completely homogeneous and laminar, it would present only minor problems for comparative, dual-beam observations. Unfortunately, the distribution of water vapor in particular is far from homogeneous, so that atmospheric emission introduces both statistical and systematic errors. Indeed, the dominant source of both statistical and systematic error in the first searches for large-scale anisotropy (e.g., Partridge and Wilkinson, 1967) was emission from the Earth's atmosphere. To minimize this source of error, all recent searches for large-scale angular variations in the CBR have been carried out at high altitude. With the exception of the two satellite experiments discussed below, all the instruments have been carried aloft, to altitudes of 20–40 km, by high-flying aircraft or, more usually, balloons. At balloon altitudes, the residual emission from the Earth's atmosphere is typically less than 1% of the CBR temperature for these microwave observations, and in addition little of that residual emission is from water vapor, since its scale height is less than that of other atmospheric gases.

6.4.2 *Sources of systematic error, especially Galactic foreground*

Emission from the Earth's atmosphere varies with the zenith angle, with $T_{atm}(z) \propto \sec z$ (Section 4.3.3). This angle dependence can in principle introduce systematic errors into the differential measurements employed in searches for large-scale anisotropy in the CBR. If the zenith angles of the two beams differ by Δz, an offset in antenna temperature of

$$\Delta T_{atm} = \frac{dT_{atm}(z)}{dz} \, \Delta z = T_{atm}(0) \, \tan z \, \sec z \, \Delta z$$

will be recorded, as we can see from eqn. (4.5). For this reason, all experiments on large-scale structure carried out beneath part or all of the Earth's atmosphere have been designed to have beams symmetrical about the vertical, so that Δz is close to zero. In experiments where the beam separation is 90°, z is then 45°, and $\Delta T_{atm} = \sqrt 2 \, T_{atm}(0) \, \Delta z$; to reduce ΔT_{atm} to a negligible level of, say, $\lesssim 10^{-5} \, T_0$, then requires any misalignment Δz to be $\lesssim 2 \times 10^{-5}/T_{atm}(0)$ rad or $\lesssim 4$ arcsec/$T_{atm}(0)$. Even at balloon altitudes, $T_{atm}(0)$ can be as large as, say, 10^{-2} K, so that the zenith angles of the two beams need to be kept equal to about 7′.

There is another feature of these experiments which helps observers control systematic error introduced by atmospheric emission. Most searches for large-scale anisotropy last many hours or longer, so that the celestial sphere moves appreciably with respect to the local vertical. This allows a separation of zenith-angle-dependent signals from any celestial anisotropy.

Another potential source of systematic error in searches for large-scale anisotropy is the radiation of the Earth or of bright celestial sources into the antennas used, either directly or by side-lobe pickup (Sections 3.3.2 and 4.4.1).

Given the small collecting area of the antennas in such experiments, or equivalently their low angular resolution, it is easy to show that celestial radio sources, with the exception of the Sun, the Moon and the Galactic plane, will have negligible effects on such measurements. The effect of the Sun can be and is avoided either by making observations only at night or, as in the case of the U.S.A. and Soviet satellite experiments, by pointing the instrument away from the Sun. Signals from the Moon may be avoided by making the observations near the time of new moon or may be removed after the fact from the data (see Section 6.6 below). Emission from the Galaxy is treated below.

We still need to consider pickup from the side- and back-lobes of the antennas used in these measurements (see Section 4.4.1). Radiation from the Earth (or from strong celestial sources) may enter the side-lobes of the antennas (see fig. 4.5 for an example). If either the side-lobe pattern of the antennas or emission from the Earth is asymmetrical, angle-dependent offsets will be introduced into the measured intensity differences, contributing to the systematic error in such measurements. Proper design of the horn antennas, in particular the use of corrugations or wave traps (Section 4.4.1), can reduce side- and back-lobes to a low level, say one part in 10^6 or less of the response of the antenna on the optical axis. We must recall, however, that the Earth acts as a 250–300 K blackbody filling nearly 2π steradians of the back- and side-lobes of these instruments, so that emission into back- and side-lobes can be as large as a few percent of the CBR temperature, even for well-designed antennas. Therefore, most groups have found it expedient to put ground screens in place as well; these large reflective surfaces shield the instruments from direct emission from the Earth. An example of the use of ground screens is shown in fig. 6.2. Well-designed ground screens can reduce side-lobe errors by a factor of about 100. It should also be noted that microwave radiometers are sensitive to man-made radio signals; the use of ground screens and proper choice of observing frequency make this problem manageable.

A final source of systematic error, which cannot be avoided, is radio emission from our Galaxy. This is a particularly troublesome source of error, because the Galaxy is manifestly anisotropic. In the microwave band, down to a wavelength of about 3 mm, the emission of the Galaxy is dominated by synchrotron radiation and bremsstrahlung or free–free radiation (discussed more fully in Section 4.4.2). The former is loosely confined to the Galactic plane, and falls off rapidly as the wavelength of observation decreases (approximately as $T(\lambda) \propto \lambda^{2.8}$). The latter is more sharply confined to the plane and falls off more slowly ($T(\lambda) \propto \lambda^{2.1}$).

At wavelengths less than about 3 mm, the dominant form of emission from the Galaxy is thermal emission from small interstellar dust particles. Infrared observations made from the IRAS satellite have shown that the interstellar dust is unfortunately far from uniformly distributed (Low *et al.*, 1984). The presence of thermal emission from dust has already proved to be a major problem to those attempting measurements of the large angular scale distribution of the CBR at wavelengths of order 1 mm (see Weiss, 1980, for a review; and Melchiorri *et al.*, 1981).

As fig. 4.10 suggests, these sources of emission from the Galaxy leave open a narrow 'window' in wavelength, centered at about 3 mm, which has been exploited for several recent measurements of the large-scale anisotropy. Even observations made in this wavelength window, however, may need corrections for Galactic emission. Such corrections for synchrotron and bremsstrahlung emission are usually based on radio maps of the Galaxy made at longer wavelengths. These maps are then extrapolated to the wavelength used for the CBR observations using an appropriate combination of the two power law dependences given above (see Conklin, 1969; Witebsky, 1978; Fixsen *et al.*, 1983, for examples of this procedure). The most useful map now available is one made at $\lambda = 1.5$ cm (Boughn *et al.*, 1992b); the extrapolation to wavelengths in the 3 mm Galactic 'window' is less uncertain than it is for the older, longer wavelength surveys of Haslam *et al.* (1981), and Reich and Reich (1986), for instance. The 1.5 cm map is shown in fig. 6.4. Note that the Galactic emission is most intense within a few degrees of the Galactic plane, but there is patchy emission elsewhere. Given such a map, the observations obtained from dual-beam instruments can be corrected point by point. Note the implicit assumption that the overall spectral index does not vary from one direction to another in the Galaxy. The long wavelength radio maps of the Galaxy themselves show that this assumption does not hold exactly. The spectral index does vary from place to place. As Bennett *et al.* (1992) point out, this variation must be taken into account in calculating corrections for Galactic emission.

Observations at $\lambda \leqslant 3$ mm also require correction for thermal emission from dust. The 100 μm maps made by IRAS (fig. 4.9) may be used to estimate the effect of dust emission at longer wavelengths. There is now good evidence, however, that the calibration of the IRAS maps for the large angular scales of interest here is in error (see Hauser *et al.*, 1991). Consequently, it will be preferable to use the maps generated by the FIRAS instrument aboard COBE (see Section 4.9.1), once these have been published in final form. An important preliminary analysis by Wright *et al.* (1991) has already appeared. In this paper, dust emission is modeled by first assuming that the dust and synchrotron spectra do not vary with position within the Galaxy, then mapping the dust intensity derived from the FIRAS observations as a function of Galactic coordinates (l, b), as shown in fig. 6.4(*b*). This preliminary map, like the IRAS 100 μm map, shows that the signal from dust is clearly not confined to the Galactic plane. Well away from the plane, say at $b \gtrsim 30°$, the regions of lowest dust emission have a surface brightness corre-

(a)

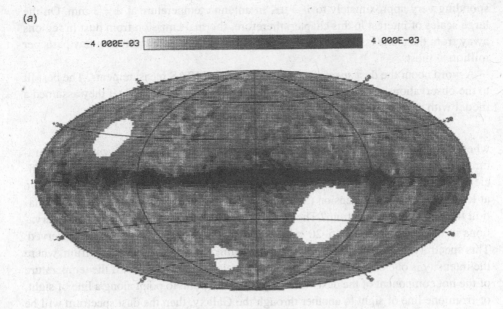

(b)

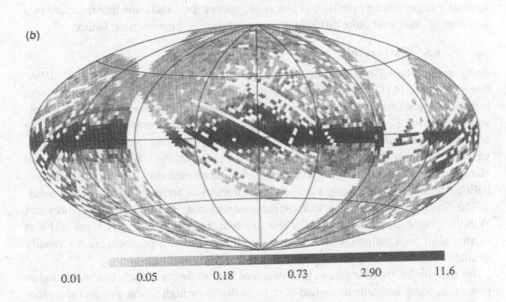

Fig. 6.4   (*a*) Map of the sky, in Galactic coordinates, made at $\lambda = 1.5$ cm (Boughn *et al.*, 1992b). The dipole signal has been removed, leaving only Galactic emission concentrated along the Galactic plane. Areas not observed appear as blanks. The range of the grey scale used is –4 mK to +4 mK, or ~0.3% of T_0 overall. (*b*) A preliminary map of Galactic dust emission derived from FIRAS measurements (Wright *et al.*, 1991). The grey scale corresponds approximately to 0.8–900 µK antenna temperature for $\lambda = 3$ mm, or 0.15–160 µK at $\lambda \sim 9$ mm.

sponding very approximately to 1–3 μK in antenna temperature at $\lambda = 3$ mm. On the large scales of interest in this chapter, therefore, thermal emission from dust in regions away from the plane is likely to produce CBR fluctuations of amplitude a few parts per million at most.

A word about the dust *spectrum* derived from the FIRAS measurements. The best fit to the observations was obtained by Wright and his colleagues (1991) if they assumed a model with dust of two different temperatures present:

$$[aB \, (\nu, 20.4 \text{ K}) + bB \, (\nu, 4.77 \text{ K})]\nu^2,$$

where $B(\nu, T)$ is the Planck function. The presence of quite cool dust, with a temperature < 2.5 K above T_0 is surprising. For the dust to remain this cool, it must be an efficient radiator at millimeter and submillimeter wavelengths, and therefore must also have at least one extended dimension (Section 5.3.1). May I suggest an alternative explanation for the observed spectrum? The cool dust component is required to fit the observations in the frequency range 5–20 cm⁻¹, where a slight excess of emission is observed. This spectral region is in the Rayleigh–Jeans region of the thermal dust spectrum, where the intensity is one or two orders of magnitude below the peak value. If the temperature of the hot component of the dust varies slightly from point to point along a line of sight, or from one line of sight to another through the Galaxy, then the dust spectrum will be broadened. This effect will show up most strongly at low intensity levels, away from the peak. Thus I would suggest that a model with small variations in the temperature of the warmer component be investigated before we accept the conclusion that there are two varieties of dust, with quite different physical properties, present in the Galaxy.

6.5   Early results

Early ground-based observations (Partridge and Wilkinson, 1967; Conklin, 1969; Boughn *et al.*, 1971), while adequate to show that the CBR was very isotropic, were insufficiently accurate to measure the dipole or higher-order moments. Conklin's observations at $\lambda = 3.8$ cm did reveal a statistically significant anisotropy component, but the observations required a large correction for Galactic emission. After 1970, observers gave up attempts to measure large-scale anisotropies in the CBR from the ground; Henry's (1971) attempt to measure the dipole component was the first to employ a balloon-borne radiometer, and a dual-beam, symmetrical arrangement of two antennas.

The first reliable detection of a dipole moment was that reported in 1976 by Corey and Wilkinson (see also Corey, 1978). Their instrument operated at $\lambda = 1.5$ cm and was carried aloft by a balloon. Unfortunately the details of this experiment are not readily available (but see Cheng *et al.*, 1979).

By the late 1970s, four groups had measured the dipole component to a much higher precision, using instruments carried aloft by balloons or high-flying aircraft (Muehlner and Weiss, 1976; Smoot, Gorenstein and Muller, 1977; Cheng *et al.*, 1979; Smoot and Lubin, 1979; Fabbri *et al.*, 1980; Boughn *et al.*, 1981). Two of these groups also reported evidence for a quadrupole component T_2 of amplitude about $T_1/4$. That result is now known to be in error (see following sections). The work of Muehlner and Weiss (1976) and of Melchiorri and his colleagues (Fabbri *et al.*, 1980) represented pioneering efforts to make use of the high sensitivity of wide-band bolometric detectors in searches for anisotropy in the CBR. Smoot and Lubin (1979) were the first to extend such measurements into the southern hemisphere to increase sky coverage.

Table 6.1 *Dipole and Quadrupole coefficients of CBR measured at $\lambda = 3$ mm (Lubin et al. 1985).*

Coefficient	Value (mK*)	Statistical error (mK*)	Total error (mK*)
T_x	-3.37	0.09	0.17
T_y	0.63	0.09	0.09
T_z	-0.49	0.09	0.09
Q_1	0.21	0.09	0.09
Q_2	0.27	0.09	0.10
Q_3	0.16	0.10	0.11
Q_4	-0.10	0.07	0.09
Q_5	0.05	0.06	0.08

*Values are given in thermodynamic temperature.

6.6   Recent measurements of large-scale anisotropy from balloons

In the 1980s, results from a second generation of more sensitive and more sophisticated balloon experiments became available. Three groups were involved in these efforts, which I will loosely classify as the California group (Lubin, Epstein and Smoot, 1983; Lubin and Villela, 1985), the MIT group (Halpern *et al.*, 1988), and the Princeton group (Fixsen *et al.*, 1983; Boughn *et al.*, 1990, 1992a). The published results of these groups' efforts are shown in summary form in table 6.2 later. Note that earlier reports of a significant quadrupole component are not borne out by these later measurements, which all show that the quadrupole amplitude is small, $T_2/T_0 < 10^{-4}$. The significant features of each program are described in the remainder of this section.

6.6.1   Features of the California experiments

The instrument used by Lubin is shown in fig. 6.2. It operated at $\lambda = 3$ mm, at the center of the 'window' in Galactic microwave emission; thus the results require smaller corrections for Galactic foreground emission than the results of other groups. The receiver was sensitive enough to detect a roughly 3 mK dipole signal in a single 100 s rotation about the local vertical. In addition, it is interesting to note that these observations reached a sensitivity adequate to detect the yearly motion of the Earth around the Sun, which produces a 3×10^{-4} K modulation in the signal. Values for the dipole and quadrupole coefficients are displayed in table 6.1.

The instrument was flown in both hemispheres, so 85% of the sky was observed (Lubin *et al.*, 1985). The last of four flights was launched in Brazil; at the end of the flight, the instrument disappeared in the Brazilian interior. It was later found by a jungle poacher, who added some of the more intriguing bits to the decor of his favorite bar.

Given the fate of their instrument, it is understandable that Lubin and his colleagues turned their attention to a different sort of measurement: searches for anisotropy on intermediate angular scales of about 1°. These results, obtained from both balloon-borne telescopes and similar instruments operated at the South Pole, will be discussed in Section 7.7.

6.6.2 Features of the Princeton experiments

The group originally based at Princeton (Fixsen *et al.*, 1983; Boughn *et al.*, 1990, 1992a) has flown two different instruments, both employing receivers based on maser amplifiers which have particularly low noise temperatures ($T/(\Delta v)^{1/2} = 1.6$ mK Hz$^{-1/2}$, giving a sensitivity of 4 mK/Hz$^{-1/2}$ in flight, with Dicke switching). Thus these results have lower statistical errors than the Berkeley measurements.

The first instrument to be flown had the conventional, symmetrical, dual-beam arrangement with two horn antennas, and operated at $\lambda = 1.2$ cm (Fixsen *et al.*, 1983). It thus produced differential signals, which had to be converted to a map of the sky (Section 6.2.3). In addition, because of the wavelength employed, the 1.2 cm observations required prior correction for bremsstrahlung and synchrotron emission from the Galaxy.

As the entries in line 6 of table 6.2 show, the value of T_1 derived from this experiment is smaller than the Berkeley value by about twice the combined errors. Each group has analyzed the other's data, leading to the conclusion that calibration uncertainty (up to 5%) in both experiments is largely responsible for the discrepancy.

That calibration and not some other systematic effect is at issue is suggested by the good agreement in *direction* of the dipole maximum, and by the general agreement in values of dipole and quadrupole coefficients between the two experiments.

The second Princeton experiment (Boughn *et al.*, 1990) operated at $\lambda = 1.5$ cm. It differed from all the instruments previously described in being a single beam instrument, so that it made absolute, not differential, measurements of the temperature of the sky as a function of position. The 3° beam was formed by a single horn antenna which rotated at 1 r.p.m. about the local vertical. The antenna was connected to one input of a cooled Dicke switch; the other input was connected to a liquid-helium cooled calibration source, carefully maintained at a fixed temperature of a few kelvin to match T_0.

The wavelength was chosen to enable the observers to map Galactic emission as well as to look for large-scale anisotropy. The use of a single beam rather than the usual dual-beam arrangement allowed the observers to construct a map of the sky directly, without the intermediate steps described in Section 6.2.3. From this map, values of T_1 and limits on T_2 were derived (Boughn *et al.*, 1992a); these appear in table 6.2. Boughn *et al.* (1992a) also used the observations to set limits on fluctuations on smaller angular scales. For the range $\theta = 1°-30°$, the 95% confidence limits are about $(2.0-0.4) \times 10^{-4}$ in $\Delta T/T$ for Gaussian fluctuations (see Chapter 8).

6.6.3 Features of the MIT experiments

The instrument flown by Halpern *et al.* (1988) is shown in fig. 6.3; the detectors employed were solid-state bolometers operating in several broad bands at wavelengths in the millimeter and submillimeter range (see Halpern *et al.*, 1988, for details). Six flights were made; the results cited here were obtained in the sixth flight when measurements were made in four spectral bands. Only the two longest wavelength bands, at effective wavelengths of about 1.7 and 0.8 mm, were used in the search for CBR anisotropies. The results from the 1.7 mm observations appear in table 6.2. In the 0.8 mm band, the measurement of T_1 was less accurate. The dipole component was found to be 4.7±1.4 mK (thermodynamic temperature), and the direction of the maximum of the

Table 6.2 Results (expressed in thermodynamic temperature) of recent measurements of the large-scale distribution of the CBR.

Group Reference	Berkeley Lubin et al. (1985)	MIT–UBC Halpern et al. (1988)	Moscow Strukov et al. (1987) Klypin et al. (1987)	Princeton Fixsen et al. (1983)	Princeton Boughn et al. (1992a)	COBE Smoot et al. (1992)
Wavelength (mm)	3	≃ 1.7	8	12	15	3.3–9.5
Vehicle	balloon	balloon	satellite	balloon	balloon	satellite
Detector	heterodyne	bolometric	heterodyne	heterodyne	maser amplifier	heterodyne
Dipole amplitude, T_1 (mK)	3.4 ± 0.2	3.4 ± 0.4	3.16 ± 0.12	3.1 ± 0.2	3.36 ± 0.10[a]	3.36 ± 0.10
Direction of solar motion, R.A. and Dec.	11.2 ± 0.1 h −6° ± 1.5°	12.1 ± 0.24 h[b] −23° ± 5°	11.3 ± 0.16 h −7.5° ± 2.5°	11.2 ± 0.05 h −8° ± 0.7°	11.0 ± 0.1 h −6° ± 1.5°	11.2 ± 0.1 h −7° ± 1°
Quadrupole moment, T_2 (mK)	≲ 0.4[c]	–	≲ 0.08[d]	≲ 0.19	≲ 0.07	0.013 ± 0.004

[a] Includes estimated systematic error.
[b] Incomplete sky coverage.
[c] Calculated from table I of Lubin et al. (1985) by the present author, by taking the appropriate quadrature sum as given by eqn. (6.2b).
[d] Taking the more conservative, model-independent upper limit.

dipole was at a lower declination than found for other measurements, $\delta = -38° \pm 21°$. A more negative δ implies that the maximum of the observed dipole lies closer to the Galactic plane; Halpern *et al.* (1988) suggest that Galactic dust emission may be contaminating their 0.8 mm observations.

A number of experimental difficulties more fully discussed by Halpern and his colleagues made calibration of the detectors in the sixth flight difficult. The MIT group therefore relied on an astronomical calibration using the Moon; the average temperature assumed for the lunar disk at 1.7 mm was 222 K. Since the temperature of the lunar surface at submillimeter wavelengths is not known, the shorter wavelength bands were calibrated by scaling from the absolute lunar calibration of the 1.7 mm band.

At the short wavelengths employed in the MIT experiment, the conversion from observed antenna temperature to true thermodynamic temperature (see Section 3.2.2) is important. As eqn. (3.8) shows, that conversion depends on the value assumed for T_0, the thermodynamic temperature of the CBR. The same is true for anisotropy in the CBR, such as the dipole component; the relation between a dipole signal measured in antenna temperature and the same dipole component expressed in thermodynamic temperature depends on T_0. So sensitive is the calculated value of the dipole component to T_0 that a comparison of the dipole amplitude measured at millimeter wavelengths with the amplitude at longer wavelengths may be used to constrain T_0, as noted in Section 4.11. As may be seen by differentiating eqn. (3.8), the relation between a dipole signal ΔT_A measured in antenna temperature and the dipole component T_1 in the CBR is

$$T_1 = \Delta T_A \frac{(e^x - 1)^2}{x^2 \, e^x}$$

or

$$T_1 / T_0 = \frac{\Delta T_A}{T_A} \frac{e^x - 1}{x e^x}, \tag{6.6}$$

where $x = h\nu/kT_0$. The MIT measurements at 1.7 mm wavelength are in acceptable agreement with the longer wavelength measurements displayed in table 6.2 if T_0 lies in the range 2.86 ± 0.26 K (see fig. 6.5). The value of T_1 appearing in table 6.2 was calculated assuming $T_0 = 2.74$ K.

The eqn. (6.6) and fig. 6.5 are based on the assumption that the dipole component in the CBR is produced by a Doppler shift in a purely thermal, blackbody spectrum. As noted in Chapter 4, some recent measurements suggested that the CBR spectrum might rise above a Planck curve at $\lambda \lesssim 1$ mm (see fig. 4.22). To allow for the possibility that the equivalent temperature of the CBR changes with frequency, eqn. (6.6) must be modified as follows:

$$T_1 = \Delta T_A \frac{(e^x - 1)^2}{x^2 e^x} \left(1 - \frac{\nu}{T(\nu)} \frac{dT(\nu)}{d\nu} \right)^{-1}. \tag{6.7}$$

This result is derived by Halpern *et al.* (1988); note that their eqn. (7) and eqn. (A5) contain a typographical error, which we have corrected in eqn. (6.7) above. We may now use this result to discover whether the antenna-temperature dipole component ΔT_A measured by Halpern *et al.* at $\lambda \simeq 0.8$ mm is consistent with the submillimeter excess discussed in Section 4.8.4 (Matsumoto *et al.*, 1988). If we assume that the submillimeter excess is cosmic in origin, so that it too will be Doppler shifted by the motion of the local

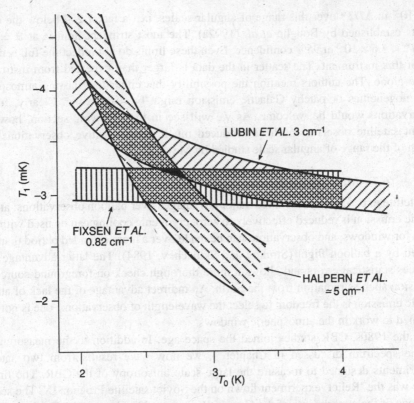

Fig. 6.5 If the dipole moment in the CBR results from the Doppler effect, T_1/T_0 is a constant. However, the value of T_1 derived from the measured anisotropy in antenna temperature, ΔT_A, will depend on both T_0 and the frequency at which the measurement is made. Values of T_1 derived from measurements of ΔT_A in three experiments (and their error bars) are displayed as a function of T_0. They agree only for a limited range of values of T_0, shown by the small, heavily shaded triangle (adapted from Lubin and Villela, 1985).

group, then T_1 can be readily calculated. Then eqn. (6.7) may be used to predict a value of ΔT_A; it turns out to be $\leq 1/2$ of the measured value for $\lambda \simeq 0.8$ mm. If the MIT–UBC results are correct, we must conclude that the submillimeter excess is either much smaller than claimed by Matsumoto *et al.* (1988) or local (perhaps Galactic) in origin. The former interpretation is strongly supported by the satellite measurements discussed in Chapter 4.

Meyer and his colleagues at MIT (Meyer, Cheng and Page, 1991) have continued bolometric observations of the CBR from balloon-borne platforms. Thus far, results from only one set of observations have been published, those made in 1989 and 1990 at $\lambda = 1.8$ mm, with a beam of 3.8° FWHM. Interestingly, the amplitude of the dipole moment was used as a calibration for the instrument. Consequently, of course, no independent measure of T_1 resulted. On the other hand, Meyer *et al.* were able to set very stringent limits on anisotropies on angular scales 3°–22°. For CBR fluctuations with a Gaussian spectrum, the 95% confidence upper limits range from 4×10^{-5} to about

2×10^{-5} in $\Delta T/T^*$ over this range of angular scales, i.e., a factor 2–4 below the upper limits established by Boughn *et al.* (1992a). The most stringent limit is at $\theta \simeq 13°$: $\Delta T/T^* < 1.6 \times 10^{-5}$ at 95% confidence. Even these limits do not reflect the full sensitivity of this instrument. The scatter in the data is larger than expected from instrument noise alone. The authors mention the possibility that either long-lasting atmospheric inhomogeneities or patchy Galactic emission might be responsible. Clearly, further observations would be welcome. As we will see in the following section, however, recent satellite observations have produced much more sensitive observations over much of the range of angular scale studied by Meyer *et al.*

6.7   Measurements from satellites

A satellite experiment offers three major advantages over balloon observations: atmospheric emission is reduced effectively to zero, cryogenic cooling can be used without a need for windows, and observations can be made over a more extended period than permitted by a balloon flight (Strukov and Skulachev, 1984). The latter advantage both reduces statistical errors and allows a more thorough check on foreground sources of emission such as radiation from the Earth. An indirect advantage of the lack of atmospheric emission is the freedom to select the wavelength of observations; one is not constrained to work in the atmospheric windows.

In the 1980s, CBR studies joined the space age. In addition to the measurements of the spectrum discussed in Chapter 4, we now have results from two satellite experiments designed to measure the large-scale anisotropy of the CBR. The first of these was the 'Relict' experiment flown on the Soviet satellite Prognoz IX. The second was one of the experiments of the Cosmic Background Explorer (COBE), described in Section 4.9.

6.7.1 Features of the Soviet 'Relict' experiment

A novel and important feature of the experiment of the Moscow group was the highly eccentric orbit of the spacecraft, which had an apogee distance of about 7×10^5 km (see fig. 6.9). The eccentric orbit ensured that the spacecraft spent most of its time far from the Earth, reducing terrestrial radiation into the side-lobes of the horn antennas employed. To reduce solar emission, the spin axis of the satellite was placed pointing away from the Sun. Two horn antennas were used to produce differential measurements of the sky brightness at $\lambda = 8$ mm. The main horn–reflector system, with a beam of about 6° width, was mounted perpendicular to the spin axis of the satellite. It thus scanned a great circle in the sky every 2 min as the spacecraft rotated. A smaller horn, with a beam aligned along the spin axis and pointed away from the Sun, provided a reference temperature. During the course of the 7 month flight in 1983 and 1984, the satellite accompanied the Earth roughly halfway around its orbit of the Sun. Thus the satellite–Sun line shifted by $\geq 180°$, allowing the main horn to survey most of the celestial sphere.

Microwave emission from the Galactic plane was clearly visible at the operating wavelength of 8 mm; to eliminate systematic error introduced by this anisotropic emission, the Soviet group dropped all measurements made within ±15° of the Galactic

---

* Here and in the following chapters, $\Delta T/T$ is the observed (or calculated) limit on CBR fluctuations expressed in thermodynamic temperature, divided by $T_0 = 2.73$ K.

plane. They do not appear to have corrected the remaining measurements at higher Galactic latitude for residual emission, which fig. 4.10 shows could be present.

Because the satellite experiment did not employ ground screens, we must consider radiation into the side- and back-lobes of the antennas employed. The side lobes of the main antenna are displayed in fig. 4.5 (the side-lobe level reaches as much as 10^{-4} of the on-axis response because a reflector was used). At 90° from the optical axis, however, the response was $\lesssim 3 \times 10^{-7}$ of the response on axis. The Sun would thus contribute a signal of about $3 \times 10^{-7} T_{\odot} (\Omega_{\odot}/\Omega_b)$, where $T_{\odot}$ is the surface temperature and $\Omega_{\odot}$ the solid angle of the Sun, respectively, and Ω_b is the solid angle of the main beam. With $\Omega_{\odot}/\Omega_b = (\theta_{\odot}/\theta_b)^2 \approx (0.5/6)^2 \approx 7 \times 10^{-3}$, and $T_{\odot} \approx 6000$ K, we find that the solar side-lobe signal is $\lesssim 0.013$ mK*.

Although the Moon's surface temperature is more than 20 times lower than the Sun's, lunar radiation introduced larger error signals into the Soviet results because the Moon was sometimes closer to the optical axis of the main horn. The preliminary results from the Relict experiment reported by Strukov and Skulachev (1984) were not fully corrected for side- and back-lobe pickup from the Moon. More recent reports (Klypin *et al.*, 1987; Strukov *et al.*, 1987) contain corrected and reanalyzed data, and it is these results that appear in table 6.2.

Once measurements within 15° of the Galactic plane and those contaminated by side-lobe pickup had been dropped, the sky coverage was about 70%. The data set was sufficiently homogeneous to allow Klypin *et al.* (1987) to search for higher order moments in the temperature of the CBR using the equivalent of eqn. (6.4). Because the coverage of the sky was incomplete, evaluation of the coefficients C_{nm} was not straightforward (see Section 6.2.3 and Klypin *et al.*, 1987, for brief descriptions of the methods used). None of the higher order moments was statistically significant, and Klypin *et al.* set 95% confidence upper limits on T_n/T_0 of $\lesssim 5 \times 10^{-5}$ for $3 \lesssim n \lesssim 10$, and $\lesssim 10^{-4}$ for $11 \lesssim n \lesssim 15$. When comparing these results with those obtained by other groups, it is important to note that the Soviet group report values for the quantity T_n given by eqn. (6.5),

$$T_n = \left(\frac{1}{4\pi} \sum_{m=-n}^{n} C_{nm}^2 \right)^{1/2}$$

rather than the coefficients C_{nm}. As noted in Section 6.1, other groups instead report or work with

$$\left(\sum_{m=-n}^{n} C_{nm}^2 \right)^{1/2}.$$

For comparison with the latter results or upper limits, the Soviet results must obviously be multiplied by $(4\pi)^{1/2}$. In table 6.2, the upper limits on the quadrupole moment are 2σ limits on T_2 itself as defined by eqn. (6.2b) or eqn. (6.5); this choice allows a more meaningful comparison with the dipole amplitude.

6.7.2 Features of the differential microwave radiometers on COBE

COBE, the first satellite devoted exclusively to studying cosmic microwave and infrared backgrounds, was launched by NASA on 18 November, 1989 (see Mather, 1982;

---

* Strukov *et al.* (1987) find an even lower value by taking the off-axis response to be $\leqslant 10^{-7}$.

Boggess *et al.*, 1992; and also Section 4.9 for a brief description of the spacecraft and its orbit and fig. 4.23 for an artist's conception). One of the three experiments aboard (the Differential Microwave Radiometer or DMR; see Smoot *et al.*, 1990) was specifically designed to search for anisotropies on all scales greater than its beam size of about 7°. As its name suggests, the instrument made differential measurements of the sky brightness at each of three wavelengths, 3.3, 5.7 and 9.5 mm, chosen to lie in the broad minimum of Galactic emission (fig. 4.10). Each of the three radiometers employed pairs of symmetrical horn antennas with optical axes separated by 60°, as shown in fig. 4.23. During the course of a year-long mission, the entire celestial sphere is scanned twice. The resulting maps (fig. 6.6) show the CBR dipole signal very clearly. Note that this signal swamps emission from the plane of the Galaxy in all the maps.

COBE's polar orbit was much closer to the Earth than the very eccentric orbit of the Relict spacecraft. Therefore, thermal emission from the Earth into the side lobes of the DMR antennas was a potentially greater problem. For this reason, COBE was equipped with ground screens to shield the instruments from the Earth, as shown in fig. 4.23. Data taken when the Earth was above or near the edge of the ground screens were dropped – see Kogut *et al.* (1992) for details. This paper also discusses the errors contributed by other foreground sources, such as Jupiter and the Moon, and by the effect of magnetic fields on the ferrite Dicke switches employed. The latter was generally one of the dominant sources of systematic error in the measurement of lower order multipoles of the CBR – but see Kogut *et al.* for a much fuller discussion. It should also be noted that the COBE data had to be and was corrected for the Doppler effect induced by the orbital velocity of the spacecraft.

We next need to consider the effect of Galactic radio and dust emission on the COBE maps. The COBE team (Bennett *et al.*, 1992) dealt with this problem by careful modeling of the Galactic foreground (see Section 6.4.2), and by using the DMR maps themselves. The availability of sensitive, complete, sky maps at three different frequencies and a map of slightly less sensitivity at $v = 19$ GHz (Boughn *et al.*, 1992b) allowed the COBE team to separate out CBR and Galactic signals on the basis of their quite different spectral properties. In addition, they dropped all data from Galactic latitude $< 10°$. Bennett *et al.* also show that the main scientific results of COBE do *not* depend strongly on the model assumed for Galactic emission or on the details of the scheme used to correct for it. Thus, despite the clear presence of the Galactic plane in the COBE maps (especially at $\lambda = 9$ mm; see fig. 6.7), cosmic anisotropy is robustly detected.

6.7.3 T_1 and T_2 from the DMR measurements

A preliminary value for the dipole moment of the CBR was reported by the COBE team in 1991 (Smoot *et al.*, 1991); $T_1 = 3.3 \pm 0.2$ mK. Note the large uncertainty in T_1 in these preliminary results. Further work on calibration of the instruments and on subtraction of the Galactic signal is needed before final values for T_1 are published. An analysis of one year's data (Smoot *et al.*, 1992) has refined T_1 to 3.36 ± 0.10 mK (see table 6.2). What guesses can we make about the eventual quality of the final results? To me, the crucial feature of the DMR experiment is not just its sensitivity but the control it offers over the major source of systematic error in searches for large-scale anisotropy, namely emission from our Galaxy. Thus I expect the *direction* of the CBR dipole given by COBE to be both more precise and more secure than previous values. On the other hand, I do not expect a major improvement in precision of the *amplitude* of the dipole moment. The DMR receivers are not particular-

Fig. 6.6   COBE maps of the microwave sky. The most prominent feature is the dipole component (peaked in the upper right quadrant of these maps in Galactic coordinates). Note the greater prominence of radio emission along the Galactic plane at the lowest frequency (from Smoot *et al.*, 1991, with permission). 'A' and 'B' are independent channels.

ly low noise (the technology was 'frozen' by NASA about a decade ago), and the in-flight calibration will not have the precision of the FIRAS calibrator. A factor of three improvement, to about 30 μK, however, should easily be achievable*.

Lest this conclusion seem too negative, it is worth remarking that we know T_1 to an accuracy of a few percent already. If the dipole anisotropy is induced by the motion of the Earth, as claimed in Chapter 8, we thus know the Earth's velocity to a few percent and its direction to an accuracy of a degree or so, comparable to the angle subtended by one's thumb held at arm's length.

Smoot *et al.* (1991) also report a preliminary upper limit on the quadrupole moment, $T_2 \leqslant 0.08$ mK. A more recent analysis by Smoot and his colleagues (1992) has revealed a non-zero quadrupole with a far smaller r.m.s. value: 13 ± 4 μK. We thus encounter for the first time a significant anisotropy in the CBR other than the familiar dipole term.

6.7.4   *Large-scale structure in the CBR detected by COBE*

COBE had one more surprise in store. Analysis of a full year of data (Smoot *et al.*, 1992) revealed apparent structure in the cosmic background on angular scales of about 7°–90°.

---

* *Note added in proof*: see Appendix C for the most recent COBE results. Kogut *et al.* (1993) report $T_1 = 3.365 \pm 0.027$ K.

The importance of this discovery cannot be overemphasized. Observational astronomers have been searching for fluctuations in the CBR since its discovery, pushing to lower and lower limits on $\Delta T/T$ over a wide range of angular scales. On the largest scales, such fluctuations have now been found. Quoting Smoot and his many colleagues:

> We observe structure in the COBE DMR maps with characteristic anisotropy of $\Delta T/T \approx$ 6×10^{-6}. The structure is larger and of a different character than known residual systematic errors and artifacts generated by the data analysis. We believe the structure is from the sky. A critical issue is whether the structure is due to Galactic or extragalactic emissions or is in the cosmic microwave background. The most economical hypothesis is to attribute the structure to the microwave background. Although we cannot rule out a heretofore undiscovered Galactic or extragalactic emission, the Galactic or discrete extragalactic origin of the measured anisotropy would require an unlikely confluence of factors. It would require the source of anisotropy to alias a thermal spectrum and, if Galactic, not have the spatial distribution associated with known components of the Galaxy. If due to discrete extragalactic sources, it would require substantially more intense sources than are now known.

The cautious tone of this paragraph nicely echoes the tone of the discovery paper by Penzias and Wilson (1965; also Appendix A).

In addition to determining the amplitude of the CBR fluctuations – roughly $\Delta T/T = 1.1 \times 10^{-5}$ smoothed to a $10°$ scale – the DMR results also established a *spectrum* for the fluctuations, that is, how ΔT depends on the correlation angle, θ. COBE found ΔT to depend only weakly on θ; in the notation used in Chapter 8, the spectral power law index is $n = 1.1 \pm 0.5$. Both the amplitude of $\Delta T/T$ and the derived value of n are of fundamental cosmological importance, as we will see in Chapter 8.

6.8 Observational programs planned for the near future

The group responsible for the Relict experiment is planning a second, multi-wavelength, experiment 'Relict-2' for launch in the 1990s (Space Research Institute, 1987). They will search for anisotropies on scales $\gtrsim 5°$, using an observing technique similar to that employed in their earlier flight.

Unlike both COBE and the earlier Relict package, Relict-2 will not orbit the Earth; instead it will follow an orbit around the outer libration point, L2, of the Earth–Sun system, at about 1.5×10^6 km from the Earth. As fig 6.8 suggests, this orbit ensures that the Earth, Sun and Moon – the major sources of possible radiation into side lobes of the antennas – are all bunched in one part of the sky, and hence can be better shielded.

To help control Galactic emission, observations will be made at five wavelengths in the range 1.6–14 mm. As the Soviet group noted, multi-wavelength observations of the dipole moment will also allow them to determine the spectrum of the CBR (Section 6.6.3).

Heterodyne receivers will be used at all five wavelengths. Those at $14 > \lambda > 3.5$ mm will be approximately a factor of two more sensitive than the corresponding DMR receivers on COBE. Unless systematic effects dominate the error budget, Relict-2 should be able to measure the lower order moments of the CBR distribution to about

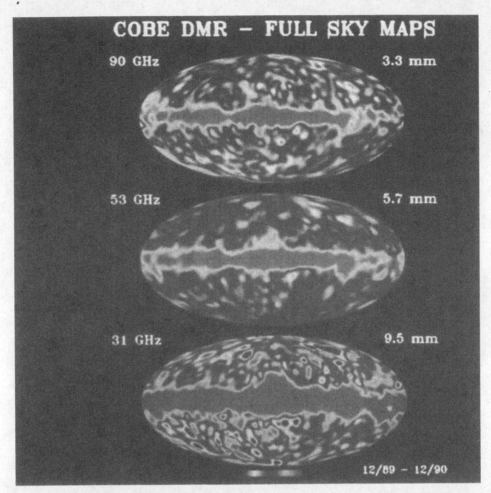

Fig. 6.7 The same COBE maps as in fig. 6.6 with the dipole signal removed, again in Galactic coordinates (from Smoot *et al.*, 1992, with permission). The non-uniformities in intensity in regions well away from the Galactic plane are the long-sought fluctuations in the CBR, with $\Delta T/T \sim 10^{-5}$.

3×10^{-6} in the course of a year, an improvement of a factor of three over the best measurements now available.

The European Space Agency also has a CBR satellite under consideration. It is designed to measure both the spectrum and the distribution of the CBR. As currently planned, the beam size would be about 2°, allowing this instrument to search for anisotropies on intermediate angular scales, a topic we defer to Chapter 7.

While observations from a satellite offer real advantages, they are also costly and take a long time to come to fruition. With the improvement in detector sensitivity – particularly the development of low-noise, cryogenically-cooled bolometers – it is now possible to equal the sensitivity of DMR in one or two nights' observations from a balloon package. Consequently, as we have seen, groups centered at MIT (S. Meyer and his col-

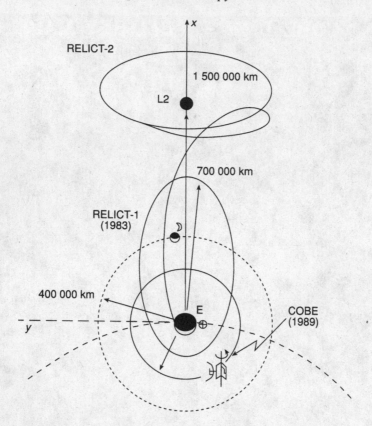

Fig. 6.8   Orbits (not to scale) of the Relict-1, COBE and Relict-2 satellite missions. Relict-2 is currently planned for launch in the mid-1990s. The lunar orbit is shown in short dashed lines.

leagues) and the University of California (Lubin, Richards *et al.*) are pushing balloon measurements forward.

The first results of these flights are just beginning to emerge (see forthcoming papers by Meyer, Cheng and Page, and by Lubin, Richards and Fischer). One example is the letter of Bernstein *et al.* (1989), which reports on *7 min* of data. The dipole was detected, but residual emission from the atmosphere introduced systematic errors into the results.

6.9   Large-scale polarization

As noted at the time of its discovery, the CBR is apparently not polarized (Penzias and Wilson, 1965; Appendix A). Nevertheless, since there are cosmological models and astrophysical processes that could produce polarization in the CBR, attempts have been made to measure or set limits on both circular and linear polarization of the radiation.

Since atmospheric emission is unpolarized, these differential experiments can be performed from the ground (see Lubin and Smoot, 1979; Nanos, 1979). However, Caderni and his colleagues (1978) have made far infrared polarization measurements from balloon altitudes.

Table 6.3 *Linear polarization of the CBR at* $\lambda = 9$ *mm (from Lubin, Melese and Smoot, 1983).*

	Linear* polarized signal (mK)		
Coefficient	Stokes parameter Q	Stokes parameter U	Error
T_0	−0.02	0.04	0.02
T_z	−0.04	0.07	0.03
T_x	−0.02	−0.03	0.04
T_y	−0.04	0.04	0.04
Q_1	−0.04	0.07	0.05
Q_2	−0.04	−0.04	0.03
Q_3	−0.06	−0.01	0.03
Q_4	0.06	0.03	0.04
Q_5	−0.01	0.12	0.04

*Limits on *circular* polarization are 20 mK on T_0 and 12 mK on $T_x \ldots Q_5$.

The ground-based instruments employed in searches for large-scale polarization are microwave radiometers with polarized inputs. Differential measurements are made not by beam switching but by changing the polarization of the input to the receiver. To control instrumental effects, the entire instrument is periodically rotated about its optical axis (generally the vertical); any instrumentally induced polarization signal will then appear with that period and can be subtracted.

The most recent and accurate sets of observations are those of Lubin, Melese and Smoot (1983) made at $\lambda = 9$ mm with a single vertical horn antenna. The results are given in table 6.3. The lower accuracy of the limits on circular polarization arises because the instrumental offsets are difficult to subtract and because of the difficulty in constructing a precise, circularly polarized, calibration source.

6.10 Summary*

The observations discussed in this chapter show clear and consistent evidence for a dipole component in the CBR temperature, with $T_1/T_0 = 1.2 \times 10^{-3}$. The dipole amplitude is essentially the same at all wavelengths from about 1 to 12 mm; this fact in turn can be used to show that the temperature of the CBR is 2.86 ± 0.26 K, in agreement with the results described in Chapter 4.

While the dipole moment is obvious, there was until 1992 no evidence for existence of any higher order moment in the CBR. Very recently, COBE has measured a quadrupole moment with $T_2/T_0 = (5-6) \times 10^{-6}$ (Smoot *et al.*, 1992). Thus the quadrupole moment is $\lesssim 0.5\%$ of the well-determined dipole moment. Limits on the octopole and higher order moments are roughly comparable in amplitude.

The DMR instrument on COBE also detected CBR fluctuations on a range of angular scales about 7°–90° and determined both their amplitude ($\Delta T/T = 1.1 \times 10^{-5}$ at 10°) and their angular spectrum, corresponding to $n = 1.1 \pm 0.5$.

The CBR is not polarized on a large angular scale. Limits on the dipole and quadrupole moments of linear polarization are $\lesssim 7 \times 10^{-5}$ of T_0 (Lubin, Melese and Smoot, 1983).

* See Appendix C for recent results.

References

Reports of the COBE Science Team are indicated below by '(COBE)' following the citation.

Bennett, C. L. *et al.*, 1992, *Ap. J. (Lett.)*, **396**, L7 (COBE).

Bernstein, G. M., Fischer, M. L., Richards, P. L., Peterson, J. B., and Timusk, T. 1989, *Ap. J. (Lett.)*, **337**, L1.

Boggess, N. W. *et al.*, 1992, *Ap. J.*, **397**, 420 (COBE).

Boughn, S. P., Fram, D., and Partridge, R. B. 1971, *Ap. J.*, **165**, 439.

Boughn, S. P., Cheng, E. S., and Wilkinson, D. T. 1981, *Ap. J. (Lett.)*, **243**, L113.

Boughn, S. P., Cheng, E. S., Cottingham, D. A., and Fixsen, D. J. 1990, *Rev. Sci. Instrum.*, **61**, 158.

Boughn, S. P., Cheng, E. S., Cottingham, D. A., and Fixsen, D. J. 1992a, *Ap. J. (Lett.)*, **391**, L49.

Boughn, S. P., Cheng, E. S., Cottingham, D. A., and Fixsen, D. J. 1992b, private communication.

Caderni, N., Fabbri, R., Melchiorri, B., Melchiorri, F., and Natale, V. 1978, *Phys. Rev.* D, **17**, 1908.

Cheng, E. S., Saulson, P. R., Wilkinson, D. T., and Corey, B. E. 1979, *Ap. J. (Lett.)*, **232**, L139.

Conklin, E. K. 1969, *Nature*, **222**, 971.

Corey, B. E. 1978, Ph.D. Thesis, Princeton University.

Corey, B. E., and Wilkinson, D. T. 1976, *Bull. Amer. Astron. Soc.*, **8**, 351.

Fabbri, R., Guidi, I., Melchiorri, F., and Natale, V. 1980, *Phys. Rev. Lett.*, **44**, 1563.

Fixsen, D. J., Cheng, E. S., and Wilkinson, D. T. 1983, *Phys. Rev. Lett.*, **50**, 620.

Halpern, M. 1983, Ph.D. Thesis, M.I.T.

Halpern, M., Benford, R., Meyer, S., Muehlner, D., and Weiss, R. 1988, *Ap. J.*, **332**, 596.

Haslam, C. G. T., Klein, U., Salter, C. J., Stoffel, H., Wilson, W. E., Cleary, M. N., Cooke, D. J., and Thomasson, P. 1981, *Astron. Astrophys.*, **100**, 209, and references therein.

Hauser, M. G. *et al.* 1991, in *After the First Three Minutes*, eds. S. S. Holt, C. L. Bennett and V. Trimble, American Institute of Physics, New York (COBE).

Henry, P. S. 1971, *Nature*, **231**, 561.

Klypin, A. A., Sazhin, M. V., Strukov, I. A., and Skulachev, D. P. 1987, *Soviet Astron. Lett.*, **13**, 104.

Kogut, A. *et al.* 1992, *Ap. J.*, **401**, 1 (COBE).

Low, F. J. *et al.* 1984, *Ap. J. (Lett.)*, **278**, L19.

Lubin, P. M., and Smoot, G. F. 1979, *Phys. Rev. Lett.*, **42**, 129.

Lubin, P. M., and Villela, T. 1985, in *The Cosmic Background Radiation and Fundamental Physics*, ed. F. Melchiorri, Editrice Compositori, Bologna.

Lubin, P. M., Epstein, G. L., and Smoot, G. F., 1983, *Phys. Rev. Lett.*, **50**, 616.

Lubin, P., Melese, P., and Smoot, G. 1983, *Ap. J. (Lett.)*, **273**, L51.

Lubin, P., Villela, T., Epstein, G., and Smoot, G. 1985, *Ap. J. (Lett.)*, **298**, L1.

Mather, J. C. 1982, *Optical Engineering*, **21**, 769.

Matsumoto, T., Hayakawa, S., Matsuo, H., Murakami, H., Sato, S., Lange, A. E., and Richards, P. L. 1988, *Ap. J.*, **329**, 567.

Melchiorri, F., Melchiorri, B., Ceccarelli, C., and Pietranera, L. 1981, *Ap. J. (Lett.)*, **250**, L1.

Meyer, S. S., Cheng, E. S., and Page, L. A. 1991, *Ap. J. (Lett.)*, **371**, L7.

Muehlner, D., and Weiss, R. 1976 in *Infrared and Submillimeter Astronomy*, ed. G. Fazio, D. Reidel, Dordrecht.

Nanos, G. P. 1979, *Ap. J.*, **232**, 341.

Partridge, R. B., and Wilkinson, D. T. 1967, *Phys. Rev. Lett.*, **18**, 557.

Penzias, A. A. and Wilson, R. W. 1965, *Ap. J.*, **142**, 419.

Reich, P., and Reich, W. 1986, *Astron. Astrophys. Suppl.*, **63**, 205, and references therein.

Scaramella, R., and Vittorio, N. 1990, *Ap. J.*, **353**, 372.

Smoot, G. F., and Lubin, P. M. 1979, *Ap. J. (Lett.)*, **234**, L83.

Smoot, G. F., Gorenstein, M. V., and Muller, R. A. 1977, *Phys. Rev. Lett.*, **39**, 898.

Smoot, G. F. *et al.* 1990, *Ap. J.*, **360**, 685 (COBE).

Smoot, G. F. *et al.* 1991, *Ap. J. (Lett.)*, **371**, L1 (COBE).

Smoot, G. F. *et al.* 1992, *Ap. J. (Lett.)*, **396**, L1 (COBE).

Space Research Institute of the Soviet Academy, 1987, *Relict-2*, a booklet published in
    Moscow.
Strukov, I. A., and Skulachev, D. P. 1984, *Soviet Astron. Lett.* **10**, 1.
Strukov, I. A., Skulachev, D. P., Boyarskii, M. N., and Tkachev, A. N. 1987, *Soviet Astron.
    Lett.*, **13**, 65.
Weiss, R. 1980, *Annual Rev. Astron. Astrophys.*, **18**, 489.
Witebsky, C. 1978, COBE Report No. 5013, unpublished.
Wright, E. L. *et al.* 1991, *Ap. J.*, **381**, 200 (COBE).

7

Searches for anisotropy in the CBR on small angular scales

Soon after the discovery of the CBR it was recognized that measurements of, or upper limits on, its anisotropy on scales of degrees or less would provide unique information about the origin and development of structure within the Universe. Such observations are of particular value because they probe cosmic times well before the appearance of any luminous objects in the Universe such as galaxies or stars. The surface of last scattering from which the CBR photons reach us is more distant than any QSO or galaxy yet detected (these lie at $z < 5$), and the CBR thus carries information about the state of the Universe at earlier times. It is quite likely, in fact, that the CBR photons we study reach us from the first few hundred thousand years of the history of the Universe; and, as we shall see in Chapter 8, the CBR may encode information from the epoch of inflation nearly 50 orders of magnitude earlier still.

With so much to learn, astronomers have worked hard to detect fluctuations in the temperature of the CBR. Indeed, there have been as many searches for structure in the background on scales $\leqslant 10°$ as measurements of the dipole and of the spectrum combined. Despite 25 years of effort by many individuals and groups, however, no anisotropy in the CBR on any scale except the dipole had been definitively and unambiguously detected until 1992, when the results of the COBE satellite were announced (Smoot *et al.*, 1992). The constraints on cosmology and on theories of galaxy formation imposed by the COBE results and by the increasingly stringent upper limits on $\Delta T/T*$ on smaller angular scales will be explored in Chapter 8; here we focus on the experimental techniques and the results of the most sensitive searches for fluctuations in the CBR on scales $\lesssim 10°$.

7.1 An outline of observational techniques

We will be concerned here with measurements on angular scales of several arcseconds to several tens of degrees; that is, a range of angular scale approaching 10^5. The wide range of angular scale makes the use of three different observational techniques necessary. For scales of a degree or more, specially designed instruments quite similar to those described in Chapters 4 and 6 have been employed. At intermediate scales, particularly $1' \leqslant \theta \leqslant 10'$, large, conventional radio antennas (such as the ones shown in fig. 3.2) are suitable. In both these cases, beam switching as described in Sections 3.3.3 or 6.2 has been used to make differential measurements between neighboring patches of

---

* Recall that I give $\Delta T/T$ in thermodynamic temperature with $T \equiv T_0 = 2.73$ K.

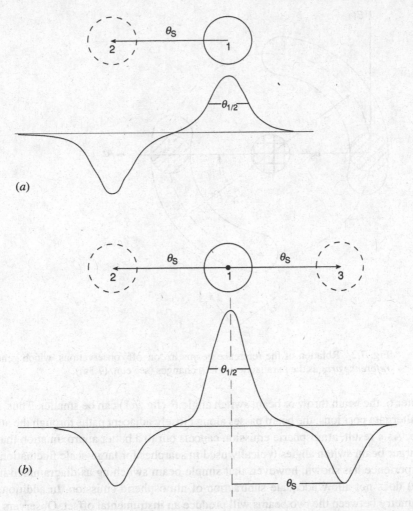

Fig. 7.1   (*a*) Beam pattern resulting from beam switching through an angle θ_s. The source is centered at location 1; 2 is the reference position. (*b*) Beam pattern resulting from 'on–off' observations; again the source is at 1, and 2 and 3 are reference positions.

the sky. Thus the greater sensitivity of differential measurements may be realized. To probe angles below an arcminute, that is smaller than the resolving power of most conventional antennas, requires a new technique: aperture synthesis (described in Section 3.5 and more fully in Verschuur and Kellermann, 1988, Chapter 10). Aperture synthesis, as we shall see, requires the use of relatively narrow band heterodyne receivers (typically $\Delta\nu < 100$ MHz for instruments now available). At larger angular scales, both heterodyne and bolometric receivers have been successfully employed.

Unlike the searches for large-scale anisotropy described in Chapter 6, most attempts to measure fine-scale anisotropy have been conducted from the ground – an obvious necessity if conventional radio telescopes are employed. What allows us to work beneath the Earth's atmosphere when making measurements on these smaller scales? Since the angular scales of interest are much smaller than those considered in

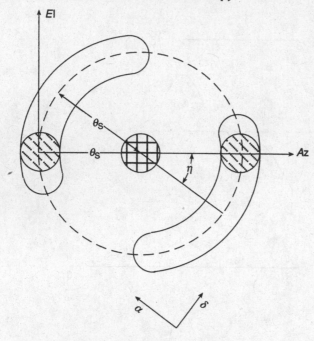

Fig. 7.2   Rotation of the reference beams in 'on–off' observations, which generates *reference arcs* as the parallactic angle η changes (see eqn. (7.3a)).

Chapter 6, the beam throw or beam switch angle θ_s (fig. 7.1) can be smaller. Thus, in its two alternate positions, the beam passes along closely adjacent paths through the atmosphere. As a result, atmospheric emission cancels out to a better approximation than for the larger beam switch angles typically used in searches for large-scale fluctuations.

Experience has shown, however, that simple beam switching as diagrammed in fig. 7.1(*a*) does not allow adequate subtraction of atmospheric emission. In addition, any asymmetry between the two beams will produce an instrumental offset. Observers have therefore devised more complicated schemes, which effectively allow subtraction of atmospheric emission to successively higher orders. The first is a standard technique of radio astronomy called, descriptively, 'on–off' observing (fig. 7.1(*b*)). In 'on–off' observing, the region of the sky under study is placed alternately in one beam or the other, usually by moving the entire telescope or instrument slightly. If this small motion, like the beam switching itself, is made in azimuth only, emission from the Earth's atmosphere will be subtracted out to high order. This technique will also subtract to first order any systematic offset between the two beams employed (see Section 7.2.2 below).

For observations not conducted at one of the Earth's poles, beam switching purely in azimuth means that the two reference positions will rotate on the plane of the sky as the hour angle of the region under study changes. Hence the reference beam will trace out arcs on the plane of the sky as the observations go on, as shown schematically in fig. 7.2.

Searches for fluctuations in the CBR have generally consisted of measurements of the antenna temperature of a set of such patterns in the sky. We then record only the temperature *difference* between the central region and the more extended reference arcs on either side of it to make our differential measurement. Note also that two angular scales

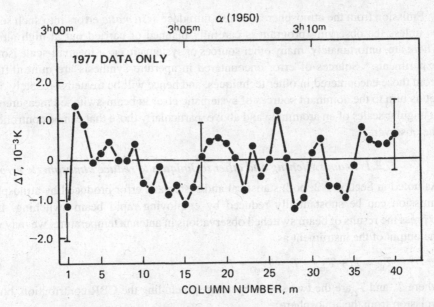

Fig. 7.3   Results of an early search for CBR fluctuations (Partridge, 1980). Clearly there are no fluctuations from one point on the sky to another exceeding $\sim 1.5 \times 10^{-3}$ K, so $\Delta T/T \lesssim 5 \times 10^{-4}$. Given the error bars on the individual measurements, however, do these data provide evidence for fluctuations greater than those expected from instrumental noise alone? Using the techniques of Section 7.4, we find that the answer is 'no,' in this case.

are involved; the FWHM of the beam employed, $\theta_{1/2}$, and the beam switch angle, θ_s.

Once measurements of a set of sky regions have been made, various statistical tests are applied to the results to see whether there is real variation in the temperature of the microwave sky from one point to another. Since each of the individual measurements has statistical error associated with it, estimating the amplitude of possible fluctuations in the CBR temperature is not a trivial task. As an instance, consider fig. 7.3, the results of a search for fluctuations made more than a decade ago. Given the size of the 1σ error bars shown for each point, is there evidence for significant point-to-point fluctuation? In this case, the answer provided by the statistical tests discussed in Section 7.4 was negative.

This illustration also underlines the importance of a careful treatment of error in such measurements. We turn to that subject next.

7.2   Local sources of statistical and systematic error (and some ways to beat them)

None of the ground-based attempts to measure small-scale fluctuations in the CBR has yet succeeded, thus error swamps any anisotropy signal, and we clearly need to look carefully at the sources of error in such observations. The two major sources of statistical error in searches for fine-scale anisotropy are, as for the observations described in Chapter 6, detector noise and fluctuations in the Earth's atmospheric emission. In several of the observational programs to be described below, receiver noise has dominated the error budget. Some more recent observations, especially those at wavelengths below 3 cm, have run into the barrier of atmospheric noise.

Emission from the atmosphere can also introduce *systematic* errors into such searches unless the observing program is carefully planned or carried out at high altitude. There are, unfortunately, many other sources of systematic error in small-scale isotropy experiments.* Sources of error encountered in aperture synthesis are quite different from those encountered in other techniques, and hence will be treated separately. Here, let us turn to the dominant sources of systematic error in beam-switched measurements at angular scales of an arcminute and above, particularly those that can be controlled by the observer.

7.2.1 Beam switching and other techniques to reduce atmospheric error

As noted in Section 6.2, both statistical and systematic error produced by atmospheric emission can be substantially reduced by employing rapid beam switching. If we express the results of beam switched observations in antenna temperature, we may write the output of the instrument as

$$T_{BS} = T_1 + T_{atm1} - T_2 - T_{atm2} + T_{offset},$$ (7.1)

where T_1 and T_2 are the two sky temperatures including the CBR contribution, but not emission from the atmosphere.

As eqn. (7.1) shows, atmospheric emission is at least partially subtracted by beam switching, providing the beam switch is in azimuth, so that the zenith angle of the two beam paths is the same. Unfortunately, the Earth's atmosphere is not perfectly uniform and laminar, so that the two atmospheric contributions, $T_{atm\,1}$ and $T_{atm\,2}$, will not cancel exactly. Water vapor in particular is clumped and poorly mixed in the atmosphere. As a consequence, there is a randomly varying atmospheric component, which adds to the statistical error of such measurements.

The offset term in eqn. (7.1) can arise from a variety of causes. One is differential ground pickup, to be discussed below. Another is a small difference in efficiency of the instrument between the two beam switched positions. For some large radio telescopes, T_{offset} can be as large as several kelvin, and may vary with the azimuth and elevation of the telescope.

'On–off' observing offers a real improvement on both fronts. Taking the difference between 'on' and 'off' readings removes to first order the beam-switch offset, as well as improving the subtraction of atmospheric emission. Expressed in antenna temperature again, the resulting signals will be

$$(T_{BS})_{on} = T_1 + T_{atm\,1} - T_2 - T_{atm\,2} + T_{offset},$$

$$(T_{BS})_{off} = T_3 + T_{atm\,3} - T_1 - T_{atm\,1} + T_{offset}.$$

We thus find,

$$\delta T_{on-off} \equiv (T_{BS})_{on} - (T_{BS})_{off} = 2T_1 - T_2 - T_3 + 2T_{atm\,1} - T_{atm\,2} - T_{atm\,3}.$$

Assuming that the atmospheric emission cancels on average if all three beams are at the same elevation, we have simply

---

* The distinction between systematic and statistical error is hard to make when we are considering attempts to measure fluctuations in a quantity, as we are here. I will use 'systematic' to refer to errors that can increase the point-to-point scatter in plots such as fig. 7.3, and 'statistical' to refer to effects that increase the size of the individual error bars on the points in such a plot.

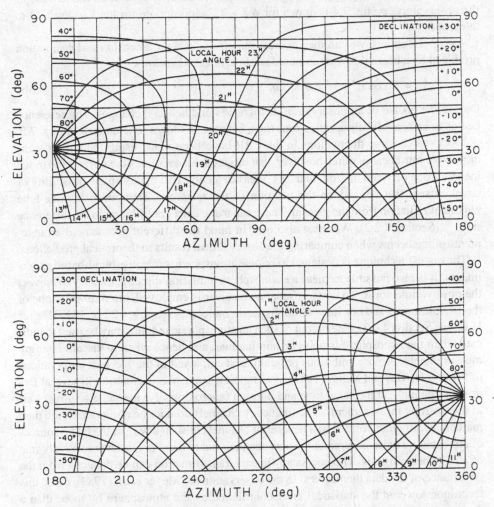

Fig. 7.4   As eqn. (7.3) shows, the connection between (Az, El) and (α, δ) depends on the latitude at which the observations are made. This figure is appropriate for $\phi = +32°$, the latitude of Kitt Peak, Arizona.

$$\delta T_{on-off} = 2T_1 - T_2 - T_3. \tag{7.2}$$

As noted above, the two reference positions (2 and 3 in fig. 7.1(*b*)) rotate about the central position (1) as the hour angle of the region under study changes, thus tracing arcs in the plane of the sky. The location and length of the refernce arcs can be deduced from plots of right ascension and declination in the azimuth–elevation plane, as shown in fig. 7.4, or in terms of the parallactic angle η, defined by

$$\eta = \tan^{-1}\left(\frac{\sin H / \cos \delta}{\tan \phi - \tan \delta \cos H}\right). \tag{7.3a}$$

In (7.3a), δ is the declination of the source, and H its hour angle, equal to $t - \alpha$, where t is the sidereal time of the observation and α is the right ascension of the source. Note that both the relations graphed in fig. 7.4 and the value of η depend on the latitude of the observatory, ϕ; fig. 7.4 is drawn for $\phi = + 32°$, appropriate for the Kitt Peak 12 m telescope.

As fig. 7.2 shows, we can then express the offsets in right ascension and declination produced by a beam throw of amplitude θ_s in azimuth as follows:

$$\Delta\alpha = \theta_s \cos \eta, \quad \Delta\delta = \theta_s \sin \eta. \tag{7.3b}$$

During the course of an observation, H increases and hence η changes. The reference beams in an 'on–off' observation thus trace out arcs as shown schematically in fig. 7.2.

Most searches for fluctuations in the CBR have been made using the on–off technique. We must bear in mind, however, that this technique requires the observer to move the telescope between the 'on' and 'off' measurements. This opens up the possibility of small offsets that depend on the position of the telescope, introduced by side-lobe pickup of radiation from the ground. These can make a net contribution to the difference $\delta T_{\text{on–off}}$ (Section 7.2.2). We must also bear in mind the different geometrical arrangements of the beams when comparing the observational results to theoretical predictions.

The on–off technique is designed to *reduce* atmospheric error in ground-based experiments. It is also possible to measure atmospheric emission directly, then try to *correct* the observations for it. To do so, a separate receiver is employed, often tuned to one of the troublesome bands or lines of atmospheric molecules (e.g., $\nu = 22$ or 183 GHz for H_2O; see Section 4.4.3). The output of such an atmospheric monitor may be used to indicate when the atmosphere is noisy or non-laminar, and hence when to discard isotropy measurements. A more ambitious program, first employed in the 1–3 mm observations of Meyer, Jeffries and Weiss (1983), is to monitor atmospheric emission in several frequency channels, at the same time and through (approximately) the same column of air as the isotropy measurements. The frequency channels used in the observations are then mathematically transformed to a new set of channels in such a way as to isolate sources of atmospheric noise in some of the new channels but to leave at least one relatively free of atmospheric signals. The necessary (linear) transformation can be deduced from the covariance of the data themselves. In the observations of Meyer *et al.* (1983), using this technique lowered the statistical error contributed by the atmosphere by more than a factor of five; the effect on possible systematic error is not stated, but is presumably comparable. Multi-frequency observations were also used to subtract atmospheric fluctuations from the Soviet observations described in Section 7.5 (Kaidanovski *et al.*, 1982; Pariiskii and Korol'kov, 1987).

Finally, it is worth making the obvious point that atmospheric error can be minimized by making the observations from high, dry and cold sites. If we use a conventional radio telescope, we have no choice in its location, but telescopes designed to work at $\lambda \lesssim 3$ cm are usually located in high, dry places to minimize atmospheric emission in any case. The smaller instruments constructed to measure $\Delta T/T$ on about 0.5°–10° scales may be moved at will. Davies *et al.* (1987), for instance, chose a dry mountain top; Mandolesi *et al.* (1986) and Timbie and Wilkinson (1990) worked at cold, dry sites in the Arctic. There are already tentative indications from unpublished work by Dragovan, Lubin and Peterson that the Amundsen–Scott Station at the South Pole may provide the best combination of altitude (about 3000 m), and low temperature and humidity for such observations (see Meinhold and Lubin, 1991). It is already known that the column density of water vapor above the South Pole can be as low as 0.3–0.5 mm.

Atmospheric emission can be bypassed almost entirely by making observations from high flying aircraft or balloons. These platforms are particularly compatible with searches at short wavelengths ($\lambda \lesssim 3$ mm); first, high altitude observations are *needed* because the atmosphere is more emissive, and second they are *possible* becuase the size of antenna required to achieve a given resolution scales with wavelength. High altitude $\Delta T/T$ measurements were pioneered by the Italian group now centered in Rome (Fabbri *et al.*, 1980; Melchiorri *et al.*, 1981; de Bernardis *et al.*, 1992). Ambitious programs to measure $\Delta T/T$ on degree scales from stabilized balloon platforms are also underway by Lubin and his Santa Barbara colleagues (see Meinhold and Lubin, 1991), by the MIT group (e.g., Meyer *et al.*, 1991), and by a Berkeley group (e.g., Fischer *et al.*, 1992; Alsop *et al.*, 1992). There is, of course, a price to be paid for reducing atmospheric error: the observing time is reduced to $\lesssim 1$ day for a balloon flight or a few hours for an aircraft. Hence very low noise receivers are required.

7.2.2 Side-lobe pickup of emission from the ground

In our description of on–off observations, we have implicitly assumed that the radio telescope employed is tracking – that is, that it is moving to cancel out the rotation of the Earth – as the observations are made. Thus the telescope moves relative to its surroundings as each pair of observations is made. This in turn raises the possibility that radiation from the ground into the side lobes of the antenna will vary in the course of the observations, hence contributing a signal to $\delta T_{\text{on–off}}$. Side- and back-lobe pickup of radiation from the ground, telescope support structure, etc. is a particularly severe problem when large, conventional radio telescopes are employed in searches for CBR fluctuations. A variety of techniques are used to make this problem manageable.

First, to reduce the side lobe amplitude, the full geometrical aperture of the primary reflecting surface of the telescope is not used. Instead, the small feed horns employed at the prime focus or Cassegrain focus are designed to have reduced response to radiation striking the outer zone of the primary reflector. This technique, similar to apodizing in optical instruments, is called tapered or graded illumination (see Section 3.3.3 or Rohlfs, 1986). Since the full geometrical aperture of the reflector is not employed, tapering causes some increase in the width $\theta_{1/2}$ of the main beam. Most large radio telescopes are designed with some taper to make the best compromise between side-lobe pickup and loss of resolution. In addition, it is frequently the practice to illuminate the primary reflector in such a way as to give an essentially Gaussian shape to the beam pattern $P(\theta, \phi)$ of the antenna.

Even a carefully designed and constructed telescope may end up with back- and side-lobe pickup of several kelvin, very roughly two orders of magnitude worse than the most carefully designed corrugated horn antennas. In addition, conventional radio telescopes are too large to shield using reflecting surfaces or ground shields (Section 4.4.1).

To first order, however, beam switching cancels out ground pickup in the side lobes. Careful adjustment to symmetrize the beam switch pattern can reduce the *difference* in side-lobe pickup to a few tenths of a kelvin. As noted above, making on–off measurements helps still further.

Motion of the telescope and consequent time variation in side-lobe pickup may be avoided altogether by keeping the telescope fixed, and allowing the sky to drift through the beam. As time goes on, the telescope will thus scan a small strip at fixed declination, and any anisotropies in the CBR will produce a time varying output from the receiver. These observations are called *drift scans*. The drift speed due to the rotation of the earth is $0.25 \cos \delta$ arcmin s^{-1}, where δ is the declination of the strip observed; hence such observations are generally made at high declinations to increase the integration time on each point observed. A variant of this technique was introduced by Partridge (1980), who made repeated drift scans over a small strip of the sky by moving the telescope back to the starting point of the region under consideration after each 20 min drift scan. Thus many scans of the same small strip could be made in a single night of observing. In searches for fine-scale anisotropy, the drift scan technique has generally been combined with on–off measurements.

A problem with the use of drift scans arises in the analysis of the observations: adjacent measurements are not statistically independent because of the finite size of the telescope beam. Thus some deconvolution is required – see Partridge (1980) or Lasenby (1981) for details.

7.2.3 *Local sources of systematic error: summary*

The antenna temperature of the Earth's atmosphere varies from roughly 1 to 10 K in the microwave region, depending on the altitude of the observatory and the wavelength used. Variations of $\geqslant 0.1$ K are certainly possible during the course of an observation. In the case of ground-based observations, we must then rely on beam switching and the on–off technique to reduce this figure by a further three orders of magnitude or more to reach the level of sensitivity desired in searches for CBR fluctuations.

Searches for anisotropies in the CBR are made with equipment at about 300 K located on or near the surface of the Earth also at about 300 K. The temperature of the equipment or the ground can easily vary by 10 K during an observation. Thus we expect about 3% swings in any side- or back-lobe pickup from the ground. As noted above, careful design of the telescope, careful illumination, and careful symmetrizing of the two beams can reduce the difference in side-lobe pickup to a few tenths of a kelvin in beam switched operation; thus variations in side-lobe pickup during the course of an observing run might typically be of order 0.01 K. We rely on use of the on–off technique (or drift scans) to reduce this number by a further factor of 100.

If nothing else, I hope that the last few pages will give the reader some sense of the difficulties involved in searches for small-scale anisotropies in the CBR. It never hurts to remember that we are looking for temperature differences of the order of 0.1 mK with instruments at 300 K, often through a turbulent atmosphere contributing something like 10 K to the telescope input.

7.2.4 *Statistical error*

In the best and most recent searches for CBR fluctuations on arcminute and degree scales, the systematic errors described above have in fact been reduced to the 0.1 mK level. The final barrier is statistical error, dominated by receiver noise and atmospheric noise. If we take 20 K and 1 GHz as representative figures for the best system temperature and bandwidth achievable with heterodyne receivers (Section 3.4), we discover that the minimum detectable temperature in 1000 s of observation is 0.02 mK. If beam switched observations are employed, so that only half the observing time is spent on the the region of interest, the sensitivity is reduced by a further factor of two. To reach $\Delta T/T \simeq 10^{-5}$ in, say, ten independent regions of the sky would require many hours of observation even with the best receiver and even if atmospheric noise were negligible.

Will temporal fluctuations in the Earth's atmospheric emission in fact allow us to push our integrating times to 10^4–10^5 s for ground-based experiments? The power spectrum of atmospheric fluctuations is not as well understood as the absolute temperature of the atmosphere. We expect most of the temporal variations to be due to fluctuations in the density of H_2O, since oxygen is well mixed in the atmosphere. There is some evidence that atmospheric fluctuations agree with a Kolmogorov turbulence model, at least for the scales of meters to tens of kilometers of interest to us (Tatarski, 1961; Dravskikh and Finkelstein, 1979). Thus the power spectrum of atmospheric fluctuations P^2 is proportional to d^α, where d is the linear separation and the index α is expected to be about 5/3 for $d \lesssim h$, the thickness of the atmosphere, and about 2/3 for $d > h$. This dependence on d explains why switching the beam through a smaller angle more effectively cancels out atmospheric fluctuations; for $d \lesssim h$, the r.m.s. noise is proportional to $d^{5/6}$.

We are also interested in the time variation of atmospheric emission. If we combine the Kolmogorov model with a simple model for the atmosphere in which the wind velocity is constant with altitude (Armstrong and Sramek, 1982), we may, following Smoot *et al.* (1986), write

$$\left.\begin{array}{l} P^2(t) \propto t^{8/3} \quad \text{for } d < h \ \text{ or } t < h/\bar{v} \\ P^2 \propto t^{5/3} \quad \text{for } t > h/\bar{v} \end{array}\right\}, \tag{7.4}$$

where $\bar{v}$ is the mean wind velocity. The r.m.s. fluctuation in H_2O column density then may be found from (7.4) by dividing by t and taking the square root:

$$\left.\begin{array}{l} \sigma(H_2O) \propto t^{5/6} \quad \text{for } t < h/\bar{v} \\ \sigma(H_2O) \propto t^{1/3} \quad \text{for } t > h/\bar{v} \end{array}\right\}. \tag{7.5}$$

Since $h \simeq 5$ km and $\bar{v} \lesssim 50$ m s^{-1}, the steeper time dependence is expected for beam switch frequencies $\gtrsim 0.01$ Hz. Observational results summarized in Smoot *et al.* (1986) show that the exponent of t in relation (7.5) is about 0.7 ± 0.2, in reasonable agreement with theory, for t up to several thousand seconds.

The constant of proportionality in (7.5) is very poorly determined, in part because the few available measurements have all been made at different sites. It appears to be about 5×10^{-2} mm precipitable, for $t = 1$ s, and is probably well below this at high altitude. We will adopt this figure to perform an illustrative calculation for observations at $\lambda = 3$ mm. From fig. 4.11, the H_2O attenuation from a high altitude site at $\lambda = 3$ mm is about $10^{-2.1}$ per millimeter, corresponding to an antenna temperature of 2.0 K mm^{-1} from H_2O. We thus expect observed temperature fluctuations of $\Delta T = (2) (5 \times 10^{-2})t^{5/6}$ or less. If the beam switch frequency is 5 Hz, the r.m.s. noise will be $\lesssim 26$ mK, as much as an order of magnitude greater than receiver noise for the best receivers with an integrating time of 0.2 s. The need for high frequency beam switching is apparent.

While this result is discouraging, there is some tentative evidence from observations made in the Arctic (Timbie and Wilkinson, 1990) and in the Antarctic (Lubin, private communication, see also Meinhold and Lubin, 1991) that at times of stable weather, the atmospheric fluctuations in the 0.3–1 cm wavelength range can be below 0.1 mK when beam-switched observations are averaged for times of the order of an hour. Further work on atmospheric fluctuations, particularly at the best sites, is needed. In particular, it would be useful to know how well atmospheric noise in the atmospheric windows can be monitored by measuring fluctuations at other frequencies centered on particular spectral lines and bands of atmospheric gases (see, e.g., Kaidanovski *et al.*, 1982).

7.3   Fluctuations from astronomical foreground sources

To some degree, the errors introduced by atmospheric emission and by ground pickup can be controlled by the observer. That is not the case for astronomical signals. As in the case of searches for large-scale anisotropies (Section 6.4), we can minimize their effect only by avoiding strong sources and by an appropriate choice of frequency. Since searches for small-scale CBR fluctuations can be made in any representative patch of the sky, obvious sources like the Sun and Moon can be avoided. Most such observations have been made at high Galactic latitudes to minimize Galactic radio emission. What we cannot avoid, at some level, is the signal from extragalactic radio sources; this fundamental limit to the sensitivity of searches for CBR fluctuations is treated in Section 7.3.2.

7.3.1   Fluctuations in Galactic radio and dust emission

First, however, let us ask what we know about *Galactic* emission, especially at high latitudes. The answer is, not nearly enough. The basic emission mechanisms are well understood (Section 3.7). The antenna temperature of Galactic emission near the Galactic pole has been measured at some microwave frequencies to about 10% accuracy; what we lack is knowledge of the small-scale structure or 'patchiness' of Galactic emission. The problem is that surveys of Galactic emission either are made with low angular resolution (e.g., the survey of Boughn *et al.*, 1992, discussed in Section 6.4, or the COBE maps, fig. 6.6), or are restricted to the Galactic plane.

Furthermore, all these surveys except COBE's have been made at wavelengths longer than most recent CBR anisotropy searches, and hence the results must be extrapolated (a risky process given the uncertainty in the spectral index of Galactic emission).

Given the absence of direct observational evidence on Galactic 'noise' at 0.01°–1° scales, our only recourse is to fall back on a model. Since this topic has received little attention elsewhere until very recently (e.g., Banday and Wolfendale, 1991; Masi *et al.*, 1991; Bennett *et al.*, 1992), I will present an illustrative model here.

Let us begin by estimating the large-scale average surface brightness of the Galaxy near the North Galactic Pole (but in regions away from the North Galactic Spur and discrete extragalactic sources). Some of the relevant measurements are given in table 7.1; see also Lawson *et al.* (1987). The total surface brightness measurements in column 2 include both the CBR brightness and a contribution from extragalactic sources. For the former, I take $T_0 = 2.73$ K; for the latter, I scale from the 178 MHz observations of Bridle (1967), which gave $T_0 + T_{extragal} = 30 \pm 7$ K. The scaling relation for the correction is thus

Table 7.1 *The surface brightness of the Galaxy near the North Galactic Pole. Values in the third column were calculated from the maps of Haslam et al. (1982) and Reich (1982) as explained in the text.*

Frequency (GHz)	Total surface brightness (K)	Surface brightness corrected for sources and CBR (K)		Reference
		Calculated	Measured	
0.408	20 ± 2	14.6 ± 2.2		Haslam *et al.* (1982)
1.42	3.35 ± 0.1	0.52 ± 0.11		Reich (1982)
2.5			0.13	Sironi and Bonelli (1986)
25			$(2.0 \pm 0.7) \times 10^{-4}$	Fixsen (1982)
90			$(4.1 \pm 1.1) \times 10^{-5}$	Lubin and Vilella (1985)

$$T(v) = [(30 \pm 7) - 2.73]\left(\frac{v}{178}\right)^n + 2.73$$

and I take $n \approx -2.8$ for typical *extragalactic* sources. The errors in the corrected brightnesses on lines 1 and 2 include a ± 0.1 uncertainty in the index n. As expected, the first four values in columns 3 and 4 of table 7.1 are in reasonable agreement with a spectral index of about 2.7 for *Galactic* emission near the pole. At $v \gtrsim 30$ GHz, say, the spectrum apparently flattens, as bremsstrahlung and possibly reemission from warm dust begin to contribute (see fig. 4.10).

With these rough figures for the *average* surface brightness in hand, let us now turn to the question of *fluctuations* in the brightness. Such fluctuations can arise in two ways: variations in the amount of emitting material or variations in the spectral index along different lines of sight. To model the former, I use the 1.42 GHz map of Reich (1982). Near the pole, the *variations* in measured brightness temperature at his resolution of about $0.6°$ are typically about 0.05 K or about 10% of the estimated 0.52 K Galactic surface brightness at the pole – see fig. 7.5. At 408 MHz (Haslam *et al.*, 1982), the fluctuation level also appears to be about 10%. To set an upper limit on the noise that Galactic emission can introduce into CBR measurements, I assume that *all* the variation in Reich's map is due to fluctuations in Galactic emission, not instrument noise, emission from background radio sources, etc. I thus adopt 10% as a conservative upper limit on spatial variations in Galactic emission on about 1° scales. Next, I assume variations in the spectral index as large as ± 0.2 (Lawson *et al.*, 1987). The results are shown in fig. 7.6; note that spectral index fluctuations could be the dominant source of Galactic noise in this model for frequencies in the 10–100 GHz range.

It is worth noting that the conservative assumptions I have made are likely to result in *overestimated* Galactic fluctuations at about 1° scale. There is some evidence that the Galactic foreground is in fact smoother than my model predicts. Pariiskii and Korol'kov (1987), for instance, report patchy emission with $\Delta T \approx 1$ mK on scales of 1°–2° at $\lambda = 7.6$ cm; at that wavelength, my model predicts ΔT closer to 2–3 mK. In addition, although the 3 cm observations of Davies *et al.* (1987) described in Section 7.7 were at a larger angular scale, the sky fluctuation level they detect falls below my model by a factor of 2–3 as well. Detailed models of Galactic emission for the particular region

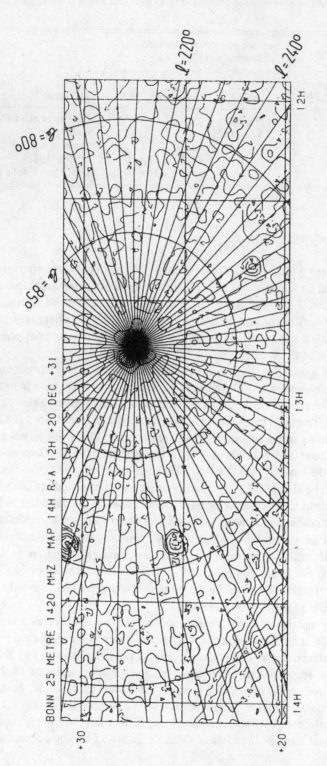

Fig. 7.5   Portion of the 1420 MHz sky map of Reich (1982, with permission). The region near the North Galactic Pole, b = 90°, is shown. The contour intervals are 50 mK in brightness temperature.

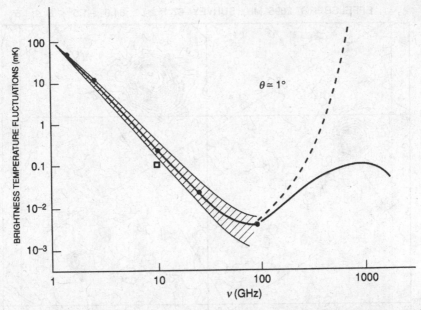

Fig. 7.6   Estimated level of fluctuations produced by patchy Galactic emission at high Galactic latitudes on a scale of ~1° (see text for description of the model). The solid line at low frequencies ($\nu < 100$ GHz) is based on the measurements of Galactic radio emission given in Table 7.1, and assumes 10% point-to-point fluctuations in intensity at 1° scale. The square is the value of ΔT measured by Davies *et al.* (1987) at 8° scale. The shaded region includes uncertainties of ± 0.2 in the synchrotron spectral index.

At high frequency ($\nu > 100$ GHz), fluctuations in dust emission dominate. The values of $\Delta T/T$ resulting from the model of Section 7.3.1 are shown in both antenna temperature (solid curve) and thermodynamic temperature (dashed curve). Note – the dust emission assumes $\Delta B = B_0$ on 1° scales; the scaling relation of Gautier *et al.* (1992) would suggest a value of $\Delta B \sim 16$ times lower.

observed by Davies *et al.* have been made by Banday *et al.* (1991); their models also predict values of $\Delta T/T \approx 1/2$ of those in fig. 7.6.

Finally, let us ask how the predictions of my model compare with the very recent results (Section 6.7.4) from the DMR instrument on NASA's COBE satellite (which, we must recall, had a beam of 7° FWHM). On the 1° scale adopted for the calculations here, we expect antenna temperature fluctuations of about 10^{-2} mK or less at the DMR frequencies, corresponding to values of $\Delta T/T$ a few times 10^{-6}. These results are consistent with both the DMR measurements and the model for Galactic emission presented in outline form by Bennett *et al.* (1992).

To model fluctuations in Galactic emission on smaller scales, we must extrapolate in solid angle as well as frequency. To get a *rough* idea of the level of optical depth variations, I use the only adequately sensitive, high resolution survey available, the 2.7 GHz survey of Reich *et al.* (1984) with a resolution $\theta_{1/2} \simeq 4.3'$. Unfortunately, it covers only a narrow strip along the Galactic equator. It is surely safe to assume that variations in emission far from the equator are smaller than those near the equator. Hence the $\lesssim 50$ mK variations in quiet regions of the 2.7 GHz map shown in fig. 7.7 are very conservative upper limits on variations on a 4.3' scale at higher Galactic latitudes. The

EFFELSBERG 2695 MHz SURVEY 67.1° ≥ L ≥ 64.9°, −1.5° ≤ B ≤ 1.5°

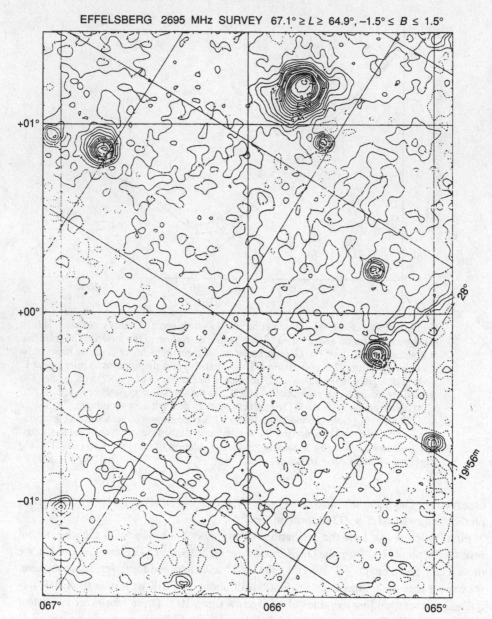

Fig. 7.7   One section of the 2.7 GHz map of Reich *et al.* (1984). Unfortunately, this survey was restricted to the Galactic plane. The lowest contours are at +0.05 K (solid) and −0.05 K (dotted). We are concerned with point-to-point variations, not the zero level of the map, about which some questions have been raised.

roughly 50 mK variations in the plane represent about 25% fluctuations in surface brightness of the faintest regions in the Reich *et al.* (1984) map. Thus I adopt a figure of 25% for variations in optical depth to construct the upper curve in fig. 7.8. Fluctuations induced by ± 0.2 variations in spectral index *n* are of smaller amplitude than 25% variations.

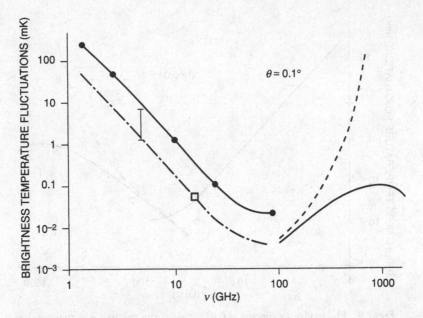

Fig. 7.8 Fluctuations expected at 0.1° scales from patchy Galactic emission at high Galactic latitudes. At low frequencies ($v < 100$ GHz), the solid line assumes 25% point-to-point fluctuations; the dot–dash curve is a more reasonable estimate determined by the observational upper limits determined at 4.85 GHz by Condon *et al.* (1989) and Readhead *et al.* (1989) at 15 GHz. At $v > 100$ GHz, ΔT from 100% fluctuations in dust emission is shown in antenna temperature (solid curve) and thermodynamic temperature (dashed curve). The scaling relation of Gautier *et al.* (1992) would suggest a value ~50 times lower.

The recent 4.85 GHz survey of Condon *et al.* (1989) allows us to check these predictions. Near the Galactic pole, they report an r.m.s. noise of about 6 mJy ≈ 7.2 mK, the largest component of which is receiver noise. Thus 7.2 mK is certainly a conservative upper limit on Galactic noise on the 0.06° scale of their observations (our curve gives about 4 mK at the same frequency). Over most of the area surveyed, however, Condon *et al.* report a confusion noise level of about 1 mJy ≈ 1.2 mK r.m.s. If we adopt this figure as an upper limit on fluctuations in Galactic emission, the lower curve shown in fig. 7.8 results. Evidence that the estimated upper limits from the 2.7 GHz map are too pessimistic is also provided by the upper limit of Readhead *et al.* (1989) discussed in Section 7.6 below – it again falls a factor of three *below* the upper curve at 15 GHz. Finally, we saw earlier in this section that region-to-region variations in surface brightness are ≲ 10% at 1° scale, and we will soon see that the variations are ≲ 2.5% at 0.01° scale. It would be surprising, therefore, if the region-to-region variations were as large as 25% at an intermediate scale of 0.1°, as inferred from the 2.7 GHz map of Reich *et al.* (1984). Hence I have added a more realistic estimate of Galactic fluctuations to fig. 7.8 (the dot–dashed line), scaled to $\Delta T = 1.2$ mK at 4.85 GHz.

On still smaller scales, say 0.01°, the only useful constraint comes from deep VLA maps at 4.86 GHz (a byproduct of the work discussed in Section 7.8 below). On about 0.01° (36") scales, no fluctuations larger than about 0.5 mK are seen. The large-scale average brightness temperature at 4.86 GHz may be interpolated from table 7.1; it is

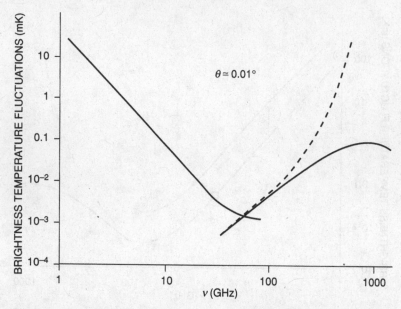

Fig. 7.9   Fluctuations expected at 0.01° scales from patchy Galactic emission at high Galactic latitudes. At low frequencies ($v < 100$ GHz), the calibration of the curve is determined by the observational upper limits of Martin and Partridge (1988) at 4.86 GHz. At $v > 100$ GHz, $\Delta T/T$ from 100% fluctuations in dust emission is shown; the scaling relation of Gautier *et al.* (1992) would suggest a value ~160 times lower.

about 20 mK. Thus no variations in optical depth $\gtrsim 2.5\%$ are seen at 4.86 GHz. This figure is used to scale fig. 7.9 for $v \lesssim 30$ GHz.

At frequencies $\gtrsim 100$ GHz, as we have noted several times, thermal emission from warm (about 20 K) dust becomes more important than synchrotron or bremsstrahlung emission. To estimate Galactic noise at high frequencies therefore requires a knowledge of the variations in dust emission on 0.01°–1° scales. Once again, we lack direct measurements at relevant wavelengths. As noted in Section 6.4, maps made at $\lambda = 100$ μm ($v = 3000$ GHz) by the IRAS satellite reveal patchy emission all over the sky, and apparently on all scales down to its angular resolution of about 0.1° (Low *et al.*, 1984). Near the Galactic pole, this patchy emission, called the infrared cirrus, has an absolute intensity of $(2–5) \times 10^6$ Jy per steradian. The lower figure is from a model of Puget (1987) or the fits of Boulanger and Perault (1988), and the upper figure includes a possible isotropic component. The discovery (Hauser *et al.*, 1991) of a calibration offset in the IRAS measurements makes the lower value more probable. The rocket-borne photometer of Matsumoto *et al.* (1988) also measured the far infrared background at 100 μm, but at a lower Galactic latitude of about 33°. Expressed as a surface brightness, their result is $B_0 = 4 \times 10^6$ Jy sr^{-1}. In the following illustrative calculations, I will adopt 4×10^6 Jy sr^{-1} for the surface brightness at the pole, which corresponds to about 1.5×10^{-5} K in antenna temperature (*not* thermodynamic temperature) at $\lambda = 100$ μm.

Since the infrared cirrus appears to be patchy on all scales down to about 0.1°, one could conservatively assume that 100% brightness fluctuations in dust emission are possible, so that $\Delta B = B_0$ and $\Delta T = 1.5 \times 10^{-5}$ K in antenna temperature at $\lambda = 100$ mm. Banday and Wolfendale (1991) find that fluctuations $\Delta B/B_0$ as large as 35% may indeed by present on scales of several degrees. A recent paper by Gautier *et al.* (1992), however, shows that the spectral power of noise in deep IRAS observations may be fitted to within a factor of two by the scaling law

$$P(\theta) = 1.4 \times 10^{-12} \left(\frac{\theta}{100} \right)^{\alpha} B_0^3$$

with θ in arcminutes and α close to 3.0; $P(\theta)$ is the spectral power in Jy^2 sr^{-1}. The angle dependence appears to hold over the range $8° > \theta > 2'$. If we adopt this result with $B_0 = 4 \times 10^6$ Jy sr^{-1}, we find $\Delta B \simeq 2.5 \times 10^5$ and 8×10^4 Jy sr^{-1} at $\theta = 1°$ and $0.1°$, respectively. Extrapolating to $0.01°$ yields $\Delta B \simeq 2.5 \times 10^4$ Jy sr^{-1}. Thus the fractional fluctuation in dust emission given by this scaling law is about 6%, 2% and 0.6% at $\theta = 1°$, $0.1°$ and $0.01°$, respectively. These values are much smaller than the 35% figure assumed by Banday and Wolfendale (1991), but the value of ΔB calculated at a $1°$ scale agrees with the estimates of ΔB made by Masi *et al.* (1991) at a $0.5°$ scale. I have chosen, however, to adopt the conservative assumption that $\Delta B = B_0 = 4$ Jy sr^{-1} in plotting the high frequency portions of the curves in fig. 7.6, fig. 7.8 and fig. 7.9. If the scaling law given above is found to be correct, the curves must be scaled downwards by factors of about 16, 50 and 160, respectively.

We need next to extrapolate these results to lower frequency. Following e.g., Matsumoto *et al.* (1988), I assume that the cool Galactic dust responsible for the cirrus has a typical temperature of 20 K and emissivity proportional to v^2. Assuming this spectrum allows us to scale the results obtained at 100 μm; the results appear in fig. 7.6, fig. 7.8 and fig. 7.9. A lower value of the emissivity index would result in larger values of ΔT at $v < 10^3$ GHz. For instance, if the emissivity were proportional to $v^{1.5}$, ΔT at 100 GHz would be about five times larger than shown. Such a large increase in ΔT is probably ruled out by COBE spectral measurements (see Wright *et al.*, 1991).

Finally, let me compare the results of fig. 7.6 with somewhat similar calcualtions performed by Banday and Wolfendale (1991). My model (fig. 7.6) predicts very similar values of ΔT at 30–200 GHz, and a value of $\Delta T \simeq 50\%$ lower than theirs at 10 GHz. Given the upper limits of Davies *et al.* (1987) at 10 GHz, my model seems more reasonable. Finally, note that Banday and Wolfendale present their results for $\Delta T/T$ in *antenna* temperature, which is related to $\Delta T/T$ in thermodynamic temperature by

$$\frac{\Delta T_A}{T_A} = \frac{xe^x}{e^x - 1} \left(\frac{\Delta T}{T} \right).$$

From fig. 7.6, fig. 7.8 and fig. 7.9, it is clear that a broad range of frequencies from about 30 to 200 GHz is available for searches for fluctuations in the CBR at a level of about $10^{-5} T_0$. The reader should also bear in mind that the model for Galactic emission on which fig. 7.6, fig. 7.8 and fig. 7.9 were based was an intentionally conservative one. In particular, fluctuations in the brightness of Galactic dust emission may be much smaller than suggested by these figures. If so, the available 'window' would extend to about 300 GHz. Some of the sensitive searches for CBR fluctuations to be described later in this chapter have already pushed below the limits shown in these figures.

Finally, it is worth remarking that the model I have developed above is generic in the sense that it applies to any region at high Galactic latitude. More specific and more precise models for Galactic emission can be constructed for a particular region of inter-

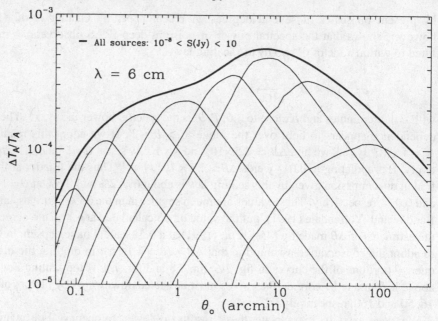

Fig. 7.10   Fluctuations resulting from foreground extragalactic radio sources (from Franceschini *et al.*, 1989, with permission). At $\lambda = 6$ cm, $\Delta T/T \gtrsim 10^{-4}$ for all angular scales from 1′ to 2°. The eight lower (lighter) curves indicate the contribution to $\Delta T/T$ from sources in eight decades of flux. Starting at the left is the curve for sources from 0.1–1.0 μJy; ending at the right is the curve for sources with $S = 1$–10 Jy.

est by using radio and IRAS maps of that region. Such projects were undertaken for the region studied by Melchiorri *et al.* (1981) and by Banday *et al.* (1991) for the $\delta = 40°$ strip studied by Davies *et al.* (1987; see Section 7.7.1); see also Masi *et al.* (1991).

7.3.2 Fluctuations from extragalactic sources

In contrast to the uncertainty in the foreground Galactic noise, the noise introduced by extragalactic radio sources is quite well understood, thanks largely to the work of the Padua group (e.g., Danese *et al.*, 1983; Franceschini *et al.*, 1989).

As Longair and Sunyaev (1969) first pointed out, the emission of radio sources introduces fluctuations in brightness into the microwave sky. As they showed, this unavoidable noise may drown out the CBR fluctuations, especially at low frequencies. The amplitude of this noise (expressed, for instance, as a fraction of T_0) will depend on both the frequency and the angular scale of the observations. To construct a figure such as fig. 7.10 showing $\Delta T/T$ as a function of scale, we need only the counts of radio sources at the frequency of interest (see Section 3.9). As shown in fig. 7.10, radio sources with different values of flux density S are dominant sources of noise at different angular scales. A very rough argument suggests why: temperature fluctuations will be a maximum when there is approximately one source per beam solid angle, i.e., when the number of sources is $N \simeq \theta^{-2}$ per steradian; but that number is related to the flux density by $N \propto S^{-\gamma}$, where $\gamma = 3/2$ for randomly distributed sources in Euclidean geometry (see Condon, 1988; or Section 3.9). There are only two complications to consider. First, the source counts at the frequency of interest may not extend to sufficiently low values of S.

For instance, at 5 GHz, direct source counts (Donnelly *et al.*, 1987; Fomalont *et al.*, 1988, 1991) peter out at $S \simeq 10$–20 μJy. Thus the estimate of $\Delta T/T$ variations on scales $\lesssim 0.5'$ depends on *extrapolation* of the source counts to lower S. At 5 GHz, only a modest extrapolation is needed. The other complication is possible clustering of radio sources. So far, we have implicitly assumed that the extragalactic sources are randomly distributed (and fig. 7.10 from Franceschini *et al.*, 1989, is based on the same assumption). If sources are *not* randomly distributed, larger fluctuations on a given scale will result. Indeed, Franceschini *et al.* (1989) use the upper limits set on fluctuations in the microwave sky by CBR measurements to establish limits on the clumping of radio sources.

Aside from these two complications, the noise contributed by extragalactic radio sources can be estimated with acceptable accuracy (say 20–30% uncertainty) for any frequency at which radio source counts have been made. Unfortunately, direct source counts are available only at $v \lesssim 5$ GHz, because of the decrease of S with increasing v (Section 3.8.3). One exception is a low sensitivity survey at 10 GHz by the Nobeyama group (Aizu *et al.*, 1987); source counts at 8.4 GHz will also soon become available (Windhorst *et al.*, 1993). To construct curves similar to fig. 7.10 for higher frequencies thus requires an extrapolation of the source counts by factors of 2–20 in frequency. Such extrapolations in turn require a knowledge of the mix of steep and flat spectrum sources, and become increasingly uncertain as we move to higher frequency and to smaller angular scale (that is, to smaller S). As the detailed calculations of Franceschini *et al.* (1989) show, however, the amplitude of the noise caused by sources also drops rapdily as v increases, so that the uncertainty in the results matters less. In fig. 7.11 are reproduced their calculations for $v = 33$ GHz; note that the equivalent $\Delta T/T$ produced by extragalactic radio sources is always $\lesssim 10^{-5}$.

Calculations for many frequencies may be combined to produce contour plots of the r.m.s. noise due to extragalactic radio sources, as shown in fig. 7.12. Note the general decrease in $\Delta T/T$ as the wavelength drops to about 0.5 cm. The sharp rise at millimeter wavelengths is due to far infrared emission from cool dust in external galaxies (like the emission from our Galaxy discussed in Section 7.3.1). The numbers and flux densities of dusty galaxies are not well known, and hence the calculations of $\Delta T/T$ for $v \gtrsim 100$ GHz must be regarded as very uncertain.

The calculations of Franceschini *et al.* (1989) take no account of another potential source of noise introduced by foreground sources: the Sunyaev–Zel'dovich (1972) effect in clusters of galaxies. This is inverse-Compton scattering of the CBR photons by hot gas in clusters of galaxies (see Section 3.7.3). For frequencies below 220 GHz, the effect produces a small *reduction* in temperature of the CBR in the solid angle subtended by a cluster; hence 'cool spots' are produced on the sky. Searches for this effect as an astrophysically interesting *signal* will be described in Section 7.10; here I mention it as a source of foreground *noise*. Estimates of the amplitude of the noise generated by the Sunyaev–Zel'dovich effect are strongly model dependent. Calculations for a simple illustrative model were published in 1981 by Rephaeli; for a specific rather extreme model by Korolev *et al.* (1986); and for a set of models by Cole and Kaiser (1988), Schaeffer and Silk (1988) and Bond *et al.* (1991). Among the common features of the results are: the maximum amplitude of the fluctuations is about $(3–5) \times 10^{-5}$ in $\Delta T/T$, is independent of frequency for $v \lesssim 100$ GHz, and occurs at an angular scale of about $1'$.

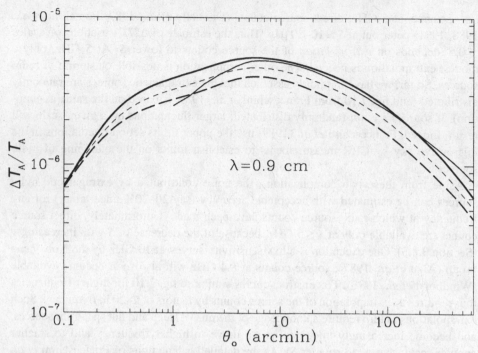

Fig. 7.11   Expected fluctuations at $\lambda = 0.9$ cm, expressed in antenna temperature, T_A, from foreground radio sources (Franceschini *et al.*, 1989). The different curves are based on different assumptions about the mean spectral index of compact radio sources at high frequencies. In no case, however, does $\Delta T_A / T_A$ exceed 6×10^{-6}.

It is interesting to ask if searches for fluctuations in the CBR have accidentally turned up evidence for the $\Delta T < 0$ 'dips' expected from the Sunyaev–Zel'dovich effect. Unpublished work by P. Hartnett at Haverford has shown that the available observations can effectively rule out the extreme model of Korolev *et al.* (1986), since that model predicts negative fluctuations in the CBR larger than those seen.

7.3.3 *Foreground sources of noise: summary*

The various sources of noise in the microwave sky that we have considered can contribute fluctuations as large as about $10^{-2} T_0$. There is a broad minimum in their amplitude at about 30–100 GHz; unfortunately, this is a difficult frequency range to work in from the surface of the Earth because of the many atmospheric bands and lines.

To date, foreground noise from radio sources has had a pronounced effect on only one class of CBR measurements: the 6 cm interferometric searches for fluctuations described in Section 7.8 below. Other searches have been conducted at shorter wavelength, where the effect of radio sources is reduced by $\lambda^{2.7}$. Until significant fluctuations are detected in the microwave sky at small angular scales, we may thus continue the past practice of ignoring foreground astronomical noise in assigning upper limits to possible CBR fluctuations. (That is, we ascribe *all* the observed fluctuations to the CBR, without correcting for foreground sources.)

In the future, as searches for CBR fluctuations are pushed to higher sensitivity, it will be necessary to correct for the effect of foreground sources. One approach is to map the

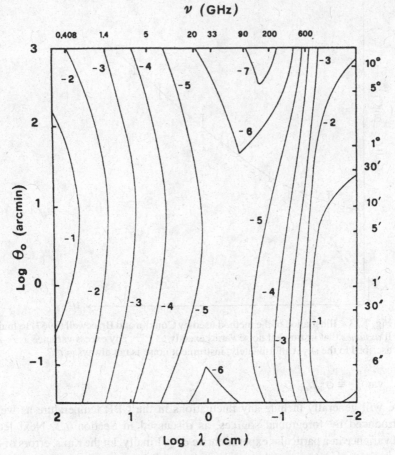

Fig. 7.12 Contour plots of the equivalent value of $\Delta T_A/T_A$ (antenna temperature) from randomly distributed foreground radio sources as a function of the angular scale and wavelength of observation (from Franceschini *et al.*, 1989, with permission).

same region of the sky at two or more lower frequencies, locate discrete radio sources, extrapolate the source flux densities to the frequency where the CBR observations were made, and finally subtract these fluxes from the data. Some variant of this approach has been used in the observational programs described in Sections 7.6, 7.7.1 and 7.8 below.

7.4 Statistical analysis of the observations (co-author: S. P. Boughn)

We turn now to the question of extracting values for, or upper limits on, CBR fluctuations from a set of measurements affected by statistical and/or systematic error. Again, use the data in fig. 7.3 as an example: given the 1σ errors associated with each measurement, is there additional, statistically significant, point-to-point variation present? If not, what upper limit may we place on point-to-point variations on the sky?

Let us begin by introducing some terminology. The quantity we seek to determine (or set limits on) is the true point-to-point variation in brightness temperature on the sky: let that variance be

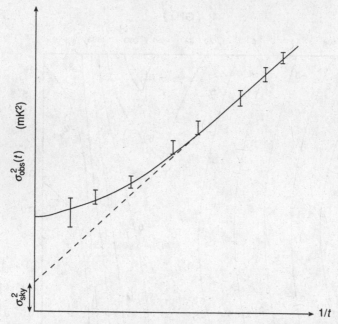

Fig. 7.13   Illustration of the method used by Conklin and Bracewell (1967) to find σ_{sky}. It *assumes* that instrument noise varies exactly as $t^{-1/2}$; any excess variance as $t \to \infty$ is ascribed to the sky. Unfortunately, instrument noise is not always $\propto t^{-1/2}$.

$$(var)_{sky} \equiv \sigma_{sky}^2.$$

This term will generally include any fluctuations in the CBR temperature as well as noise introduced by foreground sources, as discussed in Section 7.3. Next let the observed variance in a particular experiment be σ_{obs}^2. Finally, let the r.m.s. errors of each of the n individual data points be σ_i; the σ_i may all be approximately the same (as happens to be true of the data shown in fig. 7.3), but need not be.

7.4.1 *Model-dependent estimate of CBR fluctuations in the presence of noise*

The first method to extract an estimate of σ_{sky} from a set of observations was introduced by Conklin and Bracewell in 1967. It was based on the restrictive (and generally incorrect) assumption that the system noise of the radio telescope they employed scaled precisely as the inverse square root of the observing time: $\sigma_{sys} \propto t^{-1/2}$. As we saw in Sections 3.4.1 and 7.2, that assumption is unrealistic if, for instance, gain variations or atmospheric noise are present. If we were to adopt that assumption, then the observed variance could be written as

$$\sigma_{obs}^2(t) = \sigma_{sky}^2 + \frac{\sigma_{sys}^2(t=1)}{t},$$

where t is measured in seconds and $\sigma_{sys}(t=1)$ is the system noise in 1 s. If σ_{obs}^2 is plotted versus t^{-1}, the intercept at $t^{-1} = 0$ is σ_{sky}^2 (see fig. 7.13). In practice, for reasons noted above, σ_{obs}^2 often falls more gradually as t increases, as shown schematically in fig. 7.13. As a consequence, applying this method to real data would produce a misestimate of the possible amplitude of σ_{sky}, i.e., an artificially large or small upper limit on σ_{sky}.

7.4.2 *The Neyman–Pearson lemma and likelihood ratio tests*

In 1973, a more sophisticated and less model-dependent method was introduced, due largely to J. Deeter (see Boynton and Partridge, 1973). It is based on the Neyman–Pearson lemma of binary decision theory. In this method, there is no need to assume that system noise decreases inversely with integration time, but this method (and all other methods) does assume that the statistical properties of the system noise are completely characterized. That is, as with all statistical tests, the data are combined according to a prescription, usually model dependent, to form a single real number, referred to as *the statistic*. The statistic dealt with in the Neyman–Pearson lemma is the *likelihood ratio*, λ, i.e. the quotient of two *likelihood functions*, which we will now define.

Suppose that we wish to compare two alternative hypotheses: H_1, the hypothesis that there are absolutely no fluctuations in the microwave background, and H_2, that fluctuations exist with amplitude σ_{sky}. The probability density of a given set of measurements $\{\Delta T_i\}$ of the temperature fluctuations of the microwave background conditional on H_1, i.e. a universe with no fluctuations, is designated as $P(\{\Delta T_i\}|0)$, the *likehood function*. If the statistical properties of the system noise are completely characterized, then this function is straightforward to express. For example, if the data set $\{\Delta T_i\}$ is derived from a set of i statistically independent Gaussian processes, then the likelihood function is the multi-variate Gaussian distribution (see, for instance, Whalen, 1971, or another text on statistics).

$$P(\{\Delta T_i\} \mid 0) = \prod_i (2\pi\sigma_i^2)^{-1/2} \exp\left(\frac{-\Delta T_i^2}{2\sigma_i^2}\right),$$ (7.6)

where σ_i is the standard deviation of the ith measurement.

Similarly the likelihood function $P(\{\Delta T_i\} \mid \sigma_{sky})$ is the probability density of the given data set conditional on both system noise and the given model of fluctuations of the microwave background (H_2). Again this function is usually straightforward to write down (yet sometimes extremely difficult to evaluate). For the simple example of uncorrelated Gaussian fluctuations in the CBR with standard deviation σ_{sky}, the likelihood function is

$$P(\{\Delta T_i\} \mid \sigma_{sky}) = \prod_i [2\pi(\sigma_i^2 + \sigma_{sky}^2)]^{-1/2} \exp\left(\frac{-\Delta T_i^2}{2(\sigma_i^2 + \sigma_{sky}^2)}\right).$$ (7.7)

The likelihood ratio, λ, is then defined as the ratio of these two likelihood functions,

$$\lambda \equiv P(\{\Delta T_i\} \mid \sigma_{sky})/P(\{\Delta T_i\} \mid 0).$$ (7.8)

Next we evaluate λ for a given data set and then compare it with some predetermined value $\lambda^\star$. If $\lambda \geqslant \lambda^\star$ then the hypothesis H_2 is accepted. On the other hand if $\lambda < \lambda^\star$ then the hypothesis H_1 is accepted. In binary decision theory, this comparison is called a *test*.

Since the measurements are statistical in nature there exists the possibility of drawing an erroneous conclusion. The probability of rejecting hypothesis H_2 when it is true (type I error) is called the *size of the test*, α. The confidence level or efficiency of detection (i.e. the probability of accepting hypothesis H_2 when it is true) is $1 - \alpha$. The *power of the*

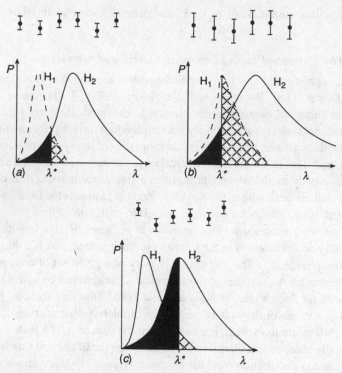

Fig. 7.14   Distribution of the likelihood ratio λ under the two hypotheses discussed in Section 7.4.2. The dark region under the curve for H_2 is the size of the test, α. The hatched area under H_1 is $1 - \beta$, where β is the power of the test. Samples of data that could produce such distributions are shown schematically. Both (*a*) and (*b*) represent situations where there is no obvious detection of fluctuations. In case (*b*), the power of the test is low, as happened in the case of the observations of Uson and Wilkinson (1984) discussed in Section 7.4.2. In both cases, σ_{sky} has been chosen so that the size of the test is 0.05 – this value of σ_{sky} is then taken as the 95% confidence upper limit on sky fluctuations. Part (*c*) represents a case where sky fluctuations are clearly detected; the size of the test is ~0.5 (Section 7.4.3), and the power of the test is large.

test, β, is one minus the probability of accepting hypothesis H_2 when hypothesis H_1 is in fact true (type II error). A large power of the test means that fluctuations have been robustly detected. In order to compute the probabilities of these two types of errors one must know the probability density function of the likelihood ratio conditional on the two hypotheses, H_1 and H_2. In principle, these can be expressed by analytic functions but it is usually the case that they are derived by Monte Carlo methods.

According to the Neyman–Pearson lemma, the likelihood ratio test is the *most powerful* test, since for any specified size of the test, the likelihood ratio maximizes the power of the test. It is this property that has enticed many of us to use the likelihood ratio to set upper limits on the amplitude of fluctuations of the microwave background.

If there is no significant detection of fluctuations, a 95% confidence level upper limit can be set by finding the amplitude of fluctuations σ_{sky} for which the size of the test is 0.05, as shown by the dark area in fig. 7.14. This means that there is a 5% chance that the fluctuations could be this large and yet result in a likelihood ratio as small as that observed.

We should remember that the likelihood ratio test is a binary test that is designed to distinguish between two alternative hypotheses neither of which, in the case of CBR observations, is likely to be completely true. That is, the amplitude of the fluctuations in the background including radio source confusion is certainly not zero; nor is it likely to be exactly some single value σ_{sky}. This does not mean that the above analysis is in error. The definition of the 95% confidence level remains a precise statistical statement. Under certain conditions, as Readhead *et al.* (1989) have shown, likelihood ratio tests may be used for composite hypotheses as well, e.g., $\sigma_{sky} < A$ versus $\sigma_{sky} \geq A$.

A more serious concern is one that is common to all methods of setting limits that make use of a statistic. Suppose that the data of a particular experiment result in a statistic, λ, which is somewhat smaller than the expected value (i.e. the most likely value in a statistical sense). This situation could be due to random statistical scatter; e.g. one out of twenty experiments will yield a value of λ in the 5% extremity of the distribution of λ values. Similarly, an overestimate of instrument noise by the experiment can easily result in a low value of λ (it should be noted that experimentalists are more often guilty of *underestimating* noise). Whatever the source, a low value of the statistic leads to what most of us would consider an artificially low 95% confidence level upper limit.

An example of this situation is the experiment of Uson and Wilkinson (1984). In that experiment the 95% confidence level upper limit was about a factor of two lower than would be expected (in the statistical sense) even if the amplitude of fluctuations in the microwave background were identically zero. We want to emphasize that this statement does not imply that the experiment had a flaw. Quite the contrary; there is every indication that observations were well understood and all the systematics were properly dealt with. The statistic λ was simply low by chance. In fact, one out of six identical experiments would yield a value of λ lower than that obtained by Uson and Wilkinson. It should also be emphasized that the 95% confidence level Uson and Wilkinson quoted was in accord with the proper statistical definition given above. The problem is more sociological than statistical. What notions does the phrase '95% confidence level upper limit' conjure up? That several identical experiments can yield significantly different 95% confidence level upper limits seems to be distasteful to most of us even though it is in perfect accord with statistics as described above.

Fortunately, it is simple to determine when such 'artificially low' limits are being set. The power of the test, β, is the probability that, under hypothesis H_1, the statistic will be less than or equal to the value determined from the data. In the case of the Uson and Wilkinson data, $\beta \simeq 0.15$, equivalent to λ being about 1σ below the median value (an entirely reasonable result as noted above); but the value of the power expected from noise alone in the absence of fluctuations in the sky is $\beta = 0.5$. Therefore, whenever $\beta < 0.5$, the 95% confidence level upper limit will be lower than expected from noise alone, the situation that confronted us in the Uson and Wilkinson experiment (see fig. 7.1(*b*)).

In order to avoid such situations, it has been suggested (Cottingham, 1987; Boughn, *et al.*, 1992) that the 95% confidence level as computed above should only be quoted when it exceeds the particular value of σ_{sky} corresponding to $\beta = 0.5$. Let us call that value $(\sigma_{sky})_{0.5}$. If the data yield a value of $\sigma_{sky} < (\sigma_{sky})_{0.5}$, then the quoted upper limit should be $(\sigma_{sky})_{0.5}$. It may seem disturbing that in this case the upper limit is determined only from the estimated noise of the experiment and not from the actual observations. However, if one considers this upper limit to be a measure of the sensitivity of the experiment, then in quoting it one is simply stating that an experiment with the quoted sensitivity failed to detect fluctuations in the microwave background.

Even though this procedure is insensitive to overestimates of errors and statistically unlikely data, it is still important to determine the extent to which the data are affected by such problems. An extremely low value of the power of the test is an indication that there may be unknown problems with the data set. If this is so, it is doubtful that the statistical characteristics of the system noise are well understood and 95% confidence level upper limits deduced from the data should be treated with due caution.

7.4.3 Estimating σ_{sky} when the power is large

The likelihood ratio test described above is useful for setting upper limits, but what about estimating the amplitude of fluctuations in the case that they are detected, i.e. when the power β is large, as shown in fig. 7.14(*c*)? Two reasonable approaches are to find the amplitude σ_{sky} in H$_2$ such that the size of the test is 0.5, or to find the value of σ_{sky} for which the measured statistic λ is the most likely. Either of these two estimates is acceptable and they will converge in the limit of large signal to noise ratio.

A more standard approach is the *maximum likelihood* method, in which we find the amplitude σ_{sky} for which the likelihood function (see eqn. (7.7)) is a maximum. For uncorrelated Gaussian noise, it is this approach that yields the familiar, weighted-least-squares method, which minimizes the χ^2 of a data set.

It is sometimes the case, however, that computing the likelihood function becomes an intractable problem. Such is the case for the 19 GHz full sky map constructed by Boughn *et al.* (1992), which has 25 000 pixels with correlated noise, and for the COBE maps. Consider a model in which fluctuations of the microwave background are described by a multi-variate Gaussian distribution (Whalen, 1971),

$$P(\{\Delta T_i\} \mid \sigma_{sky}) = [(2\pi)^N \det R]^{-1/2} \exp\left(-\sum_{j,k} \Delta T_j \Delta T_k R_{jk}^{-1}\right), \qquad (7.9)$$

where N is the numbe of pixels, ΔT_j and ΔT_k are the temperatures of the microwave background at positions j and k in the sky and $R_{jk} = \langle \Delta T_j \Delta T_k \rangle$ is the correlation of these two temperatures. In the presence of system noise n_j the distribution becomes

$$P(\{\Delta T_i'\} \mid \sigma_{sky}) = [(2\pi)^N \det \mathbf{R}]^{-1/2} \exp\left(-\sum_{j,k} \Delta T_j' \Delta T_k' \mathbf{R}_{jk}^{-1}\right), \qquad (7.10)$$

where $\Delta T_j'$ and $\Delta T_k'$ are the *measured* temperatures and $\mathbf{R}_{jk} = R_{jk} + \langle n_j n_k \rangle$. In the context of this model, the amplitude of the fluctuations is defined by $\sigma_{sky} = \langle \Delta T_j \Delta T_j \rangle^{1/2}$. It is clear that to compute the value of σ_{sky} for which this function is a maximum is out of the question, and even to evaluate it by Monte Carlo means would be an enormous effort, which includes inverting a 25 000 by 25 000 matrix in addition to generating an enormous amount of random data.

A less cumbersome approach to estimating the amplitude of CBR fluctuations is simply to choose any 'reasonable' estimator, which is unbiased in the sense that the ensemble average of the estimator equals R_{jk}. Such an approach was introduced by Lasenby and Davies (1983) who devised an unbiased estimator, which approximated the maximum likelihood solution.

As an example, consider the case in which the correlation of the microwave background temperature between two points in the sky depends only on angular separation between points and not on their relative orientation. Then the correlation matrix can be expressed in terms of a correlation function, $C(\theta) = \langle \Delta T_j \Delta T_k \rangle$, where the average is over all pairs of points in the sky separated by an angle θ. Then $R_{jk} = C(\theta_{jk})$. In the case of the full sky 19 GHz map, Boughn *et al.* (1992) made use of the following unbiased estimator of $C(\theta)$:

$$[C(\theta)]_E \equiv \sum_{j,k} \frac{\Delta T'_j \, \Delta T'_k}{\sigma_j^2 \, \sigma_k^2} \bigg/ \sum_{j,k} \frac{1}{\sigma_j^2 \, \sigma_k^2}, \tag{7.11}$$

where σ_j is the system noise associated with pixel j and the sums are over only those pixel pairs with $\theta_{jk} = \theta$. The motivation for this estimator is given below.

Now suppose the fluctuations of the microwave background are completely characterized by the amplitude of the fluctuations, $\sigma_{sky} = C^{1/2}(0)$, i.e. the functional form of the correlation function is specified. Thus we can let $C(\theta) = \sigma_{sky}^2 \, c(\theta)$ where $c(\theta)$ is specified by the model. An unbiased estimator for σ_{sky}^2 is

$$[\sigma_{sky}^2]_E = \frac{\displaystyle\sum_{j,k} \frac{c(\theta_{jk}) \, \Delta T'_j T'_k}{\sigma_j^2 \, \sigma_k^2}}{\displaystyle\sum_{j,k} \frac{1}{\sigma_j^2 \, \sigma_k^2}}. \tag{7.12}$$

In the limit of small σ_{sky} this estimator is the same as the maximum likelihood estimator and is also proportional to the likelihood ratio statistic, which motivates both its use here and, indirectly, in eqn. (7.11).

Any estimator can also be used as a statistic in order to assign confidence intervals and set upper limits. Consider the expression in (7.12). In order to find the 90% confidence interval, one must first compute (in this case by Monte Carlo methods) the probability distribution of the estimator assuming a background of given amplitude σ_{sky}. If the estimator computed from the actual data falls at the 95% point of the distribution, then that value of σ_{sky} marks the lower end of the 90% confidence interval. Similarly, if the estimator falls at the 5% point of the distribution, it marks the upper end of the 90% confidence interval. This latter value is clearly also the 95% confidence level upper limit.*

The question as to which estimator or statistic is 'best' has been the subject of much discussion and is unlikely ever to be answered. A prudent approach is to pick any 'reasonable' estimator/statistic such as the one expressed in eqn. (7.12). If different 'reasonable' choices give different upper limits, this should be taken as an indication of the 'uncertainty' in these upper limits. Experience indicates that this uncertainty is within a factor of two in most cases.

7.4.4 The Bayesian approach

A more direct approach to computing a statistic from the data is simply to compute the probability that a given model has fluctuations of amplitude σ_{sky} given the data set $\{\Delta T_i\}$, i.e. $P(\sigma_{sky}|\{\Delta T_i\})$. According to Bayes' theorem this probability density is proportional

---

* If σ_{sky} is small, and our aim is to set upper limits, the power of the test should be computed (the position of the estimator in the distribution conditional on H$_1$). If it less than 0.5, the sensitivity of the experiment, as defined in Section 7.4.2, should be quoted as the upper limit instead.

to the product of the likelihood function, $P(\{\Delta T_i\}|\sigma_{sky})$, and the probability density, $p(\sigma_{sky})$, that the microwave background has fluctuations with amplitude σ_{sky}, i.e.

$$P(\sigma_{sky}|\{\Delta T_i\}) \propto P(\{\Delta T_i\}|\sigma_{sky})\, p(\sigma_{sky}). \tag{7.13}$$

The quantity $p(\sigma_{sky})$ is called the *prior* probability density and its presence requires that we have some estimate of the probability that the microwave background has fluctuations of any given amplitude. Certainly we know that $p(\sigma_{sky}) = 0$ for $\sigma_{sky} < 0$, and also for σ_{sky} substantially greater than the upper limit set by some previous experiment, but what about intermediate values? In our ignorance, it seems reasonable to assume that $p(\sigma_{sky})$ is constant in that range. This is known as the *uniform prior* assumption. Since the likelihood function decreases rapidly for σ_{sky} much larger than typical sensitivity (defined in Section 7.4.2), $P(\sigma_{sky}|\{\Delta T_i\})$ in relation (7.13) is relatively insensitive to the upper cutoff of σ_{sky} and in most cases one can simply assume $p(\sigma_{sky})$ to be constant for all positive σ_{sky}: The constant of proportionality in (7.13) is evaluated by requiring the integral over σ_{sky} be unity. The 95% confidence level upper limit is that value of σ_{sky} above which lies 5% of the area under the distribution.

Readhead *et al.* (1989) employed the Bayesian method with uniform prior in their arcminute scale observations carried out at Owens Valley. It is interesting that the 95% confidence level upper limit determined in this way was within 10% of the upper limit derived using the likelihood ratio test – anecdotal evidence supporting the statement that all 'reasonable' methods of setting upper limits give essentially the same results.

Readhead *et al.* (1989) demonstrated that the Bayesian method is not very sensitive to overestimation of system errors, which was an objection to straightforward application of the likelihood ratio test. Recall, however, that the practice of constraining upper limits to be larger than or equal to the *sensitivity* of the measurement obviated that objection to the likelihood ratio test in any case (Section 7.4.2). In the case of statistically unlikely data the Bayesian method also fares well. For example, consider the case of twenty independent measurements of uncorrelated Gaussian fluctuations (see eqn. (7.7)). If all the measured temperatures were zero, i.e. the data were about as unlikely as possible, the 95% confidence upper level would be only a factor of two below the level set with the most likely data, i.e. $\Delta T_i^2 = \sigma_i^2$.

Even though the Bayesian method is insensitive to overestimates of errors and statistically unlikely data, it is still important to determine the extent to which the data is affected by these problems. In the case of a statistical test such as the likelihood ratio test, the power of the test provides this information. In the case of the Bayesian method, or any method for that matter, the simple χ^2 is quite informative. Any very low value of χ^2 is an indication that there may be unknown problems with the data set and that the 95% confidence level upper limits should be considered with due caution.

Perhaps the most bothersome aspect of the Bayesian method is that it requires a prior assumption of the probability that the universe we inhabit contains fluctuations of the microwave background of a particular amplitude. Formally, $p(\sigma_{sky})$ is the probability density of an ensemble of universes, not an estimate of our state of knowledge of the single existing universe. For this reason many of us would consider the function $p(\sigma_{sky})$ to be meaningless or impossible to determine. Nevertheless, the use of the Bayesian method and the assumption of a uniform prior results in 'reasonable' 95% confidence level upper limits.

Any attempt to compute a probability that a given upper limit is in error, e.g. a 95% confidence level, requires that an assumption be made as to the probability of fluctuations existing at a given level. Recall, in the case of the likelihood ratio test, that all probabilities were computed conditional on the existence of fluctuations at a given level. Such assumptions can never really be justified. Perhaps the moral of all this is the following: any 'reasonable' method of computing upper limits is acceptable, where by 'reasonable' we mean that (1) the upper limit is never significantly lower than the sensitivity level and (2) all such reasonable methods give similar results. Also these limits should be considered as rough limits (within a factor of two, say).

7.4.5   *Testing specific models*

Any model in which fluctuations of the cosmic microwave background are Gaussian is completely specified by the correlation matrix R_{jk} (see eqn. (7.9)), or correlation function $C(\theta)$, if the fluctuation amplitude depends only on separation. In the latter case the correlation function can be expanded in Legendre polynomials (e.g., Cottingham, 1987),

$$C(\theta) = \sum_n \frac{2n+1}{4\pi} C_n P_n(\cos \theta) \qquad (7.14)$$

where

$$C_n = 2\pi \int_0^\pi C(\theta) P_n(\cos \theta)\, \mathrm{d}(\cos \theta). \qquad (7.15)$$

In any real experiment the sky is observed with finite resolution with a telescope characterized by a beam pattern, $P(\theta)$ (see, e.g., fig. 7.1 and fig. 7.17 later). Since this pattern is defined on the sphere it can also be expanded in Legendre polynomials

$$P(\theta) = \sum_n \frac{2n+1}{4\pi} A_n P_n(\cos \theta). \qquad (7.16)$$

The correlation function of the microwave background smoothed by the beam can then be expressed as (Cottingham, 1987)

$$C'(\theta) = \sum_n \frac{2n+1}{4\pi} A_n^2 C_n P_n(\cos \theta). \qquad (7.17)$$

Sky subtraction techniques such as single and double beam switching are easily incorporated into this analysis. For example, in the case of single beam switching (fig. 7.1(*a*)), the difference is taken between two points on the sky separated by an angle θ_s. In this case the effective beam pattern is given by $P_e(\theta) = P(\theta) - P(\theta + \theta_s)$. For beam-switching combined with the 'on–off' technique, we see from fig. 7.1(*b*) or eqn. (7.2) that $P_e(\theta) = 2P(\theta) - P(\theta + \theta_s) - P(\theta - \theta_s)$.

The resulting model correlation function can be quite complicated but, at least in principle, the model can then be tested by any of the methods discussed above. In practice the tests, especially if they require Monte Carlo analyses, can be quite time consuming, requiring literally weeks of CPU time on respectable computers. Considering the plethora of models existing today (as described in Chapter 8), testing models from the data of a single experiment could easily consume more time than constructing the apparatus and taking and reducing the data. Luckily this is not necessary. Two generic types of models, 'power law' and 'Gaussian,' are sufficiently good approximations to virtually any model of the fluctuations of the microwave background that these two are the only ones that need be considered in most cases (but see Section 8.8.4).

A *power law* model is one in which the correlation function is a power of angular separation (at least for $\theta < 10°$ and over a limited range of angular scales). Since the fluctuations are defined on a sphere, the expansion coefficients of the correlational function are specified as powers of harmonic order n, i.e. $C_n \propto (n + 1)^\gamma$. A *Gaussian* model, on the other hand, is one in which the expansion coefficients are a Gaussian function of harmonic order n, i.e., $C_n \propto \exp(-(1/2)(n/n_0)^2)$ where n_0^{-1} is the characteristic angular scale of the fluctuations. For small θ ($\ll 1$ rad), the Gaussian model corresponds to a Gaussian correlation function as well.

To find out whether or not one's favorite cosmological model is consistent with a given experiment, simply pick the member of one of the two types described above that has a correlation function most closely approximating one's favorite model. The 95% confidence level upper limit on fluctuations for the former can then be taken as the limit on the model to be tested. It may be that there are subtle differences between one's favorite model and a power law or a Gaussian, but the inherent uncertainties in setting upper limits, which were discussed above, suggest that any more detailed comparison of a particular model with experimental data cannot really be justified.

7.5   Early results on scales about 1'–10°

With the discussion of techniques behind us, let us now turn to the observational results. In this section, I will describe some, but not all, of the searches for CBR anisotropies which led up to the best current experiments (separately described in Sections 7.6–7.8). I will emphasize in this section primarily those observational programs that broke new ground by introducing new observational techniques, by sharply reducing limits on $\Delta T/T$, or by extending the range of wavelengths used in such searches.

7.5.1 Goals and results of the first decade, 1965–1975

Within a few years of the discovery of the CBR, several groups had established upper limits on its anisotropy on scales of a degree or less. The very first published limits on CBR anisotropies on arcminute scales were those of Conklin and Bracewell (1967) and Epstein (1967).

The former measurements were made at $\lambda = 3$ cm with a beam of 10' full width at half maximum. The observed r.m.s. fluctuation from the sum of 36 days of observations was 5.6 mK. By *assuming* that the instrumental noise decreased as the square root of the observing time, the authors were able to fix a 1σ limit of 3.6 mK on σ_{sky} (but see Section 7.4.1). Hence on angular scale of 10', the CBR was stated to be 'smooth' to within about 1.3 parts per thousand (later corrected to $\Delta T/T \leqslant 1.8 \times 10^{-3}$ because of an error in the calculation of the sensitivity of the instrument). Conklin and Bracewell also noted that their measurements could be used to set limits on $\Delta T/T$ on both larger and smaller angular scales by making some assumptions about the statistical properties of the fluctuations. There are, however, direct measurements available on both larger and smaller angular scales which have set still lower limits, measurements we will consider below.

Epstein's (1967) measurements, at $\lambda = 0.34$ cm, were less sensitive but were the first to employ beam switching to reduce atmospheric noise.

These early searches were motivated both by the realization that density inhomo-geneities in the Universe could produce anisotropies in the CBR (Sachs and Wolfe, 1967; Silk, 1968) and by a desire to test non-cosmological models for the CBR. In par-ticular, in the mid- to late 1960s, several authors suggested that the CBR might be due to the summed microwave emission of many discrete radio sources (a model most fully worked out by Wolfe and Burbidge in 1969). It was quickly recognized that, in such models, fluctuations in the observed intensity of the background radiation would be expected from counting statistics alone (Hazard and Salpeter, 1969; Penzias *et al.*, 1969; Smith and Partridge, 1970). If discrete sources are to produce the *entire* background radiation and yet have no fractional fluctuations larger than some observational upper limit $\Delta T/T$ in a solid angle Ω defined by the telescope beam, their number per steradian must be large. A dimensional argument suggests $\Delta T/T \simeq n^{-1/2} \simeq (n_0\Omega)^{-1/2}$, where n_0 is the number of sources per steradian, and Ω is the solid angle of the observations. Clearly, observations at high angular resolution provide the most sensitive test of such models. Such observations were carried out by Penzias *et al.* (1969) using an 11 m diam-eter telescope at $\lambda = 0.35$ cm. They placed an upper limit on the r.m.s fluctuations in *antenna temperature* of 24 mK; this upper limit was later lowered by Boynton and Partridge (1973) using the same instrument and wavelength to 4.3 mK, or $\Delta T/T \leqslant 1.8 \times 10^{-3}$ after making the correction to thermodynamic temperature. The solid angle of the beam employed was $\Omega = 1.5 \times 10^{-7}$ sr. This upper limit on $\Delta T/T$ was sufficient to show that the number of sources required to explain both the flux and the observed 'smooth-ness' of the microwave background substantially exceeded the number of galaxies for any reasonable cosmological model and any reasonable assumption about the evolution of the sources. Since luminous radio sources are in fact a small fraction of the total number of galaxies, the discrete source explanation for the microwave background failed. It remains possible, of course, that radio sources do contribute to some extent to fluctuations in the microwave sky, a point that we have noted above in Section 7.3.2.

In 1970 appeared the first in a series of papers on $\Delta T/T$ measurements by researchers in the Soviet Union (Pariiskii and Pyatunina, 1970). These observations were made (at $\lambda = 4$ cm) with an antenna having cylindrical rather than radial symmetry; hence an asymmetrical $1.4' \times 20'$ 'fan beam' resulted. The limit set on $\Delta T/T$ on these scales, 2.3×10^{-4}, was comparable with the best upper limits then available at about $1°$ scales (Conklin and Bracewell, 1967). This work was continued by Pariiskii using instruments in both the USSR. and the USA, (Pariiskii, 1973a,b), and by Stankevich (1974) Pariiskii's upper limit of 8×10^{-5} on $\Delta T/T$ at $\lambda = 3$ cm (1973a) was later (Pariiskii, 1974) revised upwards, and is now considered to be $\leqslant 4 \times 10^{-4}$. Pariiskii (1973b) reports the detection of sky variance, with $\sigma_{sky} = (1.3 \pm 0.4) \times 10^{-4}$ K at $\lambda = 4$ cm. This value was obtained using the model-dependent method described in Section 7.4.1, and hence should be taken with some caution. In any case, as Pariiskii (1973b) points out, much of the observed sky fluctuation may be ascribed to discrete sources appearing in his $1.3' \times 40'$ fan beam. A glance at fig. 7.12 shows that discrete sources can certainly explain his results; indeed his value for σ_{sky} is surprisingly low given the wavelength and beam solid angle employed. That last remark applies even more strongly to the 11 cm observations of Stankevich (1974): the limit of 4.2×10^{-4} K that he claims on σ_{sky} cannot be correct, since it lies more than an order of magnitude below the fluctuation level predicted for radio sources alone. (Stankevich himself ascribes his σ_{sky} to discrete sources, but underestimates their contribution.) His value for σ_{sky} may also be low because he used the statistical method described in Section 7.4.1, and because he apparently failed to correct his antenna temperature measurements for the efficiency of the telescope he employed.

Finally, Carpenter, Gulkis and Sato (1973) were the first to employ a truly low-noise receiver in searches for CBR fluctuations. The system temperature of their 3.56 cm receiver was only 25 K (a remarkable achievement at that date). They made repeated drift scans of a $1°$ strip of the sky with their $2.3'$ beam. Since drift scans were used, adjacent measurements of surface brightness were not independent; the authors apply a correction factor of 0.85 for this to arrive at their final 90% confidence level upper limit on σ_{sky}: σ_{sky} < 1.93 mK. This corresponds to a 95% confidence level on $\Delta T/T$ of about 9×10^{-4}.

By 1973, then, we had reliable evidence that fluctuations in the CBR on arcminute scales were $< 10^{-3}$ of T_0.

7.5.2 Results of the second decade, 1975–1985

A number of new observational techniques were introduced in the search for CBR fluctuations in the late 1970s and early 1980s. In my view, the most important of these was the use of bolometric detectors, typically operating at $\lambda \lesssim 1–3$ mm, pioneered by the group now based in Rome (Caderni et al., 1977; Fabbri et al., 1980; Melchiorri et al., 1981) and the group at MIT (e.g., Meyer et al., 1983). To work at such short wavelengths near sea level is impossible; hence these groups also pioneered observations made from mountain altitudes, from high-flying aircraft and by balloon-borne instruments.

The work of Caderni et al. (1977) was carried out at an altitude of 3.5 km from a site in the Italian Alps. Filters were used to define the bandwidth of radiation reaching the Ge bolometer detector: $1.4 \geqslant \lambda \geq 1.0$ mm. Even at an altitude of 3.5 km, T_{atm} was still 20–50 K at these wavelengths, so the observations were made at the zenith, with a beam switch angle $\theta_s = 25'$. The two beam positions lay $12.5'$ north and south of the local zenith; the observers took care to adjust the symmetry axis to equalize the zenith angles of the two beam positions and thus to minimize systematic offsets. The individual beams were $25'$ between 20% power points, corresponding to $\theta_{1/2} = 16.4'$. if the beam profile was close to Gaussian.

The 2σ upper limit on $\Delta T/T$ is given as 1.2×10^{-4}, based on a figure for σ_{sky} of $\leqslant 6 \times 10^{-4}$ K for a single scan. In fact, I believe that the value for σ_{sky} is given in antenna temperature, not thermodynamic temperature. (All other temperature values, including T_{atm} and the calibration are expressed in antenna temperature.) Consequently, we must use eqn. (3.10) to convert 6×10^{-4} K to thermodynamic temperature: $\Delta T = 6 \times 10^{-4}$ $[(e^x-1)^2/x^2e^x]$ with $x \equiv h\nu/kT \approx 4.37$ here. The correction factor is about 4; hence I believe that Caderni et al. (1977) should quote $\lesssim 4–5 \times 10^{-4}$ for $\Delta T/T$.

Fabbri et al. (1980) flew a bolometric detector on a high-flying aircraft. At 10–15 km altitude, the antenna temperature of millimeter and submillimeter lines in the Earth's atmosphere was small enough to allow them to use a huge bandwidth, $0.5 \lesssim \lambda \lesssim 2$ mm. The system sensitivity is given as 9.8 ± 0.6 mK Hz$^{-1/2}$; the question is whether this is expressed in antenna temperature or thermodynamic temperature. Fabbri et al. assume that it is thermodynamic, and claim limits on σ_{sky} of about 9×10^{-4} K in a 20 min observation, or $\Delta T/T \lesssim 3 \times 10^{-4}$. It seems more likely to me that the system noise is, as is customary, in antenna temperature.* Then the actual upper limit on $\Delta T/T$ would be roughly an order of magnitude *higher*, the exact value depending on the shape of the very wide bandpass employed.

The third innovative experiment involved flying a bolometric detector on a balloon to an altitude of about 40 km. The main aim of this work was to measure the dipole and quadrupole moments of the CBR (see Section 6.5). Nevertheless, since the beam switch angle was only 6°, Melchiorri *et al.* (1981) could also place limits on $\Delta T/T$ on that scale (the beam size is given as 5.2°). The band pass was $0.5 \lesssim \lambda \lesssim 3$ mm; see Fabbri *et al.* (1982) for further experimental details.

Melchiorri *et al.* (1981) were well aware that emission from Galactic dust could contaminate their results and hence attempted to correct their observations by correlating them with measures of neutral hydrogen in the Galaxy, which are known (e.g., Boulanger *et al.*, 1985) to scale with the column density of dust. The corrected data show a residual anisotropy corresponding to $\sigma_{sky} = (1.1 \pm 0.2) \times 10^{-4}$ K. Once again, we must ask whether this is expressed in antenna temperature or thermodynamic temperature; the upper limit of 4×10^{-5} on $\Delta T/T$ of Melchiorri *et al.* assumes the latter. In this case, values provided for the responsivity and the throughput of the detector, and the effective wavelength of the band permit one to confirm that $\sigma_{sky} = (1.1 \pm 0.2) \times 10^{-4}$ K is expressed in thermodynamic temperature, so that $\Delta T/T \lesssim 4 \times 10^{-5}$. This upper limit is surprisingly small, barely consistent with the models of Galactic foreground emission made above.

Finally, we should recall that Meyer *et al.* (1983) were also making CBR measurements using bolometers. These observations were made from the ground (at 4000 m altitude), and hence required careful subtraction of atmospheric emission, as noted in Section 7.2.1. The final sensitivity achieved was roughly an order of magnitude poorer than the results from the balloon experiment of Melchiorri *et al.* (1981).

Both these groups, as well as groups based at Princeton University and the University of California, are extending and improving bolometric techniques. Bolometric measurements are now competitive with the best heterodyne searches for CBR fluctuations, as we will see below.

In 1977, a major new radio telescope of novel design was completed in the Soviet Union, the RATAN-600 telescope located in the Caucasus. It consists of a thin ring of 895 individual reflectors; the diameter of the ring is 600 m, and the height of the individual elements is 7.4 m (see fig. 7.15). In the late 1970s and early 1980s it was used extensively by Pariiskii and his collaborators to make counts of radio sources and to search for fluctuations in the CBR. In the latter phase of their work, a very low noise receiver at $\lambda = 7.6$ cm was employed (Berlin *et al.*, 1983); also used were ground screens to reduce side-lobe pickup to about 6 K. Repeated drift scans of a 24 h strip of the sky at $\delta = 4°54'$ were made. The observing technique and the instrument are fully described in Pariiskii and Korol'kov (1987). Because of its design, the instrument produced a fan beam, much narrower in azimuth than elevation. At $\lambda = 7.6$ cm, the beam size at the half-power points was $1' \times 10'$. The system temperature of the instrument is given as 37 K, and the bandwidth as 500 MHz. For some of the observations, the integrating time was set at 1.8 s; in this case the expected r.m.s. noise in antenna temperature is

---

* If, as Fabbri *et al.*, implicitly claim, the system noise is thermodynamic, their system equaled or bettered in sensitivity the best bolometric detectors now available.

Fig. 7.15 Photograph of the RATAN-600 telescope, located in the Caucasus. One sector of the ring reflector is shown.

1.8 mK per scan. Statistical error was further reduced by stacking many scans; if the system noise is truly random $\Delta T \propto 1/\sqrt{N}$, where N is the number of scans, 64 for the work discussed below.

According to Pariiskii and Korol'kov (1987), the RATAN data were analyzed in three different ways to produce upper limits on $\Delta T/T$ on scales of 1', 4.5'–9.0' and 1°–5°. As a first step in all cases, brief noise spikes (of duration $\leqslant 1.8$ s) were removed. Next, foreground radio sources with $S > 1$–3 mJy were removed, eliminating 10–20% of the data. In the search for fluctuations on 4.5'–9.0' scales, the next step was to apply a smoothing filter to the data, effectively smoothing out fluctuations on scales below 4.0' and subtracting baseline variations on scales $> 9.0'$. This procedure naturally results in a great reduction of the r.m.s. noise of the remaining data (to 0.083 mK). Theorists making use of these results must bear in mind that the heavy filtering of the data has smoothed away information on CBR anisotropies on all angular scales $> 9'$ or $\leqslant 4'$. At the other angular scales, 1' and 1°–5°, correlations between the 7.6 cm scans and similar scans made at 31, 3.95 and 2.08 cm were used to subtract fluctuations introduced by Galactic and atmospheric emission – see Pariiskii and Korol'kov (1987) for details.

An equation adopted from Pariiskii *et al.* (1977) was then used to convert r.m.s. values for measured antenna temperature fluctuations into values for $\Delta T/T$:

$$\frac{\Delta T}{T} = \frac{1}{(2\sqrt{N})^{1/2}} \frac{(u_p + R\sqrt{N})^{1/2}}{\eta_1 \sqrt{\eta_2}} \frac{\Delta T_{\text{measured}}}{T_0}, \tag{7.18}$$

where N is the number of scans (64), R is a correlation coefficient $\lesssim 0.1$, and $u_p = 2$ for a 2σ upper limit. The two quantities η_1 and η_2 are efficiency factors: η_1 is the aperture efficiency, given as 0.75, and η_2 is the fraction of the solid angle of the $1' \times 10'$ beam filled by a source. For the latter, Pariiskii and Korol'kov give a typical value of 0.25 in the search for fluctuations on angular scales 4.5′–9.0′. The square root of η_2 appears because of the explicit assumption that fluctuations in the CBR are randomly distributed with Gaussian statistics. If the fluctuations have non-Gaussian statistics, η_2^{-1}, not $\eta_2^{-1/2}$, is more appropriate.

Here I should note that Lasenby (1981) has raised some questions about the use of eqn. (7.18) to establish upper limits on $\Delta T/T$. As he notes, the statistical assumptions underlying (7.18) are not correct, and the effect of modifying (7.18) is to increase $\Delta T/T$. The actual increase in $\Delta T/T$ cannot be computed from the published data (Berlin *et al.*, 1983; Pariiskii and Korol'kov, 1987); Lasenby presents a heuristic argument for a factor of $\sqrt{3}$. In addition, I cannot see why the factor of $\sqrt{2}$ appears in the denominator of (7.18). Finally, if $\Delta T_{\text{measured}} \approx 0.083$ for data smoothed to a scale of 4.5′–9.0′, as given by Pariiskii and Korol'kov (1987), and if we use (7.18) with the values given by these authors, we arrive at $\Delta T/T \lesssim (2.9$–$3.4) \times 10^{-5}$, not the value of 10^{-5} claimed by them. Indeed, upper limits below about 2×10^{-5} do not appear to be consistent with the stated system noise if eqn. (7.18) is used in its unmodified form.

We need also to reexamine the question of foreground radio sources. In a beam of about 10 arcmin^2 solid angle, we expect foreground radio sources to contribute noise at a level corresponding to $\Delta T/T \simeq 5 \times 10^{-4}$ (from fig. 7.10 scaled to $\lambda = 7.6$ cm). The work of Franceschini *et al.* (1989) on which fig. 7.10 is based also shows that the foreground source confusion will be dominated by sources in the flux density range $0.1 < S < 100$ mJy; these predictions are based on direct 6 cm radio source counts, which extend to fluxes below 0.1 mJy and hence are quite secure. As noted above, the RATAN observers can detect and remove sources down to about 1 mJy. Sources with flux density below 1 mJy, however, are predicted to produce fluctuations at a level of about $(0.5$–$2) \times 10^{-4}$ in $\Delta T/T$. Thus some questions have been raised about the upper limits reported by Berlin *et al.* (1993), which are at $\theta \simeq 1'$, $\Delta T/T \leqslant 2.2 \times 10^{-4}$; and at $\theta = 4.5'$–$9.0'$, $\Delta T/T \leqslant 10^{-5}$. Confusion by sources with $S \leqslant 1$ mJy has little effect on measurements on degree scales; here the main question is the accuracy with which fluctuations introduced by patchy synchrotron emission in the Galaxy can be subtracted. The limit reported by Berlin *et al.* (1983) at $\theta = 1°$–$5°$ is $\Delta T/T \leqslant 1.5 \times 10^{-5}$, again at 2σ, or very roughly 70 times lower than the estimate of $\Delta T/T$ for Galactic emission drawn from fig. 7.6.

All three values are derived using eqn. (7.18); for reasons stated above, they may need to be multiplied by about $\sqrt{6}$, or $(6/\eta_2)^{1/2}$ for non-Gaussian CBR fluctuations.

This brief summary does not do full justice to the work of the RATAN group, which also made important contributions to the study of atmospheric and Galactic fluctuations in the microwave region, to radio source counts and to searches for small angular scale polarization in the CBR. These and other results are more fully described in the review by Pariiskii and Korol'kov (1987).

In this decade also, aperture synthesis was first brought to bear on searches for CBR fluctuations. In the late 1970s Martin, Partridge and Rood (1980) used a three element interferometer to observe the CBR at $\lambda = 11$ and 4 cm and at resolutions of $13''$ and $4''$, respectively. The data were analyzed using a variant of a technique introduced earlier by Goldstein, Marscher and Rood (1976) to correct for instrumental noise. The 11 cm observations were dominated by foreground radio sources, and in any case far better limits were soon obtained from observations at the 27-element Very Large Array described in Section 7.8 below.

The observations described so far in Section 7.5.2 were based on new technologies, new instruments or new techniques. Searches for CBR fluctuations using conventional radio telescopes continued through the early 1980s as well. For instance, I used the (then) 11 m telescope operated by the US National Radio Astronomy Observatory on Kitt Peak to search for CBR fluctuations at $\lambda = 9$ mm (Partridge, 1980). Because of known problems with side- and back-lobe pickup at this telescope, I elected to make a series of drift scans (Section 7.2.2) over two regions of the sky (beam switching in azimuth with $\theta_s = 9'$ was also employed). One set of results is shown as fig. 7.3 here. Because the angular separation of the 40 measurements shown was less than the beam width employed ($\theta_{1/2} = 3.6'$), the points are not statistically independent. I corrected for this loss of independence, but the methods I used to derive limits on $\Delta T/T$ from the observations have been criticized by Lasenby (1981). He shows that the limits I claimed were too low by a factor of 2–3.

Lasenby also contributed to the search for CBR fluctuations. With R. D. Davies, he used a conventional filled aperture telescope at Jodrell Bank to set limits on $\Delta T/T$ at $\lambda = 6$ cm on angular scales from $10'$ to $60'$ (Lasenby and Davies, 1983). The limits that they established may be affected by foreground sources, but were the most reliable limits in this interesting range of angular scales. At $\theta = 11'$, Seielstad *et al.* (1981) established comparable limits ($\Delta T/T \lesssim 2.5 \times 10^{-4}$) at $\lambda = 3$ cm.

The work of the decade 1975–1985 culminated in the careful observations of Uson and Wilkinson (1984) at $\lambda = 1.5$ cm using a 40 m telescope. They made beam-switched 'on–off' observations of a set of twelve fields distributed around a circle of constant declination near the north celestial pole. The beam size was $1.5'$ and the beam throw $4.5'$. The design of the 40 m telescope prevented Uson and Wilkinson switching the beam purely in azimuth; hence the reference beam elevation changed during the course of each 2 h observation of a single field. By selecting an appropriate range of hour angles, however, they were able (Uson and Wilkinson, 1984) to make rotation of the reference beam symmetrical about the horizontal. In that way, the elevation-dependent atmospheric (and side lobe) contribution could be modeled and subtracted.

Uson and Wilkinson took considerable pains to check other possible sources of systematic errors in their observations. They also took careful account of the efficiency of the telescope and of the conversion to thermodynamic temperature; with these corrections included, the results are as shown in fig. 7.16. Clearly, no anisotropy is present at a level of about 0.5 mK, corresponding to about 2×10^{-4} in $\Delta T/T$. As noted in Section 7.4, however, one may hope to subtract some of the instrumental noise and thereby set limits on $\Delta T/T$, which are 2–3 times lower than the r.m.s. error in the individual data points. Uson and Wilkinson (1984) employed a test for sky fluctuation based on the Neyman–Pearson lemma to show that $\Delta T/T \lesssim 2.4 \times 10^{-5}$ at 95% confidence. As we have seen, however, this test is sensitive to incorrect estimation of the errors assigned to each

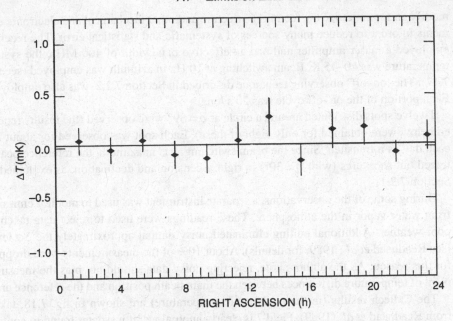

Fig. 7.16   Results of the 1984 experiment of Uson and Wilkinson, at $\lambda = 1.5$ cm (from their paper, with permission). From the size of the individual error bars (each $\sim\pm0.15$ mK), it may be shown that the reduced χ^2 of these results is < 1: the scatter from point to point is unusually low. Hence the upper limit on $\Delta T/T$ was unexpectedly low.

of the data points: a small overestimate of the size of the error bars, that is of the instrumental and atmospheric noise, can produce an erroneously small upper limit on $\Delta T/T$. The authors (Uson and Wilkinson, 1984) were aware of this point, and performed a partial test by showing that the recorded temperature differences (the quantities $T_{on} - T_{off}$ averaged over 2 h) were drawn from a uniform Gaussian distribution. On the other hand, as may be seen from fig. 7.16, the value of χ^2 for their data (with the errrors as shown) turns out to be significantly less than the number of degrees of freedom (7–8 versus 11). Such an outcome is statistically unlikely and also explains why their 95% confidence upper limit is a factor of eight below the approximate figure of 2×10^{-4} estimated above. It is now generally accepted that the observations of Uson and Wilkinson set a limit closer to 4–5×10^{-5} on $\Delta T/T$ at 95% confidence (see Section 7.4.2).

To conclude, by 1985, upper limits on $\Delta T/T$ had been pushed well below 10^{-4} on a couple of angular scales, and a range of angular scales from a few arcseconds to tens of degrees had been probed.

7.6   The best current limits on ΔT/T on arcminute scales

Readhead and his colleagues at Caltech (OVRO) adopted and modified some of the techniques introduced by Uson and Wilkinson and others, and were able to lower the upper limit on CBR fluctuations on arcminute scales to 1.7×10^{-5}, currently one of the most stringent limits on $\Delta T/T$ available at any angular scale.

Their work (Readhead *et al.*, 1989) was carried out on a 40 m telescope at $\lambda = 1.5$ cm. Since this telescope was not a multi-user, national facility, the observers had time to modify and symmetrize the feed-horn structure and other mechanical and electronic elements in order to reduce many sources of systematic and statistical error. The receiver employed a maser amplifier and had an effective bandwidth of 400 MHz; the system temperature was 40–45 K. Beam switching at 10 Hz in azimuth was employed (see fig. 7.17). The 'on–off' observing technique described in Section 7.2.2. was also employed. Each portion of the on–off cycle was 20 s long.

Twelve spots distributed around a circle at $\delta = 89°$ were observed (the results reported below were obtained for only eight of these). Each spot was observed for about 2 h each day as it transited. Since the beam switching was in azimuth, the reference beams traced out short arcs (with $\eta \simeq 30°$) in right ascension and declination, as explained in Section 7.2.1.

During some of the observations, a separate instrument was used to monitor emission from water vapor in the atmosphere. These readings were used to reject data taken in poor weather. Additional editing eliminated noisy data at approximately the 3σ level (see Readhead *et al.*, 1989, for details). About 10% of the measurements were dropped for this reason. As the authors note, rejecting noisy data should not bias the measurement of temperature differences between the main beam position and the reference arcs.

The Caltech results (in thermodynamic temperature) are shown in fig. 7.18, taken from Readhead *et al.* (1989). Field 7 is clearly anomalous; it was later found to contain a weak radio source. Hence field 7 was excluded from the remainder of the analysis, leaving seven independent measurements. We, like Readhead *et al.*, next need to ask what level of fluctuation in the CBR is consistent with the seven remaining measurements. The Caltech group employed both a Neyman–Pearson test and a Bayesian likelihood test. In both cases, the CBR fluctuations were assumed to obey Gaussian statistics. The resulting 2σ upper limits are in good agreement, and establish a 95% confidence level upper limit of 1.9×10^{-5} on Gaussian fluctuations on a coherence scale of 2.6′, at which their experiment is more sensitive (see fig. 7.19 for corresponding limits on other scales). For comparison with earlier results, the upper limit on *uncorrelated* CBR fluctuations of angular scale about equal to the beam size of 1.5′ is $\Delta T/T < 1.7 \times 10^{-5}$ at 95% confidence.

This work is marked both by the care taken to control sources of error and by completeness of statistical analysis of the results. Further results from the Caltech group should soon become available, including a lower sensitivity survey of 96 circumpolar regions (see Myers, 1990, and Appendix C).

7.7   The best current limits on $\Delta T/T$ on degree scales

As noted earlier in this chapter, searches for CBR fluctuations are increasingly plagued by noise from atmospheric emission as the angular scale examined is increased. Reduction of the statistical and systematic error introduced by the Earth's atmosphere is thus a major goal of programs designed to search for CBR fluctuations on degree scales. For such observations, fortunately, it is not necessary to use large, conventional radio telescopes. Smaller, specially designed apparatus may be used instead, allowing experiments to work at high altitudes or otherwise to minimize foreground noise from the atmosphere. As we shall see in the remainder of this section, observations have been made from mountain-tops (e.g. Davies *et al.*, 1987), dry Arctic (Mandolesi *et al.*, 1986; Timbie and Wilkinson, 1990) or Antarctic (Meinhold and Lubin, 1991; Gaier *et al.*,

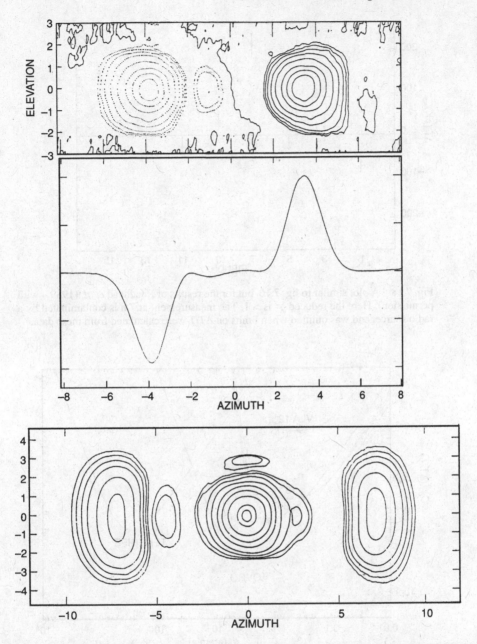

Fig. 7.17   Upper panel: the measured beam pattern employed by Readhead *et al.* (1989
– with permission). Weak side lobes are visible in both the contour map (where the
contour levels are at 0, ±2, ±3, ±5, ±10, ±20, ±40, ±60 and ±80% of peak) and in the
cross-section. Lower panel: beam pattern including on–off observations. The reference
arcs are shown for a two-hour observation (see Section 7.2.1).

1992) sites, balloons (de Bernardis *et al.*, 1990, 1992; Meyer *et al.*, 1991; Alsop *et al.*,
1992) and satellites (Smoot *et al.*, 1992).

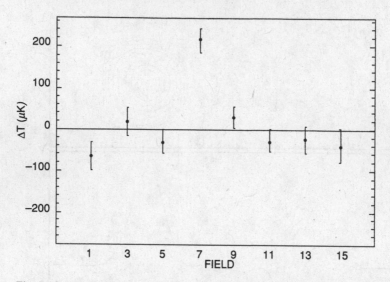

Fig. 7.18   A plot similar to fig. 7.16, but for the results of Readhead *et al.* (1989 – with permission). Here the reduced χ^2 is > 1. The measurement at 7 h is contaminated by a radio source, and was omitted when limits on $\Delta T/T$ were calculated from these data.

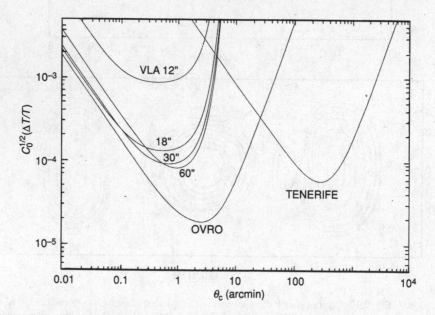

Fig. 7.19   Limits set by the observations of Readhead *et al.* (1989; OVRO), and Fomalont *et al.* (1988; VLA) assuming that CBR fluctuations have a Gaussian correlation function. Also shown are the larger scale limits of Davies *et al.* (1987; Tenerife).

7.7.1 *Ground-based observations*

The first sensitive, ground-based experiment on degree scales was mounted by Davies and his colleagues (Davies *et al.*, 1987) to search for fluctuations on a scale of 8°. They

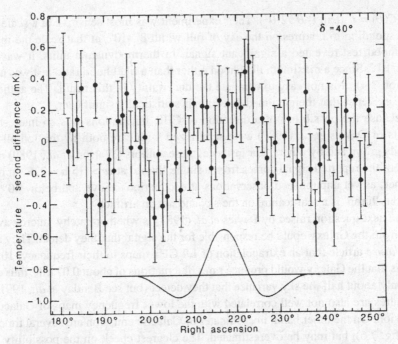

Fig. 7.20   Upper portion: measured temperature differences along a strip at $\delta = 40°$ (between 12 h and 17 h right ascension) – from Davies *et al.* (1987) with permission. A bright source moving through their beam pattern would produce a signal with the shape shown in the lower portion of the diagram. Evidence for such a 'lump' is marginally present at ~195° and at 225°, or approximately 13 h and 15 h right ascension.

observed from a dry, high altitude site in the Canary Islands (at 2300 m); chose a wavelength ($\lambda = 3$ cm) well away from water vapor lines; and used both beam switching and the 'on–off' technique, with a beam switch angle of 8.2°. A pair of matched, corrugated horn antennas was used to define the two beams, each of which had a full width at half power of 8.5°. Moving the two beams between 'on' and 'off' readings was accomplished by rotating a flat reflector placed at an angle in front of the horns. To minimize changes in the side-lobe pick-up, the apparatus was kept fixed and temperature differences were recorded as a strip of the sky at a fixed declination drifted through the beam pattern.

The receivers available to Davies and his colleagues were not particularly sensitive ($T_{sys} \simeq 100$ K), so they relied on stacking many such scans to reduce statistical noise.

At 3 cm wavelength, the emission from the Galaxy is substantial. Hence Davies *et al.* examined only a portion of their data ($12\,h \leqslant \alpha \leqslant 17\,h$) for CBR fluctuations. All these measurements were at Galactic latitude $\gtrsim 30°$.

The published results obtained by this group for one strip at $\delta = + 40°$ are shown in fig. 7.20. Given the size of the beam pattern shown it is clear that the data are substantially oversampled; there are actually only 5–10 independent measurements in their data. To set limits on $\Delta T/T$, the authors used a maximum likelihood test to see whether these data revealed a significant value of σ_{sky}. As a model for the statistical properties of the fluctuations, they took a Gaussian, so that the correlation function of the CBR signal was $C(\theta) = C_0 \exp(-\theta^2/2\,\theta_c^2)$. Their experiment was most sensitive at a scale $\theta_c = 4°$, corresponding to features on the sky of full width 8°–10°; at that scale the maximum likelihood test revealed a significant signal. In thermodynamic units, it was $\Delta T/T = 3.7 \times 10^{-5}$. Since a maximum likelihood rather than a likelihood ratio test was used (see Section 7.4), it is not easy to compute the uncertainty in this result. The authors offer some evidence that their claimed detection is statistically significant.

Because the possibility of a detection of CBR fluctuations was so interesting, the results of Davies *et al.* (1987) were scrutinized with care both by the Jodrell–Canary Islands group and by others. For instance, these authors (Watson *et al.*, 1989) note that most of the reported signal comes from a single 'lump' at $\alpha \simeq 15\,\mathrm{h} = 225°$ in fig. 7.20. Further, as yet unpublished observations at declinations above and below 40° suggest that the 'lump' is a real feature on the sky and not an artifact.

The next question raised by Davies *et al.* (1987) is whether patchy microwave emission from the Galaxy could be responsible for the signal that they detect. They argue in the *Nature* article that an extrapolation of 1.4 GHz maps to their frequency, 10.4 GHz, shows that the Galaxy would produce r.m.s. fluctuations of about 0.07 mK, thus explaining only about half the sky variance that they detect (but see Banday *et al.*, 1991). Their measures are also not well correlated with the lower frequency maps of Galactic radio emission. On the other hand, my estimates of Galactic emission are several times larger (see fig. 7.6), but may be overestimated. The clearest check on the possibility that the 3 cm signals are Galactic in origin is to re-observe the same region at the same 8° scale but at a lower frequency. Just such a program is underway at Jodrell Bank; so also are observations at a smaller angular scale (to test for discrete sources) and at $\lambda = 2$ cm (at which Galactic emission will be about three times lower). While the results of these observations are not yet available, the COBE results described in Section 7.7.3 below make it quite unlikely that Davies *et al.* have detected real fluctuations with an amplitude as large as 3.7×10^{-5} in $\Delta T/T$. Davies and his colleagues now consider this an upper limit on $\Delta T/T$ on the angular scales they surveyed.

The Antarctic plateau is higher (by nearly 1000 m), considerably drier and certainly colder than the peaks of the Canary Islands. For that reason, the Amundsen–Scott base maintained by the U.S.A. at the South Pole is rapidly becoming recognized as a premier site for ground-based CBR observations. Groups from a number of research institutions – the Bell Telephone Laboratories; the University of California at Berkeley and at Santa Barbara; Chicago; Milan; Princeton and Rome in various combinations – have made CBR measurements from the South Pole or elsewhere in Antarctica. As yet, only one set of results has been published, the letters of Meinhold and Lubin (1991) and Gaier *et al.* (1992), both from the Santa Barbara group.

This group used an off-axis telescope with a 1 m primary (fig. 7.21) to illuminate their detectors. In the case of the Meinhold–Lubin experiment, the receiver was a low-noise (about $1.6\,\mathrm{mK\,Hz^{-1/2}}$) heterodyne receiver operating at $\lambda = 3$ mm. A 1° beam switch was employed, with 0.5° FWHM beams. The telescope was pointed for about 1 min each at nine different sky positions along at strip at $\delta = -73°$; then the set of nine measurements was repeated. For each rough 9 min scan, both the average measured temperature and a gradient were subtracted to help remove time-dependent atmospheric emission (Lubin *et al.*, 1990). The residual temperature differences from many scans were then stacked

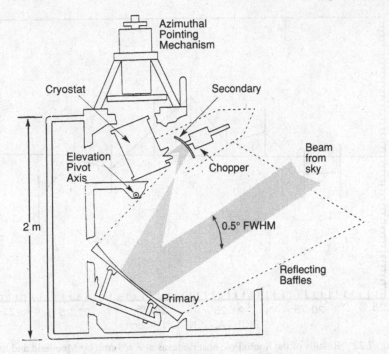

Fig. 7.21   The off-axis telescope used by Meinhold and Lubin (1991, with permission). The detectors are located in the cryostat; beam switching in azimuth is accomplished by wobbling the secondary mirror. The azimuthal pointing mechanism was used when the telescope was carried aloft by a balloon.

together. A clear linear trend was present in the stacked data; this trend in turn was subtracted to produce the final results shown in fig. 7.22. The individual error bars on each point are about ± 50 μK; the upper limits set by the experiment on various models for CBR fluctuations are explored in Bond *et al.* (1991). For comparison with other experiments, we may use $\Delta T/T \le 3.5 \times 10^{-5}$ on a scale of 0.5° (Vittorio *et al.*, 1991)

Gaier and his colleagues used the same telescope, but employed the HEMT amplifier which had a receiver noise temperature of about 30 K over a broad band of frequencies centred at 30 GHz. The lower frequency results in a broader beam of FWHM 1.65 ± 0.1° (27.7/v); the beam switch angle was 1.5°. As in the earlier work, both an offset and a gradient were subtracted. These 9 mm observations produced an upper limit on $\Delta T/T$ of 1.4×10^{-5} at 95% confidence (assuming a Gaussian autocorrelation function; see Gaier *et al.*, 1992, for further details).

As noted in Section 3.5 and in Section 7.8 below, interferometric observations cancel atmospheric noise to a significant degree. This positive feature motivated the 43 GHz observations of Timbie and Wilkinson (1990). They employed a simple, two-element interferometer consisting of two horn antennas separated by a very small distance (Timbie and Wilkinson, 1988). Thus the angular separation of the fringes on the plane of the sky was roughly comparable to the power pattern of the individual antennas, producing the beam shape shown in fig. 7.23. As a further control on sources of systematic noise Timbie and Wilkinson employed a variant of beam switching by 'wobbling' the beam pattern symmetrically about the North Celestial Pole, producing

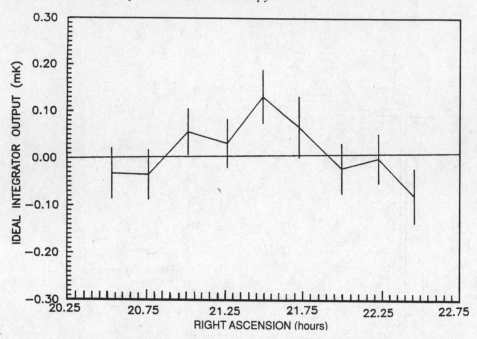

Fig. 7.22   Results of the South Pole observations at $\lambda = 9$ mm by Meinhold and Lubin (1991, with permission). Nine points along a strip at $\delta = -73°$ were observed; in this plot, a linear trend has been subtracted.

the pattern also shown in fig. 7.23. They observed from a dry, sub-Arctic site in Canada. As hoped, both the choice of site and the interferometric technique reduced atmospheric noise to a negligible level.

Their experiment has a maximum sensitivity at an angular scale of 0.5°–2°; for CBR fluctuations with a Gaussian correlation function, for instance, the 95% confidence upper limit at 1.1° scale is $\Delta T/T \leqslant 1.1 \times 10^{-4}$. This limit was obtained using a likelihood ratio test (Section 7.4.2); the power of the test was 53–62% for their data. While the limit on $\Delta T/T$ is less stringent than that of Meinhold and Lubin (1991), the use of interferometry was an important step forward, and other groups are now pursuing it.

7.7.2 Observations from balloon altitudes

Both bolometric and heterodyne receivers are now sensitive enough to allow searches to $\Delta T/T \simeq 10^{-5}$ over many independent regions of sky in the 10–20 h a high altitude balloon can stay aloft. As we saw in Chapter 6, several groups have taken advantage of this technological advance. Balloon flights impose obvious constraints on the size of the antenna; hence balloon-borne experiments have operated only at large angular scales, typically $\geqslant 1°$.

One instance is the experiment of Meyer *et al.* (1991) described in Section 6.6.3. The smallest angular scale probed was about 4°; at this scale (and at $\lambda = 1.8$ mm), $\Delta T/T \leqslant 4 \times 10^{-5}$ at 95% confidence, decreasing to $\Delta T/T \leqslant 1.6 \times 10^{-5}$ at 13° scale. Further flights are planned by this group; note that the sensitivity obtained in one 10 h flight is within a factor of about two of COBE's. Finally, in view of the detection of CBR fluctuations by

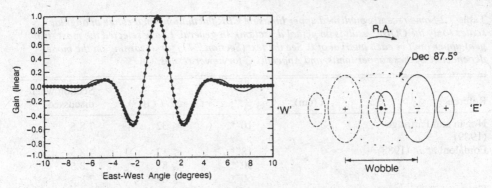

Fig. 7.23   The interference pattern of the two-element interferometer of Timbie and Wilkinson (left). In the right-hand portion, the beam pattern on the sky is shown, including the 3° beam switch (from Timbie and Wilkinson, 1990, with permission).

COBE at about the 10° scale, it is worth noting that Meyer *et al.* (1991) clearly did measure a non-zero value for σ_{sky}, at about twice the calculated level of instrument noise. Whether the measured fluctuations are truly CBR temperature fluctuations is not yet determined; Meyer *et al.* present some other possibilities, including Galactic emission.

Lubin and his students from Santa Barbara have joined forces with Richards and his students and colleagues at Berkeley to make a series of balloon flights to probe scales $\theta \geqslant 0.5°$. Both heterodyne and low-temperature bolometric receivers have been employed. To date, only a few, preliminary results have appeared (e.g. Alsop *et al.*, 1992; see also Fischer *et al.*, 1990). The latter paper reports the second flight of a balloon package dubbed MAX (for Microwave Anisotropy Explorer). The detectors employed were bolometric, operating at $\lambda \simeq 1.7$, 1.1 and 0.8 mm; $\theta_{1/2} = 0.5°$ and $\theta_{s} = 1.3°$. A region of the sky near the star γ UMi was observed. As in the case of Meyer *et al.* (1991), statistically significant sky fluctuations were detected, in this case at a level corresponding to $\Delta T/T \gtrsim 2 \times 10^{-5}$. Once again, Galactic emission may be implicated.

One speculative remark: both of these balloon experiments that detected non-zero values of σ_{sky} operated at short wavelengths (1–2 mm). Is the Galaxy 'noisier' at short wavelengths than expected (or hoped)? The estimates that I have made in Section 7.3.1. of the effective $\Delta T/T$ of Galactic dust emission, while nominally very conservative, are roughly (to within a factor of two) consistent with the level of fluctuations detected by these balloon-borne experiments. On the other hand, as Alsop *et al.* (1992) show, the spectral data derived from their observations quite strongly exclude emission from warm dust as a cause of the observed fluctuations. Future observations planned and underway by this team may elucidate the problem. In the meantime, let me note that the detection reported by Alsop *et al.* (1992), while somewhat puzzling, is not inconsistent with the upper limits reported by Meinhold and Lubin (1991) and Gaier *et al.* (1992).

On several occasions, I have noted the important role played by Italian scientists in the development and use of bolometric detectors for CBR observations. One instance is the balloon-borne experiment of de Bernardis *et al.* (1990). The detector had a single, very wide band pass, $\lambda = 0.4$–2 mm. Beam switching with $\theta_{s} = 1.8°$ was employed, with each beam having a FWHM of 0.4°. The limits on $\Delta T/T$ at an effective wavelength of 1.1 mm were a few times 10^{-4} on angular scales 0.2°–3°. Very recently, another experiment, dubbed ULISSE, has been flown (de Bernardis *et al.*, 1992). The upper limit on

Table 7.2 *Some recently published upper limits on CBR fluctuations on a range of angular scales (only the COBE result is an actual detection). In general, I have selected the most stringent upper limit in each interval of θ. See the text (Section 7.11) for a warning on the model-dependence of these upper limits, and Appendix C for newer results.*

Reference	λ (cm)	θ_s*	$\Delta T/T$ (10^{-5})	Section where discussed
Hogan and Partridge (1989)	2	10"*	$\leqslant 32$	7.8.5
Fomalont *et al.* (1988)	6	18"*	$\leqslant 12$	7.8.5
	6	30"*	$\leqslant 8$	7.8.5
		60"*	$\lesssim 6$	7.8.5
Uson and Wilkinson (1984)	1.5	4.5'	$\lesssim 4$	7.5.2
Readhead *et al.* (1989)	1.5	7'	$\leqslant 1.7$	7.6
Meinhold and Lubin (1991)	0.3	1°	$\leqslant 3.5$	7.7.1
Meyer *et al.* (1991)	0.18	$\simeq 4°$	$\leqslant 4$	7.7.2
Davies *et al.* (1987)	3	8°	$\leqslant 3.7$	7.7.1
COBE (Smoot *et al.*, 1992)		$\simeq 10°$	$= 1.1$	7.7.3

* θ_{sy} in the case of interferometric observations.

$\Delta T/T$ at the angular scale at which the instrument is most sensitive is $\lesssim 1.3 \times 10^{-5}$. Further flights and more complete reports of these millimeter-wave observations are planned by the Rome group; see Appendix C.

7.7.3 The detection of CBR fluctuations by COBE

Let me end this section by reminding readers that the DMR instrument aboard the COBE satellite (Section 6.7.4) has very recently detected CBR fluctuations at a level $\Delta T/T = 1.1 \times 10^{-5}$, just below the sensitivity reached by the experiments described above. Because the corrugated horn antennas employed in DMR were small, the angular resolution was limited to $\geqslant 7°$. Nevertheless, there is every reason to suppose that the fluctuations detected by the DMR at $\theta \simeq 10°$ will also be present at smaller angular scales. Indeed, as we shall see in Chapter 8, the DMR experiment provides tighter limits on $\Delta T/T$ on scales of about 1° than do direct measurements on that scale, at least for many conventional models of CBR fluctuations. Hence I have chosen to list the COBE results here and in Table 7.2 as well as in Section 6.7.4 (where they are more fully described).

7.8   Aperture synthesis observations of the CBR

The use of the interferometric or aperture synthesis technique described in Section 3.5 permits us both to study the angular distribution of the CBR on scales < 1' and to make two-dimensional maps of the microwave sky.

7.8.1 *Advantages and disadvantages of aperture synthesis observations of the CBR*

While high resolution and the ability to map the sky are the primary advantages of this technique, there are some additional advantages. For instance, aperture synthesis efficiently reveals discrete radio sources, whose presence can contribute to the measured sky fluctuation. In addition, because of the way signals are correlated, aperture synthesis observations are free from many of the sources of systematic errors and offsets described in Section 7.2. In particular, receiver gain variations, atmospheric noise and back- and side-lobe pickup are not significant problems in aperture synthesis.

Of course there are also disadvantages to the use of this technique. New varieties of systematic errors arise. One example is the aliasing of sources and noise into an aperture synthesis map (see Section 3.5.5 and Chapter 10 of Thompson *et al.*, 1986). As shown in the latter reference, aliasing affects only a narrow region around the boundary of a map, so this region is simply omitted from the analysis. A second disadvantage is related to one of the advantages listed above: interferometers are very sensitive to discrete radio sources. Methods used to reduce the effect of discrete sources will be discussed in Section 7.8.4 below.

Another drawback of aperture synthesis is unavoidable, however – relative lack of sensitivity to extended sources when compared with conventional radio astronomical techniques. Put simply, synthesis arrays are in a sense inefficient because only a small fraction of the total area of the array is covered by the surfaces of the antennas themselves. The efficiency is therefore reduced by a factor of order $(d/D)^2$, relative to a filled aperture of the size of the array, where d is the characteristic size of the antennas and D the size of the array. Since angular resolution is inversely proportional to characteristic size, this loss in sensitivity may be estimated as $(\theta_{sy}/\theta_{1/2})^2 \approx (\Omega_{sy}/\Omega_M)$, where θ_{sy} is the angular resolution of the array (the synthesized beam size), and $\theta_{1/2}$ is the beam width of the individual antennas of the array. The factor Ω_{sy}/Ω_M is frequently < 1%. Note, however, that a single aperture synthesis map will provide about (Ω_M/Ω_{sy}) independent samples of the sky.

For technical reasons connected with the reconstruction of the map of the sky (Section 3.5.5), the bandwidth Δv of the receivers employed in interferometry must be limited. On the other hand, each correlator connecting a pair of elements of the array effectively serves as an independent receiver of bandwidth Δv. For an array of N antennas, it is therefore possible to employ $N(N-1)/2$ receivers in each of two orthogonal polarizations. Bringing together all these factors we may express, in analogy with eqn. (3.22), the sensitivity of aperture synthesis observations as

$$T_{rms} = \frac{T_{sys}}{[\Delta v \, \Delta t \, N(N-1)]^{1/2}} \left(\frac{\theta_{1/2}}{\theta_{sy}}\right)^2 . \tag{7.19}$$

Since all the results to be discussed in this section were obtained with the Very Large Array (VLA) of the US National Radio Astronomy Observatory in New Mexico, let us pause to evaluate eqn. (7.19) for that array of 27 antennas. For a 12 h observation carried out at $\lambda = 6$ cm, where $\theta_{1/2} = 9'$, and the angular resolution is 18'',

$$T_{rms} = 1.4 \text{ mK},$$

Table 7.3. *VLA limits on CBR fluctuations. Upper limits are at 95% confidence.*

Reference	λ (cm)	Pol.	Angular Scale	Limits on $\Delta T/T$ (10^4)
Knoke *et al.* (1984)	6	I	6″–64″	< 32
Martin and Partridge (1988)	6	I	18″–80″	1.7 ± 0.5
		I	36″–160″	1.3 ± 0.2
Fomalont *et al.* (1988)	6	I	12″	< 8.5
		I	18″	< 1.2
		I	30″	< 0.8
		I	60″	< 0.6
Partridge *et al.* (1988)	6	Q	60″–160″	< 0.4
		U	60″–160″	< 0.5
		V	60″–160″	< 0.6
Hogan and Partridge (1989)	2	I	5″.3–48″	< 6.3
		I	10″–48″	< 3.2
		I	18″–50″	< 1.6
		Q	5″.3–48″	< 2.7
		U	5″.3–48″	< 1.8
		V	5″.3–48″	< 1.3
Tentative results:	3.6	I	40″–60″	≲ 0.3
Fomalont *et al.* in preparation		I	≃ 10″	≲ 1.0

where I have taken $\Delta\nu$ to be 50 MHz and $T_{sys} = 60$ K, a typical value for observations made at a wavelength of 6 cm. The r.m.s. sensitivity at an angular scale of 18″ is approximately ten times worse than that obtained at several arcminutes by Readhead *et al.* (1989) using conventional radio astronomical techniques at $\lambda = 1.5$ cm. On the other hand, a VLA map contains of order 10^3 independent samples of the sky brightness; Readhead *et al.* observed only seven patches of the sky.

It is worth remarking that a closer-packed array would reduce $(\theta_{1/2}/\theta_{sy})$ and hence increase the sensitivity of aperture synthesis observations of the CBR; just such a step is under consideration by the radio group at Cambridge University – the Very Small Array, as it is called. Also, as we saw in the previous section, a two-element interferometer with antennas nearly in contact has already been used by Timbie and Wilkinson (1990) to observe the CBR on degree scales.

7.8.2 VLA observations

Two groups have used the VLA to map the CBR at 2, 3.6, and 6 cm wavelengths (see table 7.3 for references). In the configurations in which the VLA was used, angular resolution of about 0.1′–1.0′ was achieved. The angular scales probed thus lie one or two orders of magnitude below the resolution of most conventional, single-dish, radio telescopes (the exception is the recent work at $\lambda = 1.3$ mm by Kreysa and Chini (1989) with a 11″ beam and a 30″ beam switch angle which showed $\Delta T/T \leqslant 2.6 \times 10^{-4}$).

In a typical VLA program, several 8–10 h observations of a patch of sky chosen to be free of bright discrete sources were stacked together. Bad data (e.g. noise spikes due to lightning or interference) and the output from noisy correlators were removed from the data before making the Fourier transform of the visibilities to produce a map of the sky.

Most maps were made at the full resolution of the array, which for a VLA configuration with a maximum baseline of about 1 km was

$$\theta_{sy} = 18'' \left(\frac{\lambda}{6 \text{ cm}} \right).$$

The same data, however, were also used to obtain information about sky fluctuations on larger angular scales by applying a *taper* to the data. In constructing a tapered map, one convolves the u–v plane data with a weighting function, normally a Gaussian, with unit weight at zero baseline. The weight tends to zero as the baseline increases, effectively reducing the contributions of the longer baseline spacings of the array. Thus the characteristic size of the array is reduced, increasing the synthesized beam width, θ_{sy}. Note that an increase in θ_{sy} in turn increases the sensitivity; eqn. (7.19) shows $\Delta T_{rms} \propto \theta_{sy}^{-2}$. Some of this increase in sensitivity, however, is lost because the effective number of correlators is reduced (T_{rms} is roughly proportional to N^{-1} for large N; again see eqn. (7.19)). In the case of VLA observations, this technique can be used to search for fluctuations on scales up to about five times the full resolution of the array, i.e., to $1'$–$1.5'$ for the work at $\lambda = 6$ cm described here.

7.8.3 Analysis of the VLA data; extracting an upper limit on σ_{sky}

In Fig. 7.24 is shown a VLA map of a $22' \times 22'$ portion of the sky at declination $80°$ made at $\lambda = 6$ cm with a resolution of $18''$ by Martin and Partridge (1988). Foreground radio sources dominate the map. Our interest is in the level of fluctuations in the background sky, away from the sources.

Our group (Knoke *et al.*, 1984; Martin and Partridge, 1988) therefore began the analysis of our VLA map by eliminating obvious radio sources from our maps. Our technique that we employed was simply to excise rectangular regions containing visible sources from the map. Some of the remaining 'grainy' noise in fig. 7.24 in source-free regions is instrument noise, mostly generated by the receivers and correlators.

We may use the results of Section 3.5.4 to show that instrument noise (and also sky noise) will be uniformly distributed across the entire map. This follows from the assumption that instrument noise may be modeled as an uncorrelated series of delta-function events or 'spikes' in the u–v plane. The Fourier transform of a delta-function in two dimensions is a uniform plane. (See Knoke *et al.*, 1984 and references therein for further discussion.)

Thus we expect no difference in the amplitude of instrument noise as we move from the edge to the center of an aperture synthesis map. On the other hand, real fluctuations in the microwave sky will be modulated by the primary beam pattern of the individual telescopes of the array ($P(\xi, \eta)$ in the notation of Section 3.5). Thus sky fluctuations will appear at full intensity only near the center of the map, where $P(\xi, \eta)$ is of order unity. Also, since the size of the map shown in fig. 7.24 is about three times the primary beam width, $P(\xi, \eta)$ is very small at the edge of the map. As a consequence, the array is insensitive to real sky fluctuations at the edge of the map. To a first approximation, then, the variance in the measured intensity at points near the center of the map may be expressed as

$$(\text{var})_c = \sigma_I^2 + \sigma_{sky}^2$$

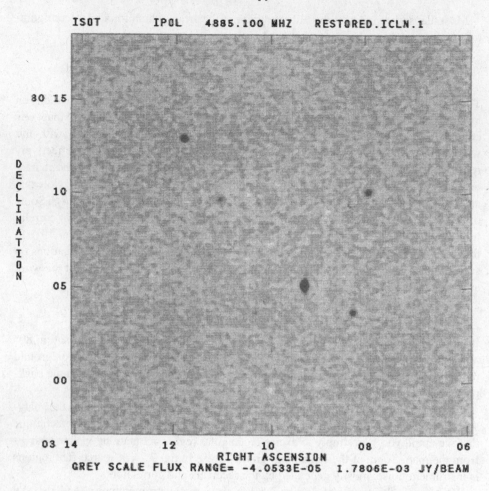

ISOT   IPOL   4885.100 MHZ   RESTORED.ICLN.1

GREY SCALE FLUX RANGE= -4.0533E-05   1.7806E-03 JY/BEAM

Fig. 7.24   VLA map of a 22' × 22' region at $\lambda = 6$ cm (from Martin and Partridge, 1988, with permission). The primary beam of the VLA antennas is ~9' FWHM, hence few sources are visible at substantially greater distances from the map center.

and at the edge as

$$(\text{var})_e = \sigma_I^2,$$

where σ_I^2 contains uncorrelated instrumental and atmospheric noise. A simple subtraction thus allows an estimate of σ_{sky}^2 alone. We assume that the instrumental noise and sky noise are independent of each other. It is also necessary to omit the very edge of such a map from the analysis, since the variance very near the boundaries can be affected by aliasing (as noted in Section 7.8.1 above).

The primary beam pattern $P(\xi, \eta)$ of the VLA antennas is known with sufficient accuracy to permit measurements from the entire map, not just the center and edge, to be used in disentangling σ_I and σ_{sky}. For instance, Fomalont *et al.* (1988) divided their VLA maps into several wide, concentric rings and calculated the average $P(\xi, \eta)$ in each. Martin and Partridge (1988), following Knoke *et al.* (1984), divided the VLA maps into an $n \times n$ grid of boxes measured the variance in each box, then minimized the sum

$$\sum_{i,j=1}^{n} [(\text{var})_{ij} - (\sigma_I^2 + P_{ij}^2 \sigma_{\text{sky}}^2)]^2,$$

where P_{ij}^2 is the mean square of the primary beam response in the i, jth box.

In both cases, these procedures allowed the observers to measure or set limits on sky fluctuations σ_{sky} substantially below the measured instrument noise σ_I (see table 7.3). It is worth noting that both groups working at the VLA (Fomalont *et al.*, 1988; Martin and Partridge, 1988) do report values of σ_{sky} differing significantly from zero. Furthermore, their values for σ_{sky} are in quantitative agreement even though the two groups looked at different regions of the sky and used different techniques to construct their 6 cm maps.

Since the measured σ_{sky} was non-zero, both groups performed checks to ensure that no systematic effects could mimic the radial dependence of sky noise. Fomalont *et al.* (1988) broke their data into two equal sets, and subtracted one map from another. As expected, the sources were absent from this difference map, but its overall noise level in regions far from the phase center was in agreement with that in a map made with all the data added together. Fomalont *et al.* did find, however, a slight increase in the noise near the center of the difference map; this they ascribe to an unknown instrumental effect, and increase their estimates of σ_I^2 near the map center accordingly. Note that this has the effect of lowering the residual value of σ_{sky}^2.

Martin and Partridge (1988) employed a different check: examining a map in Stokes parameter *V*, which employed the same visibility data as used in the total intensity map. As expected, the value of σ_I^2 was close to that found for the total intensity map. There was no statistically significant evidence for excess variance at the map center*, so no correction to σ_I^2 was made.

7.8.4 The contribution of discrete, foreground radio sources

Until very recently, the most sensitive VLA observations of the CBR were those made at a wavelength of 6 cm. At that wavelength, foreground radio sources are prominent (see Section 7.3.2 and fig. 7.10). As noted, the observers chose regions of the sky free of bright sources and in some cases excised the remaining detectable sources from the map before performing the statistical analysis outlined above. There remains the possibility that foreground radio sources too weak to be individually detected did contribute to σ_{sky}. The calculations of Franceschini *et al.* (1989) suggest a radio source contribution to σ_{sky} equivalent to $\Delta T/T \simeq (1-1.5) \times 10^{-4}$, as fig. 7.10 suggests. Our VLA maps themselves reveal the presence of subliminal sources, as may be seen by plotting a histogram of measured surface brightnesses for a patch of the sky close enough to the center of the map that $P(\xi, \eta) \simeq 1$, but not including detectable sources. Such a histogram for our 6 cm map is shown in fig. 7.25; the extended 'tail' to positive values of the brightness is evidence for faint sources.

To derive an upper limit on the amplitude of fluctuations in the CBR, some way must be found to subtract the contribution made by foreground radio sources from values of σ_{sky} calculated from our maps. Both groups modeled these contributions by extrapolating to $\lambda = 6$ cm counts of radio sources at higher flux densities and/or longer wavelength Does the variance introduced by these faint sources explain *all* the excess variance at the

---

* Note that this result allows us to place a limit on circularly polarized fluctuations in the CBR – see the following section.

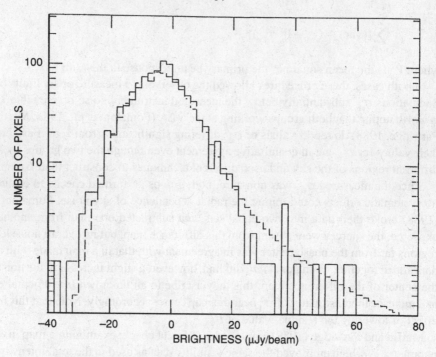

Fig. 7.25   The distribution of observed brightnesses in the central $(2.5')^2$ of the map shown in fig. 7.24 after visible sources were removed. The tail at positive values of brightness is in large part due to faint sources not directly visible in the map. An estimate of their contribution, based on an extrapolation of faint source counts, is shown by the dashed line.

map centers? Here is where the two groups disagreed. Martin and I said no (though it seemed close). Fomalont *et al.* (1988) said yes. They included an effect that we neglected – the fact that any source in an aperture synthesis map will have side lobes (in physical terms, weak secondary interference maxima). These side lobes appear with both positive and negative flux density in an aperture synthesis map. Although the amplitude of these side-lobes signals is low (as low as 1–2% of the peak response at the VLA), they do add to the variance, broadening the histogram shown in fig. 7.25 symmetrically. In our analysis, we included only positive fluctuations, and hence may have underestimated the full effect of weak sources and their sidelobes. Hence we may have overestimated the amplitude of the sky variance ascribed to the CBR.

7.8.5 *Results of the VLA observations*

Values of, or limits on, CBR fluctuations at $\lambda = 6$ cm derived from VLA aperture synthesis maps are displayed in table 7.3. Two qualitative features of these results stand out. First the higher resolution work of Knoke *et al.* (1984) is far less sensitive than later work, largely for reasons discussed in section 7.8.1. Second, there is a qualitative disagreement between Fomalont *et al.* (1988) and our group: Martin and I report statistically significant evidence for fluctuation in the microwave sky even *after* the contribution of faint radio sources has been subtracted; Fomalont and his colleagues give upper limits only. The difference almost certainly arises from the different approaches taken in modeling the contribution of foreground sources *and their side lobes*.

If the problem is indeed foreground sources, one way to resolve the apparent disagreement is to make similar observations at a shorter wavelength, since the spectra of most foreground synchrotron sources goes roughly as $T(\lambda) \propto \lambda^{2.7}$. Hogan and Partridge (1989) did just that, remapping at $\lambda = 2$ cm a small patch of the sky already studied at 6 cm. In order to match the 18" resolution of the 6 cm map, we had to taper our 2 cm map heavily. There was no clear evidence for sources in either the tapered map or the full resolution ($\theta_s \approx 6"$) map; thus there was no need to make *any* correction for foreground source contributions to σ_{sky} at 2 cm. Unfortunately, the 2 cm receivers at the VLA have much higher system noise, so we have only marginal results (table 7.3). The 95% confidence upper limit on $\Delta T/T$ on a scale of 18" does call our claim of a detection at 6 cm ($\Delta T/T = 1.7 \pm 0.5 \times 10^{-4}$) into some question.

Clearly, more work, preferably at both high sensitivity and at a wavelength < 6 cm, is needed. In 1989, the VLA was equipped with very low noise receivers operating at $\lambda = 3.6$ cm. The two groups involved in VLA mapping of the CBR have joined forces to use the VLA to search for CBR fluctuations at this wavelength. Our target sensitivity at the full resolution of the array ($\theta_s \approx 12"$) is $\Delta T/T \lesssim 5 \times 10^{-5}$; with tapering we should be able to reach $\lesssim 3 \times 10^{-5}$ at $\theta_s = 40"-60"$ (Fomalont *et al.*, in preparation). If this work succeeds, we will thus approach the sensitivity of the best work on arcminute scales (Readhead *et al.*, 1989).

Until these results are available, it would be prudent to regard the 6 cm results of table 7.3 as upper limits. I would suggest taking $\Delta T/T < 1.5 \times 10^{-4}$ and $< 1.0 \times 10^{-4}$ at scales of 18" and 60", respectively, as appropriate 95% confidence limits on CBR fluctuations on small angular scales.

These limits and the values given in table 7.3 are upper limits on the r.m.s. sky noise on an angular scale set by the size of the synthesized beam of the VLA, θ_{sy}. Thus they are not directly comparable with the upper limits set by the kind of measurements described in Sections 7.5–7.7, which involve beam-switching and hence record *differences* in temperature at points separated by the beam switch angle. Any comparison between interferometric limits on $\Delta T/T$ and limits obtained from beam-switched experiments will depend on the correlation function $C(\theta)$ or other model assumed for the CBR fluctuations. For instance, in the case of a Gaussian model for $C(\theta)$, we see from fig. 7.19 that the measurements of Readhead *et al.* (1989) provide tighter upper limits than the 6 cm VLA measurements at all angular scales. If the 3.6 cm measurements now underway at the VLA meet or exceed their target sensitivity, they will better the Readhead limits at all scales $\theta_c \lesssim 36"$.

Finally, let me note that interferometric observations place direct limits on the spatial-frequency power spectrum of sky fluctuations. This quantity, expressed in terms of the coefficients C_n defined by eqn. (7.15), is frequently employed in comparison of CBR observations and theoretical models (see Section 8.4). For small angles θ (or large harmonic orders n), C_n is the Fourier transform of $C(\theta)$. It is worth remarking that, in the case of a Gaussian correlation function, C_n will also be Gaussian distributed.

To see how the C_n emerges from interferometric observations, recall that the quantities measured in aperture synthesis are the amplitudes and phases for each baseline (see Section 3.5.4). The baselines, like n, scale with θ^{-1}. Thus the square of the measured visibility amplitudes as a function of baseline gives C_n directly (Lowenthal *et al.*, 1994). We may see the same thing by recalling that the usual way of analyzing aperture synthesis data is to Fourier transform the u–v plane measurements to produce a two-dimensional map of the sky (eqn. (3.33)). Thus the u–v plane data are the Fourier transform of the sky, just as C_n is of $C(\theta)$. Unfortunately, there are complications, such as the presence of random foreground sources, sparse sampling of the u–v plane and the primary beam response. Nevertheless, an analysis of some VLA data now underway (Lowenthal *et al.*, 1994) appears to show that the u–v plane data provide limits on the amplitude of CBR fluctuations comparable with those obtained from sky maps constructed from the same data.

7.9   Polarized fluctuations in the CBR

Up to this point in this chapter, we have considered only fluctuations in total intensity of the CBR (that is, in Stokes parameter I). As Bond and Efstathiou (1987) among others have shown, small angular scale fluctuations in the polarization state of the cosmic background are predicted by many theories of galaxy formation (Sections 8.4.6 and 8.7.3). While it is the case that the fluctuations in polarization are expected to be of smaller amplitude than fluctuations in total intensity, it is also the case that searches for polarization fluctuations are free of some of the systematic errors limiting the sensitivity of searches for total intensity fluctuations. Thus it has proven possible in some cases to set limits on polarized fluctuations, written $\Delta T_p/T_0$, somewhat below the corresponding limits on $\Delta T/T$ derived from the same instrument.

7.9.1   VLA results

At the VLA, the data required for the measurement of all four Stokes parameters (Section 3.1) are recorded simultaneously. It is thus straightforward to use the observational results described in Section 7.8 to set limits on both the circular and linear polarization of the CBR on a range of angular scales from $6''$ to $160''$. The results of such a program (Partridge, Nowakowski and Martin, 1988; Hogan and Partridge, 1989) are shown in table 7.3. Note that the upper limits on polarized fluctuations are in general lower than those on total power fluctuations, in part because some sources of systematic error were smaller and in part because foreground sources were much weaker in polarized emission. The relatively poor limits on polarized fluctuations at $6''$ scale were obtained from the data taken by Hogan and Partridge in 1989.

7.9.2   Indirect upper limits

As Pariiskii and Korol'kov (1987) have noted, many of the searches for total intensity CBR fluctuations described in this chapter were in fact made with radio telescopes having polarized inputs. Thus the limits such searches set on $\Delta T/T$ may also be taken as limits on $\Delta T_p/T_0$. For instance, the 7.6 cm feed of the RATAN telescope is linearly polarized; hence $\Delta T_p/T_0 \approx \Delta T/T < 10^{-5}$ at scales in the range $4.5'$–$9.0'$ (see Section 7.5.2). Unfortunately, it is not always the case that the polarization properties of the instruments are given in the observational papers cited in this chapter. As a consequence, the relationship between a published limit on $\Delta T/T$ and the corresponding limit on polarized fluctuations on the same scale cannot always be determined.

7.10 Searches for the Sunyaev–Zel'dovich effect in clusters of galaxies

The observational programs described so far were designed to search for cosmological fluctuations imprinted on the CBR at the epoch of last scattering. Hence the observers could look anywhere on the sky (and, perhaps unfortunately, no two groups have picked the same patch to observe). One class of observations does have specific 'target' directions, however – searches for fluctuations imprinted on the CBR as the 3 K radiation passes through and interacts with hot gas in clusters of galaxies.

As we have seen in Section 5.2.4, the inverse Compton scattering of CBR photons by hot $(T_e \simeq 10^8$ K) electrons in the clusters will distort the CBR spectrum: this is the Sunyaev–Zel'dovich (SZ) effect. In the Rayleigh–Jeans region, the CBR temperature will be lowered along lines of sight passing through clusters of galaxies containing hot plasma. Thus such a cluster will produce a 'dip' in the CBR temperature in the solid angle subtended by the cluster (at high frequencies, $v \gtrsim 220$ GHz, the sign of the SZ effect becomes positive).

The detection of such a 'dip' in the direction of a cluster of galaxies would confirm the SZ effect as well as confirming that the origin of the CBR lies beyond the cluster. More importantly, a measurement of the amplitude and angular size of the dip would provide important information on the distribution, density and temperature of the intergalactic plasma in clusters. Finally, as first pointed out by Gunn (1978) and Silk and White (1978), these results, combined with X-ray observations of the same clusters, would give us a way to determine the distance to clusters directly (and hence to check our values of H_0, Hubble's constant; see Section 8.9).

7.10.1 The first searches

Early attempts (e.g., Pariiskii, 1973a, b; Rudnick, 1978; Perrenod and Lada, 1979; Lake and Partridge, 1980) to measure the SZ effect – or even to detect it – produced discordant results. Among the problems encountered were insensitive receivers (and hence long integration times), inadequate beam switch angles (so that the reference beam was not free of the cluster), the presence of discrete radio sources in the clusters (Lake and Partridge, 1980), and systematic error introduced by side- and back-lobe pickup by the telescopes employed. The last mentioned problem was made worse by the need to track each cluster for extended periods to build up the necessary integration time to detect the roughly 1 mK signal expected (see Perrenod and Lada, 1979). In addition, the observations made by scientists in the USA were carried out at national facilities where the pressure for telescope time limited each experiment to a few days.

In Britain, Birkinshaw and his colleagues were able to employ a telescope at $\lambda = 3$ cm dedicated to the SZ search. Hence they were able to search out and correct some sources of systematic error, and to amass many days' worth of measurements. Unfortunately, the weather at their site was poor, so the data varied strongly in quality. To extract a measurement of the SZ dip therefore required an intricate statistical analysis (Birkinshaw, Gull and Northover, 1981; Birkinshaw and Gull, 1984). Several detections were reported but subsequent work has shown that either the amplitude of the SZ effect was overestimated or the associated experimental error was understated in these early papers.

Additional observations at wavelengths from about 3 mm (Radford *et al.*, 1986) to 6 cm (Lasenby and Davies, 1983) did little to clarify the situation. Nor did an attempt by Chase *et al.* (1987) to observe at a frequency ($v = 261$ GHz) high enough that the SZ signal was positive in sign.

7.10.2 The current status of SZ searches

The best current data on the SZ effect in clusters of galaxies come from two extended observing programs, one mounted at the Owens Valley Radio Observatory in California (Birkinshaw *et al.*, 1984; Birkinshaw, 1991), the other at the 40 m telescope in Green Bank, West Virginia (Uson and Wilkinson, 1988). Those results (and a few others) are gathered together in table 7.4.

Both groups employed beam switching and the 'on–off' technique and both had available low noise receivers operating at $\lambda = 1.5$ cm. Unlike some of the earlier observers, Birkinshaw and his colleagues and Uson and Wilkinson chose not to make drift scans of the clusters; instead they observed several nominally blank regions of sky on each side of the target cluster as well as the cluster center. In fig. 7.26 are displayed some of the results (from Birkinshaw, 1991). Some of the observations required corrections for the effect of background or cluster radio sources in either the main beam or the reference beams. For instance, Moffet and Birkinshaw (1989) made use of the great sensitivity of the VLA to map the region around three of the clusters he and his colleagues had observed. The measured 21, 6 and 2 cm fluxes of discrete sources were then scaled to 1.5 cm and subtracted from the measurements.

Both Birkinshaw and Uson (and their colleagues) are continuing to amass and analyze data. Hence only brief and presumably preliminary results have been published, and these omit many details of the observations and the analysis (see papers referenced above, plus Uson, 1985).

At about the same time that these results were first appearing, two new approaches to SZ observations were tried. First, Meyer *et al.* (1983) attempted to observe several clusters at wavelengths near and shorter than the peak of the 3 K blackbody spectrum, where the sign of the SZ effect reverses (Section 5.2.4). Their observations have already been mentioned in Section 7.5.2. While the technique holds promise, their instrumental efficiency was so low that no reliable detections were obtained. High frequency, bolometric observations were continued by Chase *et al.* (1987). Two clusters were observed with errors corresponding to about 1 mK if measured in the Rayleigh–Jeans region. Given the promise of such high frequency observations, I hope that they will be pursued.

In 1987, Partridge *et al.* reported the results of a first attempt to map the SZ effect in a cluster of galaxies using aperture synthesis. Since cluster 2218 in the Abell catalog of clusters of galaxies (Abell, 1958) is the best-observed SZ 'target,' it was selected for this attempt. The VLA was used at $\lambda = 6$ cm. Unfortunately, while a sensitivity < 1 mK was achieved, the angular scale of the cluster was not well suited to these aperture synthesis observations, so the limits on the amplitude of the SZ effect we obtained are not definitive (see table 7.4).

I believe it is safe to say, however, that the SZ effect has been detected in a few clusters of galaxies, notably numbers 665 and possibly 2218 in Abell's catalog, and a higher redshift cluster, 0016 + 16. The detections are at a level of 0.3–0.5 mK, typically, and not the $\gtrsim$ 1 mK claimed earlier by some observers. And, as table 7.4 shows, there still exist discordant results.

Table 7.4 A selection of recent measurements of the SZ effect in several clusters of galaxies. Results are not shown for all clusters – see references here and in the text for additional results. Some parameters of the beam geometry are given in the third column; in most cases, 'on–off' observations were made. In the sixth column, the measured values have been converted to a central decrement, assuming the cluster gas is isothermal.

Reference	Frequency (GHz)	Beam pattern $\theta_{1/2}$, θ_s	Cluster	Measured ΔT (mK)	Central decrement (mK)
Birkinshaw and Gull (1984)	10.7	$\simeq$ 3.3′, 20′	0016 + 16	− 0.72 ± 0.18	
			A576	− 0.14 ± 0.29	
			A665	+ 0.03 ± 0.25	
			A2125	− 0.31 ± 0.39	
			A2218	− 0.38 ± 0.19	
	20.3	1.8′, 7.2′	0016 + 16	− 0.37 ± 0.16	
			A665	− 0.55 ± 0.13	
			A2218	− 0.31 ± 0.13	
Birkinshaw et al. (1984)	20.3	1.8′, 7.2′	0016 + 16	− 0.64 ± 0.08	− 1.10 ± 0.34
			A665	− 0.34 ± 0.05	− 0.50 ± 0.07
			A2218	− 0.34 ± 0.05	− 0.50 ± 0.07
Birkinshaw et al. (1991) (revised version of value above)	20.3	1.8′, 7.2′	A665	− 0.43 ± 0.05	
Partridge et al. (1987)	4.9	0.3′ *	A2218	$\simeq$ + 0.5 ± 0.5	$\simeq$ + 0.5 ± 0.5
Radford et al. (1986)	90	1.2′, 4.3′	0016 + 16	− 0.8 ± 0.9	− 1.4 ± 1.5
	105	1.7′, 19′	A576	+ 0.80 ± 0.47	+ 1.8 ± 1.0
			A2218	+ 0.41 ± 0.32	+ 0.89 ± 0.69
Uson and Wilkinson (1988)	19.5	1.8′, 8.0′	0016 + 16	− 0.48 ± 0.12	− 0.78 ± 0.20
			A665	− 0.37 ± 0.14	− 0.51 ± 0.19
			A2218	− 0.29 ± 0.24	− 0.40 ± 0.33
Klein et al. (1991)	24.5	0.65′, 1.9′	A2218	− 0.6 ± 0.2	

* Interferometric measurement.

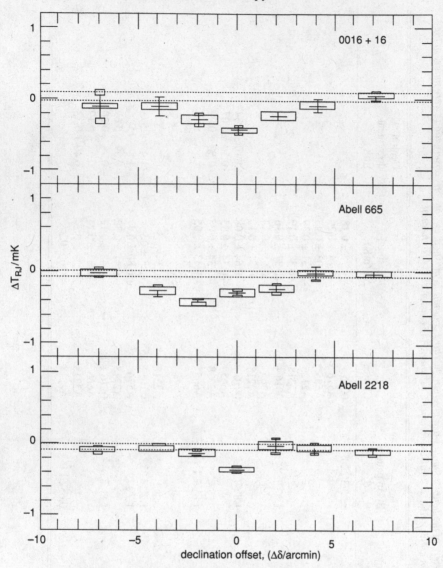

Fig. 7.26 North–south scans of three clusters of galaxies, showing the SZ effect (from Birkinshaw, 1991, with permission). The two dotted, horizontal lines indicate the uncertainty in the zero-level of these measurements – an important issue since the angular scale of the SZ signal is comparable with the length of the scan in two of the cases. Uncertainties in the corrections for radio sources lying in the main beam or reference arcs are included in the error boxes.

In order to compare different results obtained on a single cluster, we need to take into account the beam geometries employed in different experiments (e.g., Radford *et al.*, 1986; Birkinshaw *et al.*, 1991). In particular, by convolving the beam geometry with a model for the distribution of hot gas in a cluster, we can calculate the *central decrement*, the value of ΔT that would be observed in an ideal experiment with an infinitely narrow beam pointed at the cluster center. For an isotropic cluster, we can generalize eqn. (5.12) to give

$$\frac{\Delta I}{I} = \frac{\Delta T_A}{T_A} = \frac{k\sigma_T}{m_e c^2} \; f(x) \int n_e(\theta, l) T_e(\theta, l) dl, \qquad (7.20)$$

where m_e is the electron mass and σ_T the Thomson scattering cross-section. T_e is the electron temperature as a function of angular distance from the cluster center (θ), and path length through the cluster (l); n_e is the number density of electrons. The frequency dependence of the SZ signal is given by $f(x)$, where $x \equiv h\nu/kT_0$:

$$f(x) = \frac{xe^x [x \coth (x/2) - 4]}{(e^x - 1)} \qquad (7.21)$$

(Sunyaev and Zel'dovich, 1969). For small x, that is in the Rayleigh–Jeans region, $f(x) \rightarrow -2$, the same result as was obtained in eqn (5.14).

The value of the integral in (7.20) will clearly depend on both the beam geometry and the model assumed for the distribution of plasma in the cluster. One frequently used model for the plasma is the isothermal model, in which T_e is independent of l and θ, and n_e may be written as

$$n_e(\theta, l) = n_0 [1 + (\theta/\theta_c)^2 + (l/r_c)^2]^{-3\tau/2}, \qquad (7.22)$$

where θ_c and r_c are values for the cluster core radius, and τ is a parameter introduced by Cavalieri and Fusco-Femiano (1978), the ratio of the galaxy scale height to the plasma scale height in the cluster. This parameter appears to lie in the range $0.3 < \tau \leqslant 1.0$. It may easily be shown from eqn. (7.22) that the integrated SZ profile for an isotropic, isothermal cluster is

$$\Delta T(\theta) \propto [1 + (\theta/\theta_c)^2]^{1/2 - 3\tau/2}. \qquad (7.23)$$

As may be seen from eqn. (7.22), the total number (and hence mass of the cluster gas) will diverge for $\tau \leqslant 1$. Hence the isothermal model cannot hold exactly, particularly at large distances from the cluster center. The outer, low density regions, however, have only a small effect on the SZ signal, so the isothermal model remains a useful approximation. The convolution of a profile of the sort given by eqn. (7.23) with a beam pattern formed by two Gaussian beams separated by a beam-switch angle θ_s has been worked out by Radford *et al.* (1986). While the value of the central decrement does depend on parameters of the model (θ_c and τ), Radford *et al.* show that no reasonable model can account for the discrepancies between the observations listed in table 7.4.

In the last column of table 7.4, I have listed values of the central decrement calculated for one particular model, with $\tau = 0.8$. I have also given the value of the central decrement appropriate for the Rayleigh–Jeans region, where $f(x) = -2$; i.e., the frequency dependence in eqn. (7.20) has been factored out.

The agreement between values for the central decrement obtained by different groups for two clusters, 0016 + 16 and Abell 665, is heartening. SZ decrements of −1.0 and −0.5 mK for these two clusters seem fairly secure and further support is provided by fig. 7.26. The tabulated results for Abell 2218 are not as convincing. In addition, the scan of the cluster made by Birkinshaw and his colleagues (and shown in fig. 7.26) is more centrally 'peaked' than would be expected for a cluster with $\theta_c \simeq 1'$. One known

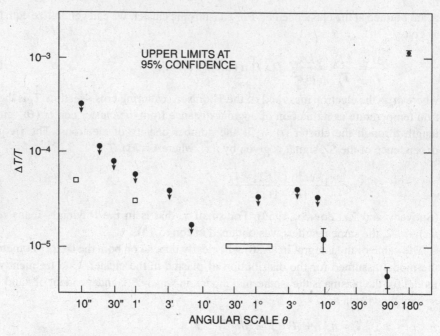

Fig. 7.27   Summary of the best current limits on $\Delta T/T$ (see Section 7.11 for cautionary advice on the use of this figure). The small squares at 10″ and 1′ represent target sensitivities of new work underway at the VLA (Section 7.8.5). The rectangle at ~30′–2° represents the probable sensitivity to be achieved by balloon-borne instruments now being flown or being readied for flight (Section 7.7.2). The values at $\theta \geqslant 10°$ are the new COBE measurements (Section 7.7.3).

problem encountered in measurements of Abell 2218 is the presence of several radio sources within a few arcminutes of the nominal cluster center (Partridge *et al.*, 1987; Birkinshaw, 1991; Klein *et al.*, 1991); these mask and interfere with observations of the SZ effect.

In the case of Abell 2218 and other clusters containing radio sources, observations at higher frequencies would help; the SZ signal changes slowly with frequency (eqn. (7.21)), and the antenna temperature of typical discrete sources as $v^{-2.7}$. In addition, clusters known to be free of bright radio sources, and well matched to the angular scale of the beam geometry, can be selected. These and other avenues are being pursued by both Birkinshaw and his colleagues, and Uson and his, and by other groups as well.

The implications of the few reliable measurements of the SZ effect already available will be considered in Section 8.9; see also Appendix C for more recent results.

7.11   Summary

Using a variety of observational techniques, astronomers have mounted searches for CBR fluctuations on angular scales from about 6″ to the large scales described in Chapter 6 – that is, over a range of angular scale approaching 10^5. The present status of a quarter-century's work is summarized in tables 7.2 and 7.3, and in fig. 7.27.

With the very important exception of the recent COBE results on scales $\geqslant 10°$ (Smoot *et al.*, 1992), there is no unambiguous and undisputed evidence for the detection of CBR anisotropies on any angular scale. The exact limit set by each of the observations listed in table 7.2 and shown in fig. 7.27 will depend to some degree on the model assumed for the CBR fluctuations. That model-dependence must be borne in mind when comparing one measurement with another and when comparing the experimental upper limits to theoretical predictions. Hence fig. 7.27 is schematic only and must be used with caution.

We may nevertheless draw some semi-quantitative conclusions from fig. 7.27. On arcminute scales, the upper limits on $\Delta T/T$ are approaching 10^{-5} at 95% confidence. Just below $1'$ and above $1°$ in scale, there are limits of a few times 10^{-5}. One could wish for tighter limits in the $10'–60'$ range, but even there, $\Delta T/T < 10^{-4}$. My guess is that upper limits on (or, more optimistically, detections of) CBR fluctuations on about $0.5°$ scale will soon be reported at the 10^{-5} level (Gaier *et al.*, 1992, for instance, find $\Delta T/T \leqslant 1.4 \times 10^{-5}$). I have indicated this level on fig. 7.27, as well as the target sensitivities of the interferometric work now underway at $10''–60''$ scale (Section 7.8.5). Newer results are summarized in Appendix C.

The remarkable smoothness of the CBR on all scales probed, up to and including the quadrupole scale, is a central fact and a crucial constraint in cosmology. In Chapter 8, we will see just how valuable these measurements are.

References

Reports of the COBE Science Team are indicated below by '(COBE)' following the citation.

Abell, G. O. 1958, *Ap. J. Suppl.*, **3**, 211.

Aizu, K., Inoue, M., Tabara, H., and Kato, T. 1987, in *I.A.U. Symposium 124, Observational Cosmology*, ed. A. Hewitt, G. Burbidge, and L. Z. Fang, D. Reidel, Dordrecht.

Alsop, D. C., *et al.* 1992, *Ap. J.*, **395**, 317.

Armstrong, J. W., and Sramek, R. A. 1982, *Radio Science*, **17**, 1579.

Banday, A. J., and Wolfendale, A. W. 1991, *Monthly Not. Roy. Astron. Soc.*, **252**, 462.

Banday, A. J., Giler, M., Szabelska, B., Szabelski, J., and Wolfendale, A. W. 1991, *Ap. J.*, **375**, 432, and references therein.

Bennett, C. L. *et al.* 1992, *Ap. J. (Lett.)*, **396**, L7 (COBE).

Berlin, A. B., Bulaenko, E. V., Vitkovsky, V. K., Kononov, V. K., Pariiskii, Yu. N., and Petrov, Z. E. 1983, in *I.A.U. Symposium 104, Early Evolution of the Universe*, eds. G. Abell and G. Chincarini, D. Reidel, Dordrecht.

Birkinshaw, M. 1991, in *Physical Cosmology*, eds. A. Blanchard *et al.*, Editions Frontiers, Gif-sur-Yvette.

Birkinshaw, M., and Gull, S. F. 1984, *Monthly Not. Roy. Astron. Soc.*, **206**, 359.

Birkinshaw, M., Gull, S. F., and Northover, K. J. E. 1981, *Monthly Not. Roy. Astron. Soc.*, **197**, 571.

Birkinshaw, M., Gull, S. F., and Hardebeck, H. 1984, *Nature*, **309**, 34.

Birkinshaw, M., Hughes, J. P., and Arnaud, K. A. 1991, *Ap. J.*, **379**, 466.

Bond, J. R., and Efstathiou, G. 1987, *Monthly Not. Roy. Astron. Soc.*, **226**, 655.

Bond, J. R., Efstathiou, G., Lubin, P. M., and Meinhold, P. R. 1991, *Phys. Rev. Lett.*, **66**, 2179.

Boughn, S. P., Cheng, E. S., Cottingham, D. A., and Fixsen, D. J. 1992, *Ap. J. (Lett.)*, **391**, L49.

Boulanger, F., and Perault, M. 1988, *Ap. J.*, **330**, 964.

Boulanger, F., Baud, B., and van Albada, G. D. 1985, *Astron. Astrophys.*, **144**, L9.

Boynton, P. E., and Partridge, R. B. 1973, *Ap. J.*, **181**, 243.

Bridle, A. H. 1967, *Monthly Not. Roy. Astron. Soc.*, **136**, 219.

Caderni, N., De Cosmo, V., Fabbri, R., Melchiorri, B., Melchiorri, F., and Natale, V. 1977, *Phys. Rev. D*, **16**, 2424.

Carpenter, R. L., Gulkis, S., and Sato, T. 1973, *Ap. J. (Lett.)*, **182**, L61.

Cavalieri, A., and Fusco-Femiano, R. 1978, *Astron. Astrophys.*, **70**, 677.

Chase, S. T., Joseph, R. D., Robertson, N. A., and Ade, P. A. R. 1987, *Monthly Not. Roy. Astron. Soc.*, **225**, 171.

Cole, S., and Kaiser, N. 1988, *Monthly Not. Roy. Astron. Soc.*, **233**, 637.
Condon, J. J. 1988, in *Galactic and Extragalactic Radio Astronomy*, 2nd edn., eds. G. L. Verschuur and K. I. Kellermann, Springer-Verlag, Berlin.
Condon, J. J., Broderick, J. J., and Seielstad, G. A. 1989, *A.J.*, **97**, 1064.
Conklin, E. K., and Bracewell, R. N. 1967, *Nature*, **216**, 777.
Cottingham, D. A. 1987, Ph.D. thesis, Princeton University.
Danese, L., De Zotti, G., and Mondolesi, N. 1983, *Astron. Astrophys.*, **121**, 114.
Davies, R. D., Lasenby, A. N., Watson, R. A., Daintree, E. J., Hopkins, J., Beckman, J., Sanchez-Almeida, J., and Rebolo, R. 1987, *Nature*, **326**, 462.
de Bernardis, P., *et al.* 1990, *Ap. J.* (*Lett.*), **360**, L31 (see also de Bernardis *et al.* 1990, in *I.A.U. Symposium 139, Galactic and Extragalactic Background Radiation*, eds. S. Bowyer and C. Leinert, D. Reidel, Dordrecht.
de Bernardis, P., Masi, S., Melchiorri, F., Melchiorri, B., and Vittorio, N. 1992, *Ap. J.* (*Lett.*), **396**, L57.
Donnelly, R. H., Partridge, R. B., and Windhorst, R. A. 1987, *Ap. J.*, **321**, 94.
Dravskikh, A. F., and Finkelstein, A. M. 1979, *Astrophys. Space Sci.*, **60**, 251.
Epstein, E. E. 1967, *Ap. J.* (*Lett.*), **148**, L157.
Fabbri, R., Guidi, I., Melchiorri, F., and Natale, V. 1980, *Phys. Rev. Lett.*, **44**, 1563.
Fabbri, R., Guidi, I., Melchiorri, F., and Natale, V. 1982, in *Proceedings of the Second Marcel Grossmann Meeting on General Relativity*, ed. R. Ruffini, North Holland, Amsterdam.
Fischer, M. L. *et al.* 1990, in *After the First Three Minutes*, eds. S. S. Holt, C. L. Bennett and V. Trimble, American Institute of Physics, New York.
Fischer, M. L. *et al.* 1992, *Ap. J.*, **388**, 242.
Fixsen, D. J. 1982, Ph.D. Thesis, Princeton University.
Fomalont, E. B., Kellermann, K. I., Anderson, M. C., Weistrop, D., Wall, J. V., Windhorst, R. A., and Kristian, J. A. 1988, *A.J.*, **96**, 1187.
Fomalont, E. B., Windhorst, R. A., Kristian, J. A., and Kellerman, K. I. 1991, *A. J.* **102**, 1258.
Fomalont, E. B., Partridge, R. B., Lowenthal, J. D., and Windhorst, R. A. 1993, *Ap. J.*, **404**, 8.
Franceschini, A., Toffolatti, L., Danese, L., and De Zotti, G. 1989, *Ap. J.*, **344**, 35.
Gaier, T., Schuster, J., Gunderson, J., Koch, T., Seiffert, M., Meinhold, P., and Lubin, P. 1992, *Ap. J.* (*Lett.*), **398**, L1.
Gautier, T. N., Boulanger, F., Pérault, M., and Puget, J. L. 1992, *A.J.*, **103**, 1313.
Goldstein, S. J., Marscher, A. P. and Rood, R. T. 1976, *Ap. J.*, **210**, 321.
Gunn, J. E. 1978, in *Observational Cosmology*, ed. A. Maeder, L. Martinet, and G. Tammann, Geneva Observatory, p. 3.
Haslam, C. G. T., Salter, C. J., Stoffel, H., and Wilson, W. E. 1982, *Astron. Astrophys. Suppl.*, **47**, 1.
Hauser, M. G. *et al.* 1991, in *After the First Three Minutes*, eds., S. S. Holt, C. L. Bennett, and V. Trimble, American Institute of Physics, New York (COBE).
Hazard, C., and Salpeter, E. E. 1969, *Ap. J.* (*Lett.*), **157**, L87.
Hogan, C., and Partridge, R. B. 1989, *Ap. J.* (*Lett.*), **341**, L29.
Kaidanovski, M. N., Korolkov, D. V., and Stotski, A. A. 1982, *Astrophys. Space Sci.*, **82**, 317.
Klein, U., Rephaeli, Y., Schlickeiser, R. and Wielebinsky, R. 1991, *Astron. Astrophys.*, **244**, 43.
Knoke, J. E., Partridge, R. B., Ratner, M. I., and Shapiro, I. I. 1984, *Ap. J.*, **284**, 479.
Korolev, V. A., Sunyaev, R. A., and Yakubtsev, L. A. 1986, *Soviet Astron. J. Lett.*, **12**, 141.
Kreysa, E., and Chini, R. 1989, in *Third ESO/CERN Symposium, Astronomy, Cosmology and Fundamental Physics*, eds. M. Caffo *et al.*, Kluwer Academic, Dordrecht.
Lake, G., and Partridge, R. B. 1980, *Ap. J.*, **237**, 378.
Lasenby, A. N. 1981, Ph.D. Thesis, University of Manchester.
Lasenby, A. N., and Davies, R. D. 1983, *Monthly Not. Roy. Astron. Soc.*, **203**, 1137.
Lawson, K. D., Mayer, C. J., Osborne, J. L., and Parkinson, M. L. 1987, *Monthly Not. Roy. Astron. Soc.*, **225**, 307.
Longair, M. S., and Sunyaev, R. A. 1969, *Nature*, **223**, 719.
Low, F. J., *et al.* 1984, *Ap. J.* (*Lett.*), **278**, L19.
Lowenthal, J. D., Fomalont, E. B., Partridge, R. B., and Windhorst, R. A. 1994, in preparation.
Lubin, P. M., Meinhold, P. R., and Chingcuanco, A. O. 1990, in *The Cosmic Microwave Background: Twenty Five Years Later*, eds. N. Mandolesi and N. Vittorio, Kluwer Academic, Dordrecht.

Lubin, P. M., and Villela, T. 1985, in *The Cosmic Background Radiation and Fundamental Physics*, ed. F. Melchiorri, Editrice Compositori, Bologna.

Mandolesi, N., *et al.* 1986, *Nature*, **319**, 751.

Martin, H. M. and Partridge, R. B. 1988, *Ap. J.*, **324**, 794.

Martin, H. M., Partridge, R. B., and Rood, R. T. 1980, *Ap. J.* (*Lett.*), **240**, L79.

Masi, S., de Bernardis, P., De Petris, M., Epifani, M., Gervasi, M., and Guarini, G. 1991, *Ap. J.* (*Lett.*), **366**, L51.

Matsumoto, T., Hayakawa, S., Matsuo, H., Murakami, H., Sato, S., Lange, A. E., and Richards, P. L. 1988, *Ap. J.*, **329**, 567.

Meinhold, P., and Lubin, P. 1991, *Ap. J.* (*Lett.*), **370**, L11.

Melchiorri, F., Melchiorri, B., Ceccarelli, C., and Pietranera, L. 1981, *Ap. J.* (*Lett.*), **250**, L1.

Meyer, S. S., Jeffries, A. D., and Weiss, R. 1983, *Ap. J.* (*Lett.*), **271**, L1.

Meyer, S. S., Cheng, E. S., and Page, L. A. 1991, *Ap. J.* (*Lett.*), **371**, L7.

Moffet, A. T., and Birkinshaw, M. 1989, *A.J.*, **98**, 1148.

Myers, S. T. 1990, Ph.D. Thesis, California Institute of Technology.

Pariiskii, Yu. N. 1973a, *Ap. J.* (*Lett.*), **180**, L47.

Pariiskii, Yu. N. 1973b, *Soviet Astron. J.*, **17**, 291.

Pariiskii, Yu. N. 1974, *Ap. J.* (*Lett.*), **188**, L113.

Pariiskii, Yu. N., and Pyatunina, T. B. 1970, *Soviet Astron. J.*, **14**, 1067.

Pariiskii, Yu. N., and Korol'kov, D. V. 1987, *Soviet Scient. Rev. E*, **5**, 40.

Pariiskii, Yu. N., Petrov, Z. E., and Cherkov, L. N. 1977, *Soviet Astron. J. Lett.*, **3**, 263.

Partridge, R. B. 1980, *Ap. J.*, **235**, 681.

Partridge, R. B., Perley, R. A., Mandolesi, N., and Delpino, F. 1987, *Ap. J.*, **317**, 112.

Partridge, R. B., Nowakowski, J., and Martin, H. M. 1988, *Nature*, **311**, 146.

Penzias, A. A., Schraml, J., and Wilson, R. W. 1969, *Ap. J.* (*Lett.*), **157**, L49.

Perrenod, S. C., and Lada, C. J. 1979, *Ap. J.* (*Lett.*), **234**, L173.

Puget, J.-L. 1987, in *Comets to Cosmology*, ed. A. Lawrence, Springer-Verlag, Heidelberg.

Radford, S. J. E., Boynton, P. E., Ulich, B. L., Partridge, R. B., Schommer, R. A., Stark, A. A., Wilson, R. W., and Murray, S. S. 1986, *Ap. J.*, **300**, 159.

Readhead, A. C. S., Lawrence, C. R., Myers, S. T., Sargent, W. L. W., Hardebeck, H. E., and Moffet, A. T. 1989, *Ap. J.*, **346**, 566.

Reich, W. 1982, *Astron. Astrophys. Suppl.*, **48**, 219.

Reich, W., Furst, E., Steffen, P., Reif, K., and Haslam, C. G. T. 1984, *Astron. Astrophys. Suppl.*, **58**, 197.

Rephaeli, Y. 1981, *Ap. J.*, **245**, 351.

Rohlfs, K. 1986, *Tools of Radio Astronomy*, Springer-Verlag, Berlin.

Rudnick, L. 1978, *Ap. J.*, **223**, 37.

Sachs, R. K., and Wolfe, A. M. 1967, *Ap. J.*, **147**, 73.

Schaeffer, R., and Silk, J. 1988, *Ap. J.*, **333**, 509.

Seielstad, G. A., Masson, C. R., and Berge, G. L. 1981, *Ap. J.*, **244**, 717.

Silk, J. 1968, *Ap. J.*, **151**, 459.

Silk, J., and White, S. D. M. 1978, *Ap. J.* (*Lett.*), **226**, L103.

Sironi, G., and Bonelli, G. 1986, *Ap. J.*, **311**, 418.

Smith, M. G., and Partridge, R. B. 1970, *Ap. J.*, **159**, 737.

Smoot, G. F., Levin, S. M., De Amici, G., and Witebsky, C. 1986, *Radio Sci.*, **22**, 521.

Smoot, G. F., *et al.* 1992, *Ap. J.* (*Lett.*), **396**, L1 (COBE).

Stankevich, K. S. 1974, *Soviet Astron. J.*, **18**, 126.

Sunyaev, R. A., and Zel'dovich, Ya. B. 1969, *Astrophys. Space Sci.*, **4**, 301.

Sunyaev, R. A. and Zel'dovich, Ya. B. 1972, *Comments Astrophys. Space Sci.*, **4**, 173.

Tatarski, V. I. 1961, *Wave Propagation in a Turbulent Medium*, Dover, New York.

Thompson, A. R., Moran, J. M., and Swenson, G. W. 1986, *Interferometry and Synthesis in Radio Astronomy*, J. Wiley, New York.

Timbie, P. T., and Wilkinson, D. T. 1988, *Rev. Sci. Instr.*, **59**, 914.

Timbie, P. T., and Wilkinson, D. T. 1990, *Ap. J.*, **353**, 140.

Uson, J. 1985, in *Observational and Theoretical Aspects of Relativistic Astrophysics and Cosmology*, eds. J. L. Sanz and L. J. Goicoechea, World Scientific, Singapore.

Uson, J. M., and Wilkinson, D. T. 1984, *Ap. J.*, **283**, 471.

Uson, J., and Wilkinson, D. T. 1988, in *Galactic and Extragalactic Radio Astronomy*, 2nd edn., eds. G. L. Verschuur and K. I. Kellerman, Springer-Verlag, Berlin.

Verschuur, G. L., and Kellermann, K. I. 1988, *Galactic and Extragalactic Radio Astronomy*, 2nd edn., Springer-Verlag, Berlin.

Vittorio, N., Meinhold, P., Muciaccia, P. F., Lubin, P., and Silk, J. 1991, *Ap. J. (Lett.)*, **372**, L1.

Watson, R. D., Rebolo, R., Beckman, J. E., Davies, R. D., and Lasenby, A. N. 1989, in *Large Scale Structure and Motions in the Universe*, eds. G. Giuricin *et al.*, Kluwer Academic, Dordrecht.

Whalen, A. D. 1971, *Detections of Signals in Noise*, Academic Press, New York.

Windhorst, R. A., Fomalont, E. B., Partridge, R. B., and Lowenthal, J. D. 1993, *Ap. J.*, **405**, 498.

Wolfe, A. M., and Burbidge, G. R. 1969, *Ap. J.*, **156**, 345.

Wright, E. L., *et al.* 1991, *Ap. J.*, **381**, 200 (COBE).

8

What do we learn from the angular distribution of the CBR?

In the standard Big Bang model (introduced in Chapter 1), the Universe is assumed to be exactly homogeneous and isotropic, and only gravitational forces between particles are considered. Under these conditions, all particles in the Universe are at rest in comoving coordinates and the CBR is completly isotropic.

The actual Universe we observe clearly differs from this idealized model in a number of obvious ways; for instance, the Universe is visibly inhomogeneous on all length scales up to many megaparsecs. Any departure from the idealized conditions assumed in the Big Bang model can introduce anisotropy into the CBR.

8.1 Sources of anisotropy in the CBR

In this section, we list briefly some cosmological processes which can introduce anisotropy into the CBR on various angular scales, starting with the largest. Suppose, for instance, that we allow large-scale magnetic fields, and not just gravity, into our cosmological model. Then the CBR will become anisotropic (Thorne, 1967). Likewise, long wavelength gravitational waves added to an otherwise homogeneous Universe will induce a quadrupole moment (Dautcourt, 1969; Burke, 1975).

A more fundamental source of anisotropy in the CBR is the anisotropic expansion of the Universe as a whole (see, e.g., Barrow *et al.*, 1983). The cosmological equations derived from General Relativity permit a wide variety of anisotropic solutions as well as the isotropic Robertson–Walker models discussed in Chapter 1. In such anisotropic models, the rate of expansion of the Universe will not be the same in all directions. As a consequence, the observed temperature of the CBR will also be angle dependent. Rotation as well as anisotropic expansion is possible; it too will produce anisotropy (Hawking, 1969).

Even in an entirely isotropic, non-rotating Universe, large-scale anisotropy can be introduced by inhomogeneities. The simplest and perhaps most important example is that a relatively local 'lump' of matter or void can accelerate the Galaxy gravitationally. The induced velocity of the Galaxy, with us in it, will introduce a dipole component into the CBR.

Inhomogeneities in the material contents of the Universe will induce smaller scale anisotropies as well (e.g., Sachs and Wolfe, 1967; Silk, 1968). Since the Universe is manifestly inhomogeneous on all scales up to and beyond the scale of clusters of galaxies, fluctuations in the CBR temperature *must* be present at some level. The connection between cosmic structure and theories of its formation, on the one hand, and angular fluctuations in the CBR, on the other, has been subject to more research than any other topic in the field – even though $\Delta T/T$ fluctuations on any angular scale save the dipole were not discovered until 1992!

In the remainder of this chapter, we will outline the various processes responsible for anisotropies in the CBR. Given the huge range of research on CBR anisotropies – there are hundreds of papers on $\Delta T/T$ fluctuations and cosmic structure alone – it is neither possible nor profitable to treat each process in detail. Instead, we will look primarily at the physical basis of each cause of anisotropy in the CBR, and at the constraints set by various searches for anisotropy on important cosmological parameters and theories. Because the literature is so rich, there are good review articles which present the material of this chapter more technically and in more detail. Specifically, the reader may wish to consult a series of papers by Bond and Efstathiou (e.g., Bond, 1990; Efstathiou, 1990) or by Silk and Vittorio (e.g., Vittorio and Silk, 1992). Because the literature on the theory of CBR anisotropies is so rich, I make no claim to have covered it completely or even fairly in this chapter. Appendix C contains more recent references.

8.2   The dipole component

Since the dipole moment in the CBR is the most convincing anisotropy yet detected (with an amplitude $T_1/T_0 = 1.2 \times 10^{-3}$), we begin by examining processes which can induce such a pattern in the CBR. A wide variety of cosmological and astrophysical processes have been offered as explanations of some or all of the observed dipole moment, but only one – the Doppler effect – has gained much support, as we shall see.

8.2.1   What is responsible for the dipole component?

Let us start by remarking that counts of nearby galaxies made in both the optical (Lahav, 1987) and the far infrared (Meiksin and Davis, 1986; Yahil *et al.*, 1986; Strauss and Davis, 1989) show that they are anisotropically distributed about us. The direction of the dipole moment of that distribution is rather closely aligned with the direction of the CBR dipole given in table 6.2. Is it possible that microwave emission from these anisotropically distributed galaxies is responsible for the CBR dipole signal of about 3 mK? The answer is no; Lahav and Partridge (1988) have shown that the radio emission from nearby galaxies (at distances $r \lesssim 300$ Mpc) can contribute no more than 1%, and probably less than 0.1%, of the observed dipole amplitude. The argument depends on the fact that the microwave emission of galaxies and other radio sources has a very different spectrum to the CBR (see Section 3.8). If nearby galaxies were responsible for the dipole signal observed by Lubin *et al.* (1983) at $\lambda = 3.3$ mm, measurements of the CBR at much longer wavelengths would reveal a much larger dipole (and quadrupole) component. For instance, T_1/T_0 would be about 10 at $\lambda = 12$ cm – and no such anisotropy is observed.

Next, let us consider anisotropic expansion or rotation (or vorticity) as causes of the observed dipole. Both can induce a dipole moment in the CBR (Hawking, 1969; Barrow *et al.*, 1983). In the case of anisotropic expansion, however, the quadrupole component will generally be larger than the dipole (see Section 8.3 below). Since the observations show $T_2/T_1 \lesssim 0.005$, anisotropic expansion cannot explain all of the roughly 3 mK dipole component.*

---

* See Paczynski and Piran (1990) for a special counterexample.

The generally accepted explanation of the dipole component is the Doppler effect. An observer in motion with respect to a frame in which the CBR is isotropic will see the photons of the CBR blueshifted to higher energy or shorter wavelength in the direction of his or her motion. In the opposite direction, the photons will be redshifted to lower energy. A dipole anisotropy results. Since the observed anisotropy is small, our velocity must be small. In the non-relativistic case, and assuming a blackbody spectrum for the CBR, Peebles and Wilkinson (1968) showed that

$$\frac{T_1(r)}{T_0} = \frac{v \cdot \hat{r}}{c} \tag{8.1}$$

with a maximum in the direction of motion. The resulting anisotropy is an almost pure dipole moment; there is a small quadrupole term of $(1/2)|v/c|^2 \, T_0$ in amplitude, that is $\lesssim 0.1\,\%$ of the dipole moment. The results displayed in table 6.2 establish that the Earth is moving at 360 ± 20 km s^{-1} in the direction $\alpha = 11.2$ h and $\delta = -7°$ (lying near the constellations Leo and Crater).

8.2.2 *What is responsible for the motion of the observer?*

In this section and Section 8.2.3, we will use the working assumption that the dipole moment in the CBR is caused by the Doppler shift. We then need to know what causes our motion. Before we turn to possible answers to that implied question, it is worth looking at a more fundamental point: does our claim that the CBR dipole results from the motion of the observer violate an important tenet of special relativity that only *relative* motion is meaningful? Are we setting up the CBR as some sort of absolute reference frame of the kind Einstein drove out of physics in 1905? Strictly speaking, the velocity that appears in eqn. (8.1) *is* the relative velocity between the observer and the matter at large distances, which last scattered the CBR photons (see fig. 1.11). One could argue that the huge shell of matter is rushing past us, but it seems more modest and less forced to assume we are the ones in motion (just as it seemed more reasonable to Aristarchus that the smaller Earth moved around the larger Sun). Thus we will continue to refer to *v* as the motion of the observer.

Mention of the Earth's yearly motion around the Sun should remind us that the Sun moves in a circular orbit around the Galactic center. Astronomers have known for many decades that the amplitude of the Sun's orbital velocity is about 220–250 km s^{-1}. Is this the motion revealed by the CBR dipole? The answer is no, since it happens that the Sun's velocity in the Galaxy lies roughly in the opposite direction from the apex of the CBR dipole (fig. 8.1).

An obvious consequence is that the speed of the center of the Galaxy relative to the CBR must be roughly twice the 360 km s^{-1} amplitude inferred from measurements of T_1.

Our Milky Way Galaxy, in turn, is one member of a group of a dozen galaxies, which are apparently bound together gravitationally: the Local Group. The Milky Way and M31 (the Andromeda spiral) are the dominant members. Gravitational forces between the members of the Local Group have induced velocities relative to the center of mass of the Local Group. The solar motion relative to the center of mass of the Local Group is thus the vector sum of the Sun's orbital velocity and the velocity of the Milky Way in the Local Group. It happens to be easier to measure the Sun's velocity relative to the center of mass of the Local Group than the Sun's orbital velocity in the Galaxy alone.

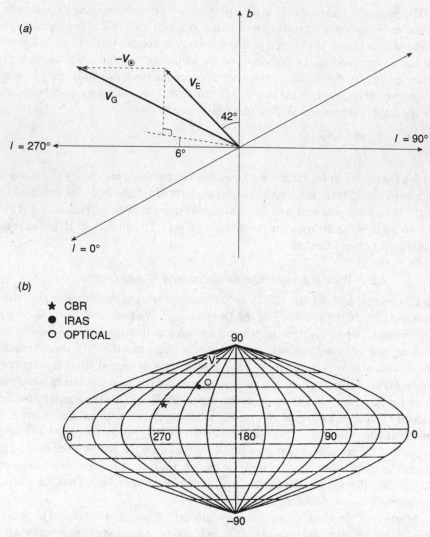

Fig. 8.1 (*a*) The composition of v_G, the velocity of the center of mass of the local group. The velocity of the Earth relative to the CBR is v_E and $v_\odot$ is the velocity of solar system relative to the local group, as discussed in the text. (*b*) Directions (again in Galactic coordinates) of the CBR dipole and to the Virgo cluster (V). Also shown are the alignments of the mass dipoles found by counting galaxies in the optical and from the IRAS catalog (see table 8.1 for references). v_G and the mass dipoles coincide to within ~40° or less.

Astronomers have determined the amplitude and direction of the former to moderate accuracy; Yahil *et al.* (1977), for instance, give $v_\odot = 300 \pm 20$ km s^{-1}, towards the celestial coordinates 23 h right ascension and +50° declination. This determination of $v_\odot$ is just one of many (see Lynden-Bell and Lahav, 1989, for a summary and discussion), which differ in direction by up to about 20°. Note that this uncertainty is many times larger than the errors in the CBR determinations. Because of this uncertainty, it has become conventional to adopt a standard value for $v_\odot$ suggested in 1976 by de

Table 8.1 *Direction of the Local Group velocity, in Galactic coordinates.*

Nature of observation	Reference	Direction of Dipole l	b
CBR	adopted here	$268° \pm 4°$	$26° \pm 4°$
IRAS sources	Meiksin and Davis (1986)	$235° \pm 10°$	$45° \pm 10°*$
	Yahil *et al.* (1986)	$248° \pm 9°$	$40° \pm 8°$
	Harmon *et al.* (1987)	$272° \pm 10°$	$31° \pm 3°*$
Optical galaxies	Lahav (1987)	$227° \pm 23°$	$48° \pm 8°$
Elliptical galaxies	Lynden-Bell *et al.* (1988)	$307° \pm 11°$	$9° \pm 8°$

* My estimates of errors.

Vaucouleurs *et al.*: $|v_\odot| = 300$ km s^{-1} towards $l = 90°$, $b = 0°$ in Galactic coordinates (corresponding approximately to $\alpha = 21.2$ h, $\delta = 48°$). We, too, will adopt that standard for the calculations in table 8.1. Some authors use $v_\odot$ from Yahil *et al.* (1977) instead.

To find the velocity of the Local Group relative to the CBR frame, we then subtract $v_\odot$ from the velocity derived from the CBR observations. In addition, the measurements are now sensitive enough to need a small correction for the Earth's orbital velocity around the Sun. When these corrections are applied, we find for the velocity of the center of the mass of the Local Group $v_G = 600 \pm 30$ km s^{-1} towards 10.5 h right ascension and $-25°$ declination. These values differ very slightly from those found by Aaronson *et al.* (1986), because we have both included recent Soviet and USA satellite results given in table 6.2, and used $T_0 = 2.73$, not 2.8 K. In Galactic coordinates, and using the standard value of $v_\odot$, v_G is directed towards $l = 268°$ and $b = 26°$ with an uncertainty of $\pm 4°$ in each coordinate. If $v_\odot$ from Yahil *et al.* (1977) is used instead, $v_G = 620 \pm 30$ km s^{-1} toward $l = 276°$, $b = 28°$.

We conclude that the CBR dipole implies more than motion of the observer; it implies motion of the entire Local Group of galaxies, which has a mass in excess of $10^{12} M_\odot$, with a speed of about $2 \times 10^{-3}c$.

The amplitude of v_G is surprisingly* large. Could the motion of the Local Group be a relic of a much earlier epoch in the Universe or even 'built in' as an initial condition? The problem with this suggestion is that random velocities of this sort, unlike velocities resulting from the expansion of the Universe, decrease with time. Hence the velocity of the Local Group would have been much larger in the past. Specifically, if we treat v_G as a velocity of the Local Group relative to the comoving coordinates introduced in Section 1.3.1, we find $|v_G(z)| = |v_G|(z + 1)$, where $v_G(z)$ is the velocity at an earlier epoch corresponding to redshift z. This relation would then imply $v_G(z) \simeq c$ at $z \simeq 1000$, the epoch of decoupling, which is not allowed by the observations described in Chapter 7 (Anile *et al.*, 1976).

If v_G is not primordial, it must have been induced; and on scales as large as $10^{12} M_\odot$ in mass and about 1 Mpc in length, only gravity can be responsible. Almost all cosmologists now assume that the velocity of the Local Group has been gravitationally induced by a large 'lump' of matter (or, more realistically, an assembly of lumps and voids) lying

---

* 'Surprising' at least in the context of some models now favored for the origin of large-scale structure in the Universe (to be discussed later in this chapter).

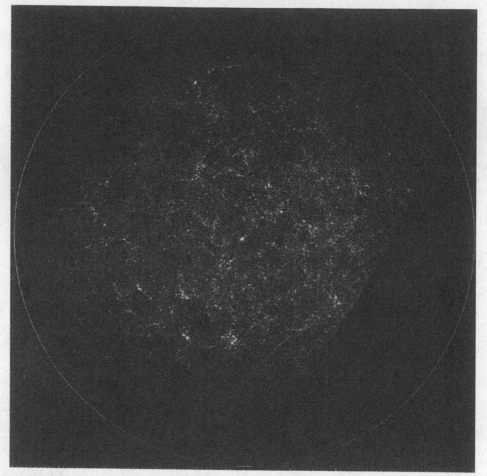

Fig. 8.2   The distribution of $\sim 10^6$ nearby galaxies in Galactic coordinates, with the North Galactic Pole at the center. Obscuration by dust is responsible for the absence of galaxies at low Galactic latitudes (reproduced with the permission of P. J. E. Peebles).

near the Local Group. Note that this explanation cannot work unless the Universe is inhomogeneous on scales in excess of 1 Mpc and $10^{12}M_\odot$.

It is intriguing that there is a region of higher than average density of galaxies centered on the Virgo Cluster at 12.5 h right ascension and $+13°$ declination. This region is clearly apparent near the center of a plot of the distribution of the brightest galaxies (fig. 8.2). Although the radius vector to the Virgo cluster and v_G lie about 45° apart (fig. 8.1), perhaps this lump or others like it are responsible for some or all of the motion of the Local Group. We will follow up the consequences of this suggestion next.

8.2.3   *The mass dipole, v_G and Ω.*

Our aim in this section is to relate the amplitude of v_G to the inhomogeneous distribution of matter near the Local Group. Since $|v_G|$ is known to better than 10% accuracy, we may then use that relationship as a source of important cosmological information. In particular, a simple linear model for the connection between v_G and a spherical 'lump' or void will allow us to set limits on Ω, the cosmological density parameter (defined in Section 1.3.4).

We shall begin, however, by comparing the value of v_G found above with the velocity of the Local Group found using other astronomical techniques, since this comparison affords some insight into the models we will be considering later.

A careful inspection of fig. 8.2 reveals not just lumpy structure but large-scale inhomogeneity; there are more bright galaxies, for instance, in the quadrant $270° < l < 360°$ than in the quadrant $0° < l < 90°$. This impression can be quantified by calculating the dipole moment of the distribution of galaxies, appropriately weighted by some measure of their mass, e.g., their luminosity. Such calculations were first made (Meiksin and Davis, 1986; Yahil *et al.*, 1986) using the catalog of extragalactic sources detected by the IRAS infrared satellite. This satellite surveyed the entire celestial sphere at four wavelengths between 12 and 100 μm. The resulting catalog of sources is quite uniform and of course has full sky coverage. In addition, obscuration by dust in the plane of our Galaxy is a far smaller problem at infrared wavelengths than at visible wavelengths (see the references above plus Lahav, 1987). On the other hand, emission from wispy clouds of dust in our Galaxy (the 'infrared cirrus') contaminates the catalog, especially at the two longer wavelengths, 60 and 100 μm (see Yahil *et al.*, 1986; Harmon *et al.*, 1987). The teams working with the IRAS data have used different methods of coping with the contamination by 'cirrus,' with the results displayed in table 8.1. The direction of the Local Group velocity is given in Galactic coordinates, since those are employed by most of the observers. Note that the largest difference* between the directions v_G derived from the IRAS dipole and from the CBR dipole is $\lesssim 40°$.

Catalogs of galaxies made at ordinary optical wavelengths may also be used to determine a dipole moment in the distribution of nearby galaxies (Lahav, 1987). In this case, the problems are obscuration by dust in the plane of our Galaxy, and the fact that no single catalog covers the entire celestial sphere, so that dissimilar catalogs have to be combined. Nevertheless, the optical dipole direction is in good agreement with the IRAS dipole direction; it lies less than 40° from the CBR apex.

The work summarized above shows both that bright galaxies are inhomogeneously distributed, and that the dipole moment of their distribution is at least approximately aligned with the microwave dipole (see also Lynden-Bell *et al.*, 1989). There is thus support for the idea that v_G is gravitationally induced. The results summarized in table 8.1, however, provide no information on the *amplitude* of the velocity induced by the inhomogeneous distribution of matter. We turn next to measurements of the velocity of the Local Group based on optical observations.

As noted above, the dipole anisotropy T_1 is a measure of our velocity relative to matter at large distances, on the surface of last scattering. It is also possible to determine the velocity of the observer relative to matter at much smaller distances by studying the distribution of redshifts (and hence recession velocities) of a set of objects having known or inferable distance. Knowing the distance to each object allows us to subtract the cosmological part of its recession velocity, leaving only its peculiar velocity. Rubin and her colleagues (1976a, b), who pioneered this technique, used a spherical shell of spiral galaxies at redshifts of order 0.01–0.02. Observations of this sort are not easy and are

---

* In the most recent study (Strauss *et al.*, 1992), this difference is reduced to about 20°.

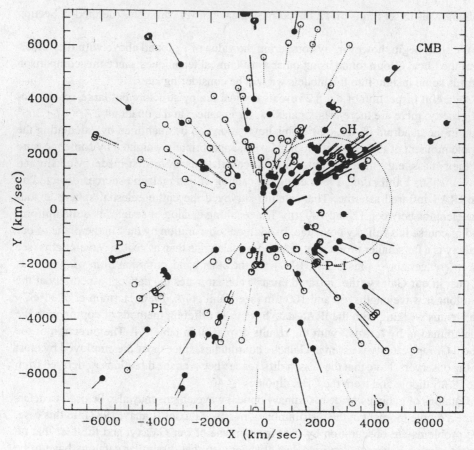

Fig. 8.3   Evidence for large-scale streaming velocities (from Lynden-Bell *et al.*, 1988, with permission). The inferred peculiar velocities of a number of elliptical galaxies lying within 45° of the supergalactic plane are plotted. Note the streaming towards the upper right corner, in the general direction of the Great Attractor.

subject to several sorts of potential systematic error (see Fall and Jones, 1976; Rubin *et al.*, 1976b; Schechter, 1977; for discussions). Nevertheless, Rubin and her colleagues were able to derive a speed of about 600 km s⁻¹ for the Local Group. Unfortunately, the direction was approximately orthogonal to that found from the CBR measurements. It is possible, as suggested by Collins *et al.* (1986), that the entire shell of galaxies observed by Rubin and her colleagues is in motion (with velocity about 1000 km s⁻¹) relative to matter at large distances. If not, the two observational results must be considered discrepant.

Some years ago, Hart and Davies (1982; see also Davies, 1983, and the useful discussion following it) re-examined this problem using a different shell of galaxies and found a value for v_G in reasonable agreement with the CBR value, as did Aaronson *et al.* (1986). On the other hand, Collins *et al.* (1986) found a value of v_G in better agreement with Rubin's original work.

Is there a way to reconcile all these results? In principle, the answer is yes – by allowing large-scale streaming motions in the Universe, as suggested by fig. 8.3. The presumption is that these large-scale motions are gravitationally induced. In the past decade, evidence for large-scale motions in the Universe has accumulated rapidly (see, for instance, Aaronson *et al.*, 1982; a review by Davis and Peebles, 1983a; and especially Lynden-Bell *et al.*, 1988, and the volume by Rubin and Coyne, 1989). Surveys of galaxies out to redshifts $z = 0.02$–0.03 suggest coherent or 'bulk' motion (with respect to the frame established by the CBR) which has an amplitude of 500–600 km s^{-1}. Specifically, Lynden-Bell *et al.* (1988) present evidence that galaxies in an approximately spherical volume of radius $80h^{-1}$ Mpc* around us are falling towards a large clump of galaxies located at $l = 307°$, $b = 9°$ in the sky, and at a distance of 40–45h^{-1} Mpc from us. This concentration of matter is now called the Great Attractor.

We thus conclude that large-scale motions of the kind needed to explain the discrepancies between local measures of the velocity of the Local Group and the CBR dipole can and do exist. Whether the various observational studies mentioned in this section (and others not mentioned here) can be brought into harmony is not yet clear. Such large-scale velocities were discovered (or recognized) only recently, and the field is growing explosively as data pour in (recent reviews are provided by Lynden-Bell *et al.*, 1989; Kaiser, 1991; and Scaramella *et al.*, 1991). Because the observational situation is changing so rapidly, and because the results bear only indirectly on the CBR measurements, let us turn back to the interpretation of the velocity of the Local Group derived from the CBR dipole component.

The gravitational acceleration produced by a clump of matter (such as the Great Attractor) may be written

$$g = G \int_V \frac{(\rho(x') - \rho_0)(x' - x)}{|x' - x|^3} \, dV', \tag{8.2}$$

where x gives the position of the observer. We write ρ_0 as the mean mass density of the Universe, as in Chapter 1, and $\rho(x')$ as the density at position x'.

We now need to calculate the magnitude of the velocity induced by g in an expanding Universe. For small induced velocities (small relative to the expansion velocity, $H_0|x' - x|$) and small differences $\rho(x') - \rho_0$, we may use the linear theory as developed by Peebles (1976; see also Chapter 2 of his 1980 book, *The Large-Scale Structure of the Universe*). We will consider only the growing mode, in which the velocity induced by gravitational inhomogeneities increases with time (unlike the case of primordial velocities discussed in Section 8.2.2). For the growing mode, the induced velocity is parallel with and proportional to the acceleration g. The relation is

$$v = \frac{2f(\Omega)}{3H_0 \, \Omega} g. \tag{8.3}$$

The function $f(\Omega)$ and a detailed derivation of (8.3) are given in Peebles' 1980 book. As Peebles shows, for values of Ω† in the approximate range $0.1 \leq \Omega \leq 1$, $f(\Omega)$ may be well approximated by $\Omega^{0.6}$. This result is quite general, as long as no large velocities are involved. On the large scales under discussion in this chapter, linear theory applies very well; no non-linear velocities or density inhomogeneities are expected or found.

---

* Here as elsewhere, h is Hubble's constant in units of 100 km s^{-1} per megaparsec.

† Recall the notation introduced in Chapter 1: $\Omega \equiv \rho_0 / \rho_c$, where ρ_c is the critical density defined in Section 1.3.4.

A more specific model may be constructed by adopting two more assumptions. The first is that the gravitational lump or void is spherical; the second is that matter lying outside the gravitating lump is homogeneous and isotropic and hence exerts no net gravitational force. Then the amplitude of g in eqn. (8.2) reduces to $G\,\delta M/R^2$ where δM is the *excess* mass inside the spherical lump of radius R. The total mass contained inside R is $M + \delta M$, where $M = \rho_0 V = \Omega \rho_c (4/3\pi R^3)$ or $H_0^2 R^3 \Omega/2G$ when we substitute $\rho_c \equiv 3H_0^2/8\pi G$. Finally, then, we may rewrite (8.3) in the linear approximation with spherical symmetry as

$$|v| = \frac{1}{3}\, H_0\, R\Omega^{0.6}\, \frac{\delta M}{M}. \tag{8.4}$$

It is important to notice that the amplitude of v depends on both the mass excess and the density parameter Ω in this model. Since we know v to high precision, we can solve eqn. (8.4) for the important cosmological parameter Ω if we can independently estimate $\delta M/M$ for the lump producing the velocity of the Local Group. As a first approximation, let us assume that the mass excess can be determined by merely counting up galaxies, so $\delta M/M = \delta N/N$, where N is the number of galaxies in a particular volume. In the region of the Great Attractor, for instance, there is an excess of galaxies with $\delta N/N \simeq 1.5$ (Dressler, 1988). The distance between the center of mass of the Local Group and the Great Attractor is about $40h^{-1}$ Mpc, so $H_0 R = 4000$ km s^{-1}. Taking $|v_G| = 600$ km s^{-1}, we find $\Omega = 0.13$. While this value of Ω is barely consistent with the value derived from cosmic nucleosynthesis (Section 1.6.4), there are strong theoretical arguments for a value $\Omega = 1$ (Section 1.5.5). Is $\Omega = 1$ consistent with the CBR dipole observations? First, let us ask if our assumptions of linearity and exact spherical symmetry could tilt our estimates of Ω. Davis and Peebles (1983a) show that dropping the assumption of spherical symmetry would change the amplitude of v derived from eqn. (8.4) by a factor $\lesssim 2$. Mild non-linearity ($\delta M/M \simeq 1$ but not > 10) may be treated by adding a correction factor $[1 + \delta M/M]^{-0.25}$ (e.g., Yahil, 1985); the modified equation is

$$|v| = \frac{1}{3}\, H_0\, R\Omega^{0.6}\, \frac{\delta M}{M}\left(1 + \frac{\delta M}{M}\right)^{-0.25}. \tag{8.4a}$$

For the numerical example given above, this modification alone would increase Ω from 0.13 to 0.20, still well below unity. The use of either eqn. (8.4) or (8.4a) requires the assumption that the Universe *outside* the spherical lump is homogeneous and isotropic. As we have already seen, however, the Universe is far from isotropic on scales of tens of megaparsecs and above, so we have good reason to question this assumption. A more crucial assumption in this sample calculation is that v is induced entirely by the Great Attractor, a lump at $H_0 R = 4000$ km s^{-1}. More recent studies (summarized by Lynden-Bell *et al.*, 1989; Kaiser, 1991; and Scaramella *et al.*, 1991) suggest that much of the velocity of the Local Group is induced by more local inhomogeneities with a characteristic value of $H_0 R$ more like 2000 km s^{-1}. Adopting a value of $H_0 R$ a factor of two lower increases Ω by a factor of about three, allowing for values much closer to the theoretically favored value of unity. Recent work by Kaiser *et al.* (1991) also favours $\Omega \simeq 1$.

Most workers in the field now use a less model-dependent approach first introduced by Peebles (1976). It is based on the use of eqns (8.2) and (8.3) directly, rather than (8.4) or (8.4a). This approach has been used, for instance, by Kaiser *et al.* (1991) and Yahil *et al.* (1991 and work in progress) to derive values of Ω close to unity. In a variant of this approach, Dekel *et al.* (1990) and Bertschinger *et al.* (1990) effectively inverted the problem by using an inversion of eqns. (8.2) and (8.3) to derive a density distribution from the observed peculiar velocities of a set of galaxies. That model density distribution may then be compared with the actual distribution of galaxies – say the IRAS galaxies – to provide an estimate for the cosmological parameter Ω. The preliminary suggestions from this group are that Ω, or more specifically the product of $\Omega^{0.6}$ and the bias factor b, is close to unity. These results suggest that the simple, 'single spherical lump' model introduced above may be a rather poor approximation to reality; I have presented it for its pedagogical value.

8.2.4 'Bias' or does the light trace the matter?

Note that Dressler and others rely on counts of galaxies to determine $\delta M/M$. Using galaxy counts in this way requires the assumption that *all* the matter in the Universe is distributed in the same way as the galaxies themselves. Recent advances in particle physics and in theories of the early Universe (discussed further in Sections 1.4 and 1.5, and in Section 8.8 below) have taught us to be cautious on this point. Many forms of 'dark matter' may exist, including particles that may be more uniformly distributed than the luminous (baryonic) matter in galaxies. Light need not trace matter. In addition, galaxies may form only in regions of higher than average density – a notion referred to as 'bias' in galaxy formation (see, for instance, the reviews by Bardeen, 1986, and Kaiser, 1986). The amount of bias is given by a factor b: $\delta M/M = b^{-1} \delta N/N$. In either of these situations, counts of galaxies would overestimate the value of $\delta M/M$. If the true value of the mass excess is smaller, it is easy to see from eqn. (8.4) that Ω can be greater; a value as large as $\Omega = 1$ can by no means be excluded.

In principle, then, the precise value for the motion of the Local Group derived from T_1 could allow us to determine Ω, a crucial and as yet undetermined parameter of cosmology. In practice, however, we cannot employ (8.3), (8.4) or (8.4a) to this end until we know that counts of galaxies do accurately measure $\delta M/M$ or we find an independent means of determining δMM, b or g directly.

8.2.5 Conclusions from v_G

The picture of the Universe that has emerged in the last few pages is quite different from the idealized, homogeneous, Big Bang model we started with in Chapter 1. A variety of recent astronomical studies has revealed the large-amplitude and large-scale velocities of galaxies in the Universe. Speeds of about 500 km s^{-1} are detected for regions as large as 10–100 Mpc (Rubin and Coyne, 1989). For reasons laid out in Section 8.2.2, these velocities cannot be primordial. If they are gravitationally induced, large-scale density perturbations (like the Great Attractor) must be present in the Universe. The required mass excesses, δM, must be large. As an example, let us estimate the extra mass of the Great Attractor assuming that light traces matter, so that no bias is present. We may then write $\delta M = 1.5M = 1.5\ \Omega \rho_c\ (4/3\ \pi\ R^3)$, where $\delta M/M = 1.5$ is taken from the work of Dressler (1988). With $\Omega = 0.13$, we find $\delta M = 2.7 \times 10^{49} h^{-1}$ g or $(1.4–2.8) \times 10^{16} M_\odot$ for $0.5 < h < 1$. We will need to keep the existence of such large density perturbations in mind when we consider the small angular scale isotropy of the CBR later in this chapter.

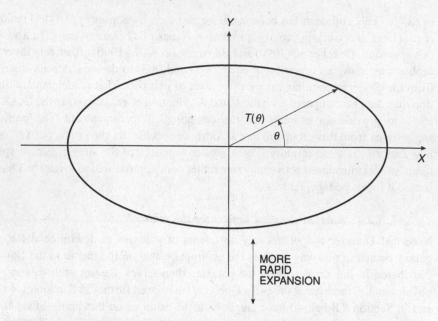

Fig. 8.4   Anisotropic expansion, in this case more rapid along the Y axis, produces a quadrupole moment in the distribution of the CBR.

8.3   Anisotropic expansion of the Universe and its effects on the CBR (especially T_2)

In Section 8.2 we asked what caused the observed dipole component in the CBR; we have good evidence that it is caused by our motion. We now focus on the quadrupole moment, T_2, and higher order moments, which must be produced by different processes.* The observational results will be used in particular to place strong constraints on one other possible cause of anisotropy in the CBR – anisotropic expansion of the Universe. Anisotropic, angle-dependent expansion may be present if the underlying geometry of the Universe is anisotropic (in which case the isotropic Robertson–Walker metric introduced in Section 1.3 is no longer valid), or if anisotropic, non-gravitational forces are present (such as a large-scale magnetic field; Thorne, 1967). Whatever its cause, anisotropic expansion will introduce anisotropy into the CBR. To see this qualitatively, consider a spatially flat geometry which has an expansion rate along one axis slightly larger than the rate along the two orthogonal axes. Then the redshift of the surface of last scattering will be larger along that axis, and hence the observed temperature of the CBR lower, since $T_0 = T_s(z + 1)^{-1}$ (eqn. (1.5)). This simple argument suggests that anisotropic expansion would produce a quadrupole component in the CBR, as fig. 8.4 suggests. In fact, the situation is more complicated, as we will soon see, when we generalize the argument from this simple illustrative example to other anisotropic geometries.

---

* Except for the very small kinematic quadrupole

$$\frac{T_2}{T_0} = \frac{1}{2}\left(\frac{v}{c}\right)^2 \quad \cos\ 2\theta \text{ or } \frac{1}{2}\left(\frac{v}{c}\right)\frac{T_1}{T_0} \text{ in amplitude.}$$

8.3.1 Anisotropic geometries

Earlier in this chapter, we encountered evidence that the Universe is far from homogeneous; now we are introducing the possibility that its geometrical structure may be anisotropic as well. The notion of an anisotropic – and hence anisotropically expanding – Universe may not be as aesthetically tidy and simple as the exactly isotropic model introduced in Chapter 1, but it is not inconsistent with the cosmological equations derived from General Relativity. Indeed, as Collins and Hawking (1973) have pointed out, the number of cosmological solutions that demonstrate exact isotropy well after the Big Bang origin of the Universe is a small fraction of the set of allowable solutions of the cosmological equations. It is therefore necessary to take seriously the possibility that the Universe is anisotropic, and to investigate what effect anisotropic expansion will have on the angular distribution of the background radiation.

I will restrict my attention here to cosmological models (or mathematical descriptions) of the Universe, which, while anisotropic, are nevertheless homogeneous on a very large scale; that is, they have the same average density at all points* (see Barrow *et al.*, 1983, for a discussion of mildly inhomogeneous cases). The geometries appropriate to anisotropic but homogeneous cosmological models were classified nearly a century ago by Bianchi (1898). Since the Universe is evidently fairly isotropic, we will initially consider only three of the ten Bianchi classes, each of which contains one of the isotropic Robertson–Walker metrics of Section 1.3.1 as a special case. One of these, Bianchi class I, may be thought of as the anisotropic generalization of the flat space, Einstein–de Sitter cosmology. Bianchi class IX is the anisotropic generalization of closed, high density models, and Bianchi class V is the generalization of open, low density, ever-expanding models (for more details, see MacCullum, 1985, and Barrow *et al.*, 1983).

If the appropriate description for the Universe is one of the Bianchi class I or IX geometries, the anisotropic expansion will produce primarily a quadrupole component in the CBR, and not higher order moments. In the simplest of all anisotropic models, an axisymmetric Bianchi class I (or cylindrically symmetric model), the metric may be written simply as

$$ds^2 = c^2 \, dt^2 - A^2(t)(dx^2 + dy^2) - B^2(t) \, dz^2, \tag{8.5}$$

where A and B are two different functions of time. In this case (Brans, 1975; Barrow *et al.*, 1983), *only* a quadrupole component and no dipole or higher order components are present; this is the case that we considered qualitatively above. In the more general case of the Bianchi class IX model, both a quadrupole and a dipole component are present (Barrow *et al.*, 1985).

If the geometry of the Universe fits into Bianchi class V, the anisotropic generalization of the $\Omega < 1$, negatively curved, Robertson–Walker metric, a more interesting pattern for the anisotropy results, as first noted by Novikov (1968). In effect, the anisotropy is squeezed into a small patch of sky, of angular radius $\theta \simeq \Omega$ rad. The resulting pattern, shown in fig. 8.5, has been referred to as a 'navel' or 'hotspot.' Note that the amplitude ΔT of the 'hotspot' maxima can be several times larger than the amplitude of the larger-scale quadrupole component itself. Note also that the detection

---

* We require essentially that the scale of any inhomogeneity be much smaller than the curvature scale of the anisotropic geometry.

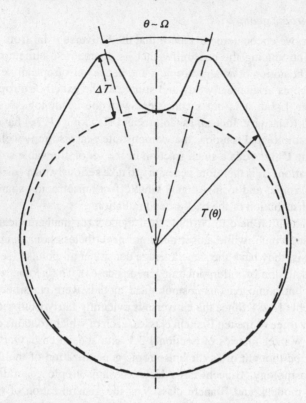

Fig. 8.5   In low-density ($\Omega < 1$), anisotropic cosmological models, the CBR anisotropy is restricted to a small region of the sky – the 'hotspot' – of angular scale $\theta \sim \Omega$. Clearly, multipole moments other than the quadrupole are present in $T(\theta)$ in this case.

and mapping of such a pattern in the CBR would provide a direct measure of the important cosmological parameter, Ω.

The pure quadrupole moment in a Bianchi class I model (with metric given by eqn. (8.5)) results from *shear* in the expansion. Shear may be present in Bianchi classes IX and V as well, but there is an additional, interesting anisotropy present in these more general classes: *vorticity*. 'By vorticity,' writes Hawking in his pioneering 1969 paper, 'one means the rotation of "nearby" matter about an observer moving with the matter, relative to an inertial frame defined by gyroscopes. (Here "nearby" should be interpreted as meaning at distances of about a hundred megaparsecs, near compared with the Hubble radius but far compared with the length scales of local phenomena such as the rotation of the Galaxy.) Thus in a sense, the whole Universe would be rotating, though, as the model is homogeneous, there would be no centre of rotation. Such rotation, if it existed, would be of great interest for the dynamical effects it could have on the Universe and on the formation of galaxies and for its relation to Mach's Principle. This states that the local inertial frame should be non-rotating with respect to distant matter.' Vorticity in turn introduces both dipole and quadrupole moments into the angular distribution of the CBR. The observational limits on T_1 and T_2 therefore allow us to check

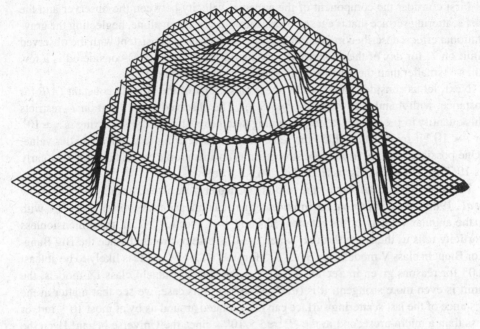

Fig. 8.6   The pattern of the CBR temperature $T(\theta, \phi)$ in a Bianchi type VII_0 model. Only one hemisphere, corresponding to $\pi/2 \leqslant \theta < \pi$, is shown. Note the 'spiral' pattern in $T(\theta, \phi)$ (from Barrow *et al.* (1985), with permission).

the validity of Mach's Principle (and indirectly, the inflationary model)* as well as to constrain simple anisotropic expansion of the Universe.

The Bianchi classes we have examined so far are not the most general anisotropic geometries. Bianchi classes I and V are in fact special cases of more complex geometries which have unique signatures in the angular distribution of the CBR. These geometrical effects have been examined by Barrow *et al.* (1985). In the most general flat space model (Bianchi class VII_0), a spiral pattern is imprinted on the quadrupole component (fig. 8.6). In the generalization of negatively curved geometry, the spiral is imprinted on the hotspot pattern shown earlier. No spiral pattern is present in positively curved anisotropic geometries.

8.3.2 *Observational constraints*

Anisotropic geometries, and the shear and vorticity they cause, thus can produce a variety of characteristic signatures in the CBR radiation. On the other hand, none of these features has been detected in the CBR. We may thus use the observational results – especially the stringent upper limits on T_2 given in table 6.2 – to fix limits on the anisotropy of the Universe. This process was initiated by Thorne (1967), Misner (1968) and Hawking (1969; see also Collins and Hawking, 1973). The most recent work is by Barrow *et al.* (1985), who used the then-available upper limit of 7×10^{-5} for T_2/T_0. In what follows, we will scale their results (and earlier ones) using $T_2/T_0 \lesssim 5 \times 10^{-6}$, the best value now available.

---

* The validity of Mach's Principle (and zero vorticity) is a prediction of inflationary cosmology.

First consider the component of the apparent velocity between the observer and the last scattering surface that is caused by anisotropic expansion alone, neglecting the gravitational effects described in Section 8.2. The largest speed consistent with the observed limit on T_2 for *any* of the anisotropic models Barrow *et al.* (1985) considered is a few km s^{-1} (smaller than the Earth's orbital velocity about the Sun!).

Next, let us consider limits on the shear anisotropy of the Hubble constant ($\dot{A}/\dot{B}$ for instance, with A and B as given in eqn. (8.5)). The observed upper limit on T_2 restricts this quantity to $1 \pm \varepsilon$ with $\varepsilon < 10^{-9}$, assuming that the redshift of last scattering is $z_s \simeq 10^3$, or $\varepsilon < 10^{-6}$ if reionization shifts the redshift of last scattering to a much smaller value. (One possible 'escape' from these strong limits on anisotropic expansion, noted as early as 1967 by Thorne, is discussed in Section 8.3.3 below).

Limits on the vorticity set by the results of Chapter 6 are also extremely tight (Barrow *et al.*, 1985). We will present the limits in terms of a dimensionless vorticity (ω/H_0), with ω the angular velocity of the rotation. Since $H_0^{-1} \approx t_0$ (section 1.2.3), the dimensionless vorticity tells us the angle through which the Universe has rotated since the Big Bang. For Bianchi class V models, $(\omega/H_0) \lesssim 10^{-8}$ for $0.05 \le \Omega < 1$ (and Ω is likely to be at least 0.05 for reasons given in Section 1.6.4). For the closed, Bianchi class IX models, the limit is even more stringent: it is $(\omega/H_0) \lesssim 10^{-13}$. In this case, we see that matter at the distance of the last scattering surface can have rotated around us by at most 10^{-13} rad, or less than a micro-arcsecond in the $H_0^{-1} \approx 5 \times 10^{17}$ s since the Universe began! Even the largest permitted value of (ω/H_0) – for a low density, Bianchi class VII model – is still $\lesssim 10^{-4}$ (Barrow *et al.*, 1985, with updated limit on T_2).

As Barrow *et al.* (1985) are careful to point out, however, these limits on shear and vorticity depend on the redshift assumed for the surface of last scattering. The very tight limits given above were based on the assumption that $z_s \simeq 1000$, i.e. that there was no rescattering of the CBR photons (see Section 1.7.3). If the surface of last scattering has been shifted to a lower redshift by reionization of the matter in the Universe, limits on anisotropic expansion would become up to 50 times less stringent.

The limits on anisotropic expansion imposed by the upper limits on T_2 are nevertheless extremely tight. Since the observed quadrupole may be produced by other mechanisms discussed below, the observations are of course also fully consistent with a completely isotropic Robertson–Walker metric.

Finally, what can we say about the smaller angular scale features such as those shown in fig. 8.5 and fig. 8.6? No such signals have been detected in maps of the CBR, which now cover most of the celestial sphere. Inspection of fig. 6.6 and fig. 6.7, for instance, suggests limits of order $(1–2) \times 10^{-3}$ on the fractional amplitude of any such hotspot. Again, the angular distribution of the CBR is completely consistent with a purely isotropic geometry; isotropic geometry in turn is consistent with inflation.

8.3.3 *Observational constraints from limits on the polarization of the CBR*

The quadrupole moment and other angular variations that we have been discussing above arise because of the differential redshift produced by anisotropic expansion. Thus, in effect, the presently observed anisotropy signal is proportional to the time-average of the anisotropic expansion rate from the epoch of last scattering to the present. One could imagine 'fine-tuning' an anisotropic model in such a way (see, e.g., Thorne, 1967) as to allow large anisotropy now or at an earlier epoch, say the last scattering, while at the same time keeping the time-averaged anisotropy small. Models with a rapidly decaying anisotropy, or the so-called mix-master models (Misner, 1968) in which the direction of maximum anisotropic expansion switches from one spatial coordinate to another, can be so arranged. As it happens, however, we have at least two further constraints on anisotropic expansion of the Universe at specific early epochs.

The first of these epochs is the time of nucleosynthesis, a few minutes after the initial state of infinite density of hot Big Bang models (see Section 1.6.4). The fraction of light nuclei, such as ^{4}He, produced at this early epoch depends on the expansion rate; anisotropic expansion thus alters the results of nucleosynthesis (Barrow, 1976, 1984; Rothman and Matzner, 1985). Measurements of the abundances of light nuclei in fact set tighter constraints on some classes of anisotropic models than the CBR observations discussed above (Barrow, 1976).

The second epoch is the epoch of last scattering of the CBR photons. Anisotropic expansion at that particular epoch will induce linear *polarization* in the CBR (Basko and Polnarev, 1980; Negroponte and Silk, 1980; Tolman, 1985). The upper limits discussed in Section 6.9 thus constrain the amount of anisotropy present at that particular early epoch. The constraints depend strongly on whether the epoch of last scattering was at $z_s \simeq 1000$ or much lower (see Negroponte and Silk, 1980, for a discussion). Given the observational results presented in Chapter 6, the authors show that direct measurements of anisotropy in the CBR temperature are generally a more useful diagnostic than polarization measurements, at least in the case that the redshift of any reionization epoch is $z_s \lesssim 75$. For further details, consult Negroponte and Silk (1980) and Tolman (1985).

8.3.4 Long wavelength gravity waves

Another kind of cosmic anisotropy can be produced by imposing long wavelength gravity waves on an otherwise isotropic, homogeneous background metric. An observer 'riding' such a gravity wave will see an induced quadrupole moment in the CBR (Burke, 1975). Since T_2 is known to be small, we can fix an interesting upper limit on the amplitude of gravity waves passing us. If we make the further assumption that such waves are present with random phases (Burke, 1975), we can go on to set limits on the energy density that long wavelength gravity waves contribute to the Universe. Expressed as a fraction of the critical mass density we find $\Omega \leq 0.3\lambda^{-2}$ where λ is the typical wavelength of the gravity waves expressed in megaparsecs. Random gravity waves of $\lambda \gtrsim 1$ Mpc cannot, therefore, play a major role in determining the dynamics of the Universe (see also Dautcourt, 1969).

What about perturbations on a much larger scale, one with λ greater than the observable radius of the Universe, ct_0? The possibility of inhomogeneities on these very large scales is raised by inflationary cosmology, the modification of the Big Bang model discussed in Section 1.5. Grishchuk and Zel'dovich (1978) have shown that the observational constraints on the large-scale anisotropy of the CBR can place limits on gravitational perturbations on scales much larger than ct_0. The most recent analysis (Mukhanov and Chibisov, 1984) shows that the scale of any appreciable perturbations must be very large indeed, $\lambda \geq 45ct_0$, or else the quadrupole moment they induce would exceed the limits given in table 6.2.

8.4　Density perturbations and their imprint on the CBR

We turn now to the single most active field of CBR studies – the investigation of models for the origin and growth of density perturbations in the Universe and the imprint of these perturbations on the CBR. This imprint is most prominent on angular scales smaller than those treated so far in this chapter, and most of the useful observational constraints are treated in Chapter 7 and summarized in table 7.2.

Despite great activity, there is as yet no single, widely accepted model for the origin and growth of density perturbations and of the structure in the Universe which arises from them (see Section 1.7). As we shall see, however, upper limits on CBR fluctuations on a range of angular scales have narrowed the choices in important ways, and the recent detection of fluctuations on scales $\gtrsim 10°$ (Section 7.7.3) promises to narrow the field still further. (That a single observational result would precipitate out a unique, widely accepted model is too much to hope for.)

Just because there is so much activity in this area – scores of papers are published each year – it is extraordinarily difficult to summarize it. On the other hand, that very activity has ensured that several good reviews are available (Kaiser and Silk, 1986; Bond, 1990; Efstathiou, 1990; Kolb and Turner, 1990). I will not try to summarize all the work in the field. Instead, I regard this and the following several sections as a sort of 'primer' to the field. Sections 8.4–8.8 will therefore be sketchier than Chapter 5 or the earlier sections of this chapter. I hope, however, that the essential physical processes will be clearly enough presented to allow readers to make some sense of the confrontation between theory and the observations described in Chapter 7. I also hope that these sections will prove to be a useful introduction to the more detailed reviews referred to above.

8.4.1　Classification of density perturbations

All theories for the origin of structure in the Universe are based on the notion that initially small density perturbations (regions of slightly higher than average density) grew in amplitude to produce the current, manifestly inhomogeneous architecture of the Universe. The initial density perturbations may well be quantum fluctuations originating at an epoch $t \simeq 10^{-32}$ (Section 1.5.3). Our interest at the moment is not in the origin of density fluctuations but in their properties.

Throughout the remainder of this chapter, we will consider only small amplitude perturbations in the density, that is, with the density at an arbitrary point represented by

$$\rho(x) = \rho + \Delta\rho(x),$$

we will take $\Delta\rho \ll \rho$. This assumption allows us to employ, explicitly or implicitly, linear perturbation theory to analyze the properties and evolution of the perturbations. While the sign of $\Delta\rho$ will vary with position, our interest is in positive $\Delta\rho$, i.e., regions of (small) overdensity. Assuming $\Delta\rho \ll \rho$ also allows us to decompose an arbitrary density fluctuation into three independent components: scalar perturbations, vector perturbations (vortices or 'whirls' in which the velocity of the matter is inhomogeneous) and tensor perturbations (gravity waves). Vortices or 'whirls' were investigated in a series of papers by Ozernoi and his colleagues some years ago (Chibisov and Ozernoi, 1969; Ozernoi, 1974; see also Anile et al., 1976). More recently, it has been recognized (e.g., by Starobinskii, 1985, and Fabbri *et al.*, 1987) that in some cases the amplitude of tensor perturbations can approach that of scalar perturbations. Until very recently, however, most research has focused on scalar perturbations in the density. We will adopt that focus as well.

In the case of scalar perturbations, we may distinguish between two classes, *adiabatic* and *isocurvature* perturbations. In the former, both matter and radiation are perturbed in such a way as to keep the specific entropy constant inside and outside the perturbed region. Thus the fractional perturbations in the radiation and the matter components are related as follows:

$$(\Delta\rho/\rho)_r = \frac{4}{3}(\Delta\rho/\rho)_m,$$

where the subscripts refer to radiation and matter components, respectively. If adiabatic perturbations are present, the density and pressure, and hence the space curvature, will vary from point to point in the Universe (since density and pressure serve as the source terms for Einstein's equations). In the case of isocurvature perturbations, just the opposite is true – the space curvature is constant but the specific entropy varies from point to point. As a consequence, this class of fluctuations is also frequently referred to as 'entropy fluctuations.' Consider a region in which there is a fractional overdensity $\Delta\rho_m$ in the matter component. To keep the curvature constant inside and outside the perturbed region, there must be a corresponding underdensity in radiation, $-\Delta\rho_r$. For most of the epochs that we will be considering, the Universe was radiation dominated with $\rho_r \gg \rho_m$. Thus it is easy to see that $|\Delta\rho/\rho|_r \ll |\Delta\rho/\rho|_m$. Hence another early (and approximate) name for this class of perturbations, 'isothermal.'

Density perturbations can also be characterized by the nature of the matter involved. As we shall see, baryon perturbations evolved differently from perturbations in cold dark matter or in hot dark matter. Recall from Section 1.6.1 that cold dark matter particles were not relativistic at the time they froze out, whereas hot dark matter particles were. As examples of the former, take WIMPS ('weakly interacting massive particles'); as an example of the latter, light neutrinos with a non-zero mass. That difference, too, has an important bearing on the evolution of perturbations of the two types of dark matter.

Perturbations may also be characterized by their *power spectrum* – how the amplitude depends on the scale, or equivalently on the wave number, k, of the perturbation. For many models it is possible to express the spectrum as a power law in k. If we represent the density field $\Delta\rho(x)$ in terms of the Fourier components $\delta(k)$, the power spectrum is then

$$P(k) = \langle |\delta(k)|^2 \rangle \propto k^n, \tag{8.6}$$

Of more physical interest is the variance in density over a particular comoving scale l. Since $l = 2\pi/k$, we may write (see Efstathiou, 1990):

$$\left(\frac{\Delta\rho}{\rho}\right)_k^2 = \left(\frac{\Delta\rho}{\rho}\right)_l^2 = \frac{V}{(2\pi)^3} \int_0^{k \sim l^{-1}} P(k')\, d^3k' \propto k^{3+n} \propto l^{-(3+n)}. \tag{8.7}$$

It follows that

$$\frac{\Delta\rho}{\rho} \propto l^{-(3+n)/2} \propto M^{-(1/2)-(n/6)} \tag{8.8}$$

for power-law spectra. Here, as in eqn. (8.7), l is the comoving scale of the perturbation. Until Section 8.8, we will assume that the spectral index n is fixed, that is, that n is independent of scale.

Two particular values of n are of special interest in the case of adiabatic fluctuations. If $n = 0$, the power spectrum is white noise. On the other hand, the inflationary paradigm favors a specific value, $n = +1$, the so-called Harrison–Zel'dovich or 'scale invariant' spectrum, described in Section 1.5.3. In this case, $\Delta\rho/\rho \propto l^{-2} \propto M^{-2/3}$. Here the reader may be tempted to ask why this spectrum is defined as 'scale invariant' if $\Delta\rho$ does depend on the mass of the perturbation. 'Scale invariance' as used here means that the amplitude $\Delta\rho/\rho$ is independent* of the mass of the perturbation when that perturbation crosses the horizon and becomes causally connected (see, e.g., Kolb and Turner, 1990, Chapter 9). Perturbations of smaller mass, however, cross the horizon at an earlier epoch (Section 1.5.3). As a consequence, they have time to grow while 'waiting' for larger mass perturbations to enter the horizon. If we now ask how $\Delta\rho/\rho$ depends on M *at a particular moment*, we will thus find that $\Delta\rho/\rho$ is larger for small M (or l). That is the dependence I have presented in eqn. (8.7) and eqn. (8.8). It is appropriate in the context of this volume because our information about $\Delta\rho(M)$ is largely derived from observations made of a particular epoch, the time of last scattering. Hence I will continue to employ that convention (especially in Section 8.4.6).

While the inflationary paradigm also favors adiabatic fluctuations (see Kolb and Turner, 1990), isocurvature perturbations are also possible under certain circumstances (see Efstathiou and Bond, 1986, for instance). In this case, of course, there are no curvature perturbations; it is the specific entropy or equation of state that varies from place to place. Scale invariant entropy fluctuations initially have a power law dependence with $n = -3$. The initial power spectrum of scale-invariant fluctuations in the total density may also be specified (e.g., Gouda *et al.*, 1989); in this case, $n = +1$. Some authors adopt one of these two conventions; some the other. I choose the latter, for consistency with Holtzman (1989), Suto *et al.* (1990) and Gouda *et al.* (1991), and for easy comparison with the results of the COBE DMR experiment, which found $n \approx 1$. Peebles, on the other hand, uses the former convention. Adding to the possible confusion, Peebles' (1987) isocurvature baryon models happen to fit the observations best if the initial spectrum of the entropy perturbations is *not* scale invariant, but is roughly $k^{-0.5\pm0.5}$, corresponding in the convention I have adopted to $n = 3.5 \pm 0.5$, strongly scale dependent and barely consistent with the COBE results.

As we will see in the following section, the spectrum of temperature fluctuations generated on the surface of last scattering will differ from the initial, pure power law spectrum. For instance, the spectrum of temperature fluctuations for an $n = 1$ adiabatic model differs from that for Peebles' (1987) isocurvature model by less than the differences in initial indices might suggest.

Finally we may ask about the statistical properties of the perturbations. For the moment (again until Section 8.8) we will assume that the phases of fluctuations of different wave number present in any volume are random and uncorrelated. In this case, perturbations of small amplitude will obey Gaussian statistics: a plot of measured values of the CBR temperature measured at many different points on the sky will approximate a Gaussian curve centered at T_0. It is important to note that this statement is not the same as assuming a Gaussian *power spectrum* for the amplitude of the fluctuations.

---

* Ignoring small, model-dependent factors of order unity – for details see Efstathiou (1990).

8.4.2 *Growth of perturbations in the expanding Universe*

By 'growth' I mean an increase in the density contrast $\Delta\rho/\rho$. The growth of density perturbations in the expanding Universe is treated in a number of papers (Lifschitz, 1946; see also Section 1.7.1, Peebles, 1980, and the useful review by Efstathiou, 1990). As Lifschitz first showed, the increase of $\Delta\rho/\rho$ in the expanding Universe is not exponential in time, as would be true for a static fluid (Jeans, 1902), but rather is a power law (there is also a decreasing mode, which is not of interest for us). The index of the power law depends on whether the expansion is radiation- or matter-dominated. For instance, during the radiation-dominated phase $\Delta\rho/\rho \propto t$. Later, during the matter-dominated phase, $\Delta\rho/\rho \propto t^{2/3}$. A comparison with Section 1.3.3. shows that the scale factor also increases as $t^{2/3}$, and consequently $\Delta\rho/\rho \propto (z+1)^{-1}$. That connection between growth rate and redshift is a generally useful relation, which we will use to follow the later evolution of density perturbations.

Two important limitations on this argument need to be noted. First, in an open, low-density, cosmological model, the growth of density perturbations ceases when the expansion becomes linear with time (see Section 1.3.3). That occurs at a redshift given by $(z+1) \simeq 1/\Omega$, where Ω is the density parameter (Sunyaev, 1971). If only baryons are present in the Universe, Ω is very roughly 0.1, so the growth of perturbations would cease at a redshift of order 10. The second point has to do with the transition from linear growth, described by the $\Delta\rho/\rho \propto (z+1)^{-1}$ law, to non-linear growth (Section 1.7.2). Roughly speaking, that transition occurs as $\Delta\rho/\rho \rightarrow 1$. Non-linear, non-gravitational processes, such as generation of shock waves, etc., can rapidly alter $\Delta\rho$. Since these non-linear processes may depend in complicated ways on the scale of the perturbation, they will also rapidly alter the initial spectrum of the perturbations. We will be concerned here only with the linear regime, a point we return to briefly in Section 8.7 below.

In the linear regime, at least, do all classes of perturbations, and perturbations of all wave numbers, grow at the same rate? The answer is no, and behind that lies much interesting physics.

First we need to remind ourselves of the argument introduced by Jeans in 1902 (and presented briefly in Section 1.7.1). In the case of adiabatic perturbations with a positive value of $\Delta\rho$, the pressure inside the perturbation is higher than in the surrounding 'fluid' of radiation and matter. This overpressure resists gravitational contraction. If the pressure dominates, the amplitude of the density perturbations will oscillate (these are compressional sound waves). If gravity dominates, the perturbations will contract and $\Delta\rho/\rho$ will increase in amplitude. Since the gravitational force on a perturbation of characteristic size l is proportional to $\rho V/l^2 \propto l$, gravity wins above a characteristic scale known as the Jeans length,

$$l_{\mathrm{J}} = c_{\mathrm{s}} \left(\frac{\pi}{G\rho} \right)^{1/2} = \left(\gamma \frac{\pi kT}{G\rho m} \right)^{1/2}, \tag{8.9}$$

where $\gamma = 5/3$ for a monatomic gas. This result is dervied, for instance, in Peebles (1971) and other texts. In eqn. (8.9), c_{s} is the sound speed in a fluid of temperature T consisting of particles of mass m. Corresponding to this characteristic length scale is a mass, the Jeans mass, M_{J}. Clearly $M_{\mathrm{J}} \propto l_{\mathrm{J}}^3 \rho \propto \rho^{-1/2} T^{3/2}$.

Let us now evaluate M_{J} at two characteristic epochs in the early Universe. First, consider a moment at $z \approx 1000$, just after the primordial plasma has recombined (Section

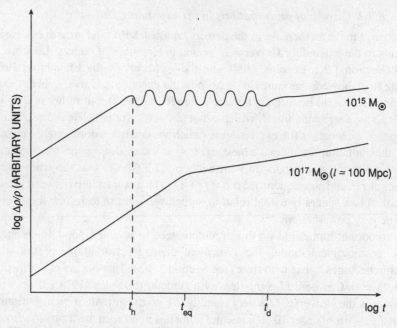

Fig. 8.7 The growth of adiabatic perturbations of two different scales is shown schematically. The initial amplitude of the more massive perturbation is smaller (corresponding to $n > 0$ in eqn. (8.6)), but its mass is $> M_J$, so it grows without interruption. The lower mass perturbation has $M < M_J$ and hence oscillates once it enters the horizon at t_h. In this case, I have selected $M > M_D$ so that the perturbation is not Silk damped; Silk damping is shown in fig. 8.8.

1.6.5). The temperature of the cosmic fluid is about 3000 K and $m \approx m_H \approx m_p$. Taking l_J as the diameter of a region of mass M_J, we find $M_J = 3.0 \times 10^{38} h^{-1}$ g $= 1.5 \times 10^5 h^{-1} M_\odot$ for $\Omega = 1.0$. Thus, after recombination, all masses $\gtrsim 10^5 M_\odot$ are free to grow gravitationally, with $\Delta\rho/\rho$ increasing as $(z + 1)^{-1}$. Note that all galactic masses exceed this value of M_J. At an earlier epoch, however, when the Universe was still radiation-dominated, M_J was much larger because $c_s \approx c/\sqrt{3}$ during radiation-dominated expansion, and was not dependent on the temperature. From Section 1.6.6, we know that radiation-dominated expansion lasted until the epoch corresponding to a redshift $z = 2.5 \times 10^4 \Omega h^2$. Let us then evaluate M_J at $z = 3 \times 10^4$; it is about $10^{16} M_\odot$ for $\Omega = 1$, $h = 1/2$. Thus adiabatic density perturbations of mass less than this (large) value will oscillate, not grow, during the radiation-dominated epoch. Note also that if we take a value of $\Omega < 1$, the calculated value of M_J will increase, both because radiation-dominated expansion lasts longer and because ρ evaluated at a particular epoch is lower. We see that masses up to about $10^{16} M_\odot$ or more, well in excess of the mass of even massive clusters of galaxies, will undergo oscillations during the radiation-dominated phase of expansion. Only very large-scale adiabatic perturbations, with $M > M_J \simeq 10^{16} M_\odot$ or more are free to grow gravitationally until matter-dominated expansion takes over. As a consequence, a break at $M \simeq 10^{16} M_\odot$ will be introduced into the intial power spectrum of perturbations: only for perturbations of larger mass will the original dependence of $\Delta\rho/\rho$ on k (and hence on M) be preserved. Higher wave number (lower mass) perturbations will end up with reduced amplitudes, as fig. 8.7 shows schematically.

The Jeans argument applies only to *adiabatic* perturbations. In the case of isocurvature perturbations, as we have seen, the radiation component is barely perturbed while the expansion is radiation-dominated. As a consequence, the pressures inside and outside the density perturbation are essentially identical. Hence isocurvature perturbations experience little growth or oscillation during the radiation-dominated phase. The subsequent growth of perturbations of this class is more complex than the growth of adiabatic perturbations. Recall that both the matter and the radiation are perturbed, but with opposite signs and very different amplitudes in such a way as to keep the curvature constant. Recall also that the two components $\rho_r(t)$ and $\rho_m(t)$ evolve differently. The consequence is that the rates of growth of $(\Delta\rho/\rho)_r$ and $(\Delta\rho/\rho)_m$ can be quite different and indeed $(\Delta\rho/\rho)_m$ can even decrease with time. The evolution of isocurvature perturbations is reviewed by Efstathiou (1990) and treated in more detail by Wilson and Silk (1981), Efstathiou and Bond (1986), Peebles (1987) and Gouda *et al.* (1991) among others. Here I want to extract just two results from these more complete treatments. First, as the expansion of the Universe shifts from radiation- to matter-dominated, the ratio of ρ_r to ρ_m drops. To maintain constant curvature, the ratio $(\Delta\rho/\rho)_r/(\Delta\rho/\rho)_m$ must increase – in a sense power is shifted to the radiation component. Thus, even though we introduced this class of perturbation as essentially 'isothermal,' perturbations in the radiation density and hence temperature fluctuations *are* present at late epochs. Second, as Efstathiou (1990) for instance shows, scale invariant isocurvature perturbations, like adiabatic ones, produce the same $P(k) \propto k^1$ spectrum at recombination, at least on large scales. Thus we expect differences between the two classes to show up only at *smaller* scales (as confirmed by fig. 8.9 later).

8.4.3 *Damping of perturbations*

We now consider mechanisms that can sharply *decrease* the amplitude of density perturbations, starting with the so-called 'Silk damping' of adiabatic baryonic perturbations, first discussed by Silk in 1968. The physical basis is not complicated – photons diffuse out of overdense regions, 'dragging' baryons with them*. Thus $\Delta\rho/\rho$ in both components is decreased. For this process to operate with any efficiency, the photons must be (1) free to random-walk out of the density perturbation, yet (2) sufficiently coupled to the baryons that they drag the baryons with them. Condition (1) has two consequences – first, that the photon–electron coupling cannot be too tight, or else no diffusion would occur, and second, that Silk damping operates only in regions small enough to allow photon diffusion in a time short relative to the expansion time scale. Hence Silk damping operates only on masses *less than* a critical value M_D. Detailed calculations (e.g., Efstathiou and Silk, 1983) show that

$$M_D \approx 2 \times 10^{12} \Omega_b^{3/2} \, \Omega^{1/4} \, h^{-5/2} M_\odot, \tag{8.10}$$

or

$$M_D \simeq 2 \times 10^{10} h^{-5/2} M_\odot$$

---

* The 'dragging' of baryons is indirect. The photons couple directly to the electrons via Compton scattering. The electrons, in turn, have a Coulomb interaction with the protons. Note, therefore, that Silk damping ceases at recombination when the free electrons vanish.

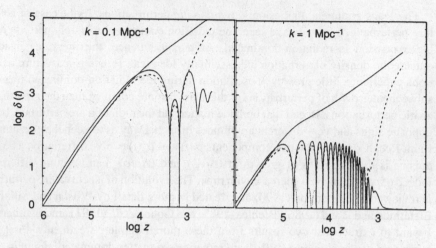

Fig. 8.8 Evolution of adiabatic perturbations in a cold dark matter model (from Efstathiou, 1990). Left panel: perturbations of comoving scale 10 Mpc (or $M \sim 10^{14}$ $M_\odot$); right panel: a scale 10 times smaller or $M \sim 10^{11}$ $M_\odot$ for $h = 1$. The heavy dark lines show the evolution of the perturbation in cold dark matter as time increases and the redshift decreases. Note that growth continues throughout. The dashed lines show the baryonic component which begins to oscillate shortly after the perturbation enters the horizon (at $z = 10^4$ for the more massive perturbation and $z = 10^5$ for the less massive one). The less massive perturbation suffers Silk damping. After recombination at $z \approx 10^3$, photon drag ceases, and the baryons rapidly fall into the gravitational potential wells of the dark matter perturbations, so $(\Delta\rho/\rho)_b$ rapidly shoots up. Our primary interest is in fluctuations in the *radiation* component (light, oscillatory, solid curves), since it is these that are revealed in anisotropy of the CBR. Note that the amplitude of $(\Delta\rho/\rho)_r$ goes essentially to zero for the smaller scale perturbation.

for $\Omega_b \simeq 0.05$ (Peebles, 1981; Efstathiou, 1990). As many authors have noted, M_D is intriguingly close to the mass of a typical galaxy.

Silk damping affects only the matter coupled to the radiation. Therefore, this mechanism does not decrease the amplitude of perturbations in any form of 'dark' matter. To investigate the consequences let us consider for a moment adiabatic perturbations of mass less than M_D. Before Silk damping has time to operate, $(\Delta\rho/\rho)_b = (\Delta\rho/\rho)_{dark}$. Silk damping will then sharply reduce $(\Delta\rho/\rho)_b$; the dark matter fluctuations will continue to evolve unaffected. Thus the strong distinction for fluctuations of mass $< M_D$ shown in fig. 8.8 can develop.

There is, however, another kind of damping, which affects non-interacting, collisionless particles such as the various dark matter candidates (see, e.g., Bond and Szalay, 1983). It is *free streaming*, migration of collisionless particles from high to low density regions. This migration clearly serves to reduce, or 'damp,' the amplitude of density perturbations. The time scale for this process depends on whether the particles are relativistic or not; slow-moving 'cold' particles do not move fast enough. For 'hot' particles, the scale over which this form of damping is effective depends on the mass of the particle. Bond and Szalay (1983) consider the specific case of hot dark matter in the form of light neutrinos of non-zero mass. In this case, the characteristic scale for free-streaming damping is

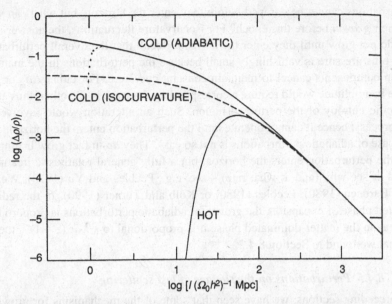

Fig. 8.9 The spectrum of density perturbations at the surface of last scattering for several assumptions about the properties of the perturbations. 'Cold' and 'hot' refer to the dark matter component, which is assumed to dominate Ω_b in these models. Note the high frequency cutoff for the hot dark matter model. The dotted line displays schematically the high frequency cutoff in the *baryonic* component caused by Silk damping (see Section 8.4.5).

$$M_F \approx 3 \times 10^{15} \left(\frac{m_v}{30 \ eV} \right)^{-2} M_\odot, \tag{8.11}$$

where m_v is the neutrino rest mass in electronvolts. Note that M_F is roughly equivalent to the mass of a cluster of galaxies or a supercluster for $m_v \approx 10$ eV.

8.4.4 *The role of the particle horizon*

Causal, physical processes (such as Silk damping) cannot operate on scales larger than the particle horizon, approximately ct at an epoch t. At an early epoch, when the particle horizon is small, the characteristic wavelength of a density perturbation may be much larger than ct. If so, Silk damping or other causal processes will not affect it. Later, as ct increases to match the size of the perturbation, causal processes can begin to operate, and thus to influence $\Delta\rho/\rho$. We call this moment the epoch at which the perturbation enters the horizon.

Clearly, perturbations of smaller scale enter the horizon earlier, and hence are affected for a longer time by the damping processes described in Section 8.4.3. The result is shown in fig. 8.8, based on detailed calculations by Efstathiou (1990). Silk damping, for instance, reduces the amplitude of both $(\Delta\rho/\rho)_m$ and $(\Delta\rho/\rho)_r$ for masses $< M_D$. Hence the initial perturbation spectrum is cut off at short wavelengths, as shown in fig. 8.9. It is worth remarking once again that the initial spectrum (eqn. (8.7)) is preserved only at *large* scales.

Perturbations cannot be damped before they enter the horizon, but what can we say about their *growth* before this epoch? For isocurvature fluctuations, the answer is easy – they do not grow until they enter the horizon. Recall that the overall perturbation in density plus pressure is vanishingly small because the perturbations in the matter and radiation components cancel to maintain constant curvature. For either $\Delta\rho_m$ or $\Delta\rho_r$ to increase in amplitude would require a corresponding decrease in the other, thus altering the specific entropy of the perturbed region. Such an alteration would involve some causal process; hence it cannot operate until the perturbation enters the horizon.

The case of adiabatic perturbations is not so easy. They *do* in fact grow in amplitude before the perturbation enters the horizon, but a fully general relativistic treatment is required to see why (and at what rate) – see, e.g., Peebles and Yu (1970), Weinberg (1972), Bardeen (1980), Peebles (1980) or Kolb and Turner (1990). In the radiation-dominated phase of expansion, the growth of adiabatic perturbations is proportional to t, whereas in the matter-dominated phase it is proportional to $t^{2/3}$ or $(z + 1)^{-1}$, the same result that we found in Section 8.4.2.

8.4.5 Perturbations on the surface of last scattering

In the preceding sections, we have seen that some of the mechanisms for growth and damping of density perturbations change precipitously at the epoch of recombination. That epoch fairly closely precedes the epoch of last scattering, which defines the surface we observe when we study the CBR. For both reasons, the amplitude of density pertur-bations just at recombination is of particular interest. Let us ask what classes, scales and amplitudes of density perturbations survive to reach the surface of last scattering at $z_s \simeq 1000$. That will be the burden of this section; in Section 8.4.6 we will see how the density perturbations introduce observable temperature fluctuations on the surface of last scat-tering.

We begin by evaluating the scale of, and the mass contained within, the particle horizon at $z = 1000$. The two are

$$l_H = ct_s \approx 100 \text{ Mpc},$$
$$M_H \approx 10^{17} M_\odot,$$
(8.12)

where l_H is expressed in comoving coordinates. That linear scale may be converted to a corresponding angular scale (see, e.g., Weinberg, 1972); it is

$$\theta_H \simeq 2°\Omega^{1/2}.$$
(8.13)

Perturbations on this scale and above reach the surface of last scattering unaffected by any causal, physical processes. Hence, perturbations on scales $> l_H$ or θ_H do preserve the initial power spectrum as expressed in eqn. (8.7) or eqn. (8.8). Hence the great impor-tance of the COBE results (Section 7.7.3) on scales of 10° and above; they probe the pri-mordial power spectrum directly.

As we have seen in the preceding portion of Section 8.4, on scales smaller than M_H, a variety of mechanisms can operate to alter the initial spectrum of perturbations. For instance, as we have seen, Silk damping introduces a high frequency (low mass) cutoff in adiabatic perturbations in the baryonic matter. Likewise (but at a larger characteristic scale), free streaming of hot dark matter particles produces a similar high frequency cutoff (as shown in fig. 8.9). That figure also displays the differences between isocurvature and adiabatic perturbations: if we consider perturbations of each class but of the same scale (say 10 Mpc), the value of $\Delta\rho/\rho$ for the former is smaller because isocurvature perturbations did not grow before they entered the horizon.

The effect of many of the physical processes described thus far in Section 8.4 is to alter the initial spectrum of density perturbations given by eqn. (8.6). As Peebles (1982) first showed, for a wide range of models these alterations can be incorporated in a transfer function with a simple analytic form. For the scale invariant case, Peebles uses

$$P(k) \doteq Ak(1 + \alpha k + \beta k^2)^{-2},$$

and Bond and Efstathiou (1984) introduce a transfer function for CDM models,

$$P(k) = Ak\{1 + [ak + (bk)^{3/2} + (ck)^2]^\nu\}^{-2/\nu},$$

with ν close to 1.13 for $\Omega_b = 0.03$. Both expressions provide good approximations to the spectrum of density fluctuations reaching the surface of last scattering.

The coefficients α, β, a, b and c are model-dependent (see Bond and Efstathiou, 1984; Peebles, 1984; or Efstathiou *et al.*, 1992; for examples). Nevertheless, we may explore the general functional forms of both transfer functions. First, for large scales ($k \rightarrow 0$), $P(k) \propto k$; the initial spectrum is unaffected, as we noted in Section 8.4.4. For large k, $P(k) \propto k^{-3}$ in the scale-invariant case, or k^{n-4} generally. This is a result of the cutoff at small scales produced by oscillation and damping as described above. In addition, the coefficients depend on Hubble's constant. In general, the model dependence determines both the overall shape of $P(k)$ between the two asymptotic values found above and the linear scale of the transition from k^1 to k^{-3} behavior.

Thus far, I have tried to isolate the basic physical processes governing growth (or damping) of density perturbations up to the epoch of last scattering. For two reasons I have not presented quantitative results. First, the calculations are complex (see, e.g., Bond, 1990; Efstathiou, 1990). Second, our real interest is in the amplitude of temperature fluctuations in the CBR, and these do not necessarily scale in any direct way with $\Delta\rho/\rho$. Thus we will save more quantitative work until Section 8.5.

Before we turn to more quantitative results, however, there is one additional process we need to consider, one that occurs *after* the epoch of last scattering. It is a crucial feature of cold dark matter models for the origin of structure in the Universe.

As fig. 8.8 and fig. 8.9 show, adiabatic perturbations of *baryons* are strongly damped on mass scales $\lesssim (10^{10}-10^{11})\ M_\odot$. The cold dark matter component, however, is not affected by photon drag. Hence $\Delta\rho/\rho$ in dark matter continues to grow (fig. 8.8). Once the primeval plasma recombines at $z \simeq 1000$, photon drag ceases abruptly, and the baryonic matter is free to fall into the potential wells created by the perturbations in the dark matter. Thus $(\Delta\rho/\rho)_b$ rapidly 'catches up' with $\Delta\rho/\rho$ in dark matter. Consequently, well after recombination, there is no observable cutoff in the spectrum of density perturbations in cold dark matter cosmologies.

It is important to remember, however, that CBR observations reveal only the amplitude of $\Delta\rho/\rho$ (and the corresponding $\Delta T/T$) exactly *on* the surface of last scattering. Hence the observable temperature fluctuations are sharply reduced for scales $< M_D$, as shown by the dotted line in fig. 8.9. The rapid growth of $(\Delta\rho/\rho)_b$ occurs after the epoch we sample when we observe the CBR.

8.4.6 How CBR fluctuations relate to density perturbations on the last scattering surface

I have frequently made the claim that an inhomogeneous distribution of matter on the surface of last scattering will introduce fluctuations into the CBR. We now need to explore that claim by considering the physical processes which can introduce fluctuations in temperature in the CBR from point to point in the sky. As we shall see, these mechanisms are model-dependent and in some cases rather complex. From the point of view of the theorist, of course, that model dependence is a virtue: different models for the growth of structure, or for the nature of a possible 'dark' component, leave different and potentially observable imprints in the CBR.

Let us begin with the easiest case, adiabatic perturbations. In this case, both radiation and matter are perturbed, and $(\Delta\rho/\rho)_r = (4/3)(\Delta\rho/\rho)_m$. Since $\rho_r \propto u \propto T^4$ (eqn. (3.3)), it is easy to see that $\Delta T/T = (1/3)(\Delta\rho/\rho)_m$. Note that the scale, l, of the perturbation does not enter this simple relationship. Thus, for adiabatic perturbations, the dependence of $\Delta T/T$ on the comoving scale l of the perturbation is the same as the dependence of $\Delta\rho/\rho$ on l. For a power law k^n spectrum, we have from eqn. (8.8) $\Delta\rho/\rho \propto l^{-(3+n)/2}$. Hence, for adiabatic perturbations,

$$\Delta T/T \propto \theta^{-(3+n)/2}. \tag{8.14}$$

A word of warning before we proceed. The magnitude of the temperature fluctuations given above is the quantity that would be measured by a 'local' thermometer. We shall see below that the observable value of ΔT at the surface of last scattering may be considerably smaller because of averaging effects.

We turn next to $\Delta T/T$ fluctuations introduced by perturbations in the gravitational potential, called the Sachs–Wolfe effect, since it was first pointed out in their important 1967 paper. The physical mechanism responsible for point-to-point temperature fluctuations is the gravitational redshift of photons leaving a gravitational potential well. If the gravitational potential ϕ varies from place to place, so will the measured temperature. Spatial variations in ϕ can be produced by density perturbations (scalar fluctuations in ϕ) or by a stochastic background of gravity waves (tensor fluctuations; see Starobinskii, 1985). Sachs and Wolfe (1967) considered both classes (but in flat space only). By integrating the equations describing the propagation of photons through a perturbed metric, they derived a relationship between the amplitude of density perturbations responsible for perturbations in the gravitational potential and the amplitude of the resulting fluctuations in the temperature of the CBR.

Here, we may use a heuristic, Newtonian argument to establish the scale-dependence of the Sachs–Wolfe effect. We begin by noting that

$$\frac{\Delta T}{T} \approx \frac{1}{3}\frac{\Delta\phi}{c^2}.$$

In turn, we may write $\Delta\phi = G\,\Delta M/l$ for the amplitude of gravitational potential fluctuations on a scale l. Thus

$$\frac{\Delta T}{T} \approx \frac{1}{3}\frac{G\Delta M}{lc^2} \propto \Delta\rho l^2,$$

and using eqn. (8.8) we find

$$\Delta T/T \propto l^{(1-n)/2} \propto \theta^{(1-n)/2}. \tag{8.15}$$

Note that, if $n = 1$, as is the case* for the Harrison–Zel'dovich spectrum favored in inflationary cosmologies, then the amplitude of Sachs–Wolfe temperature fluctuations is independent of angular scale on the sky.

The final mechanism for production of temperature fluctuations is the Doppler effect. For small amplitude perturbations we have, as in eqn. (8.1),

$$|\Delta T/T| = v/c,$$

where v is the velocity of the scattering material relative to the bulk of matter on the surface of last scattering. We may then use an argument very similar to the one developed in Section 8.2.3 to show

$$v \propto l \frac{\Delta M}{M}$$

or

$$v \propto l \frac{\Delta \rho}{\rho}$$

on a comoving scale l. Finally, then,

$$\Delta T/T \propto \theta^{-(1+n)/2} \tag{8.16}$$

for a k^n power law spectrum of perturbations.

Doppler fluctuations will be particularly pronounced for vector (vortex) perturbations. In addition, in the case of Doppler fluctuations, linear polarization of the fluctuations is possible because there is a preferred direction picked out by the velocity vector.

A quick glance at (8.14), (8.15) and (8.16) will show that each mechanism produces a different power law dependence of $\Delta T/T$ on angular scale. Hence each may dominate in some range of angular scale θ. In particular, the Sachs–Wolfe effect dominates on the largest scales, where we have measurements of, not just upper limits on, $\Delta T/T$.

8.4.7 Effect of the thickness of the last scattering surface

As noted in Section 5.1.2, recombination of the primordial plasma is not instantaneous. Thus Thomson scattering of photons does not cease instantaneously – the surface of last scattering has a characteristic 'thickness' of $\Delta z_s \approx 100$ (Jones and Wyse, 1985). The optical depth of the surface of last scattering increases gradually from about 0 to about 1 as we look back through this thickness. It follows that perturbations of linear scale less than the thickness of the last scattering surface will be partially transparent. To calculate the value of the CBR temperature along a line of sight, we must solve the transport equation through the surface of last scattering. Such calculations have been performed by Peebles and Yu (1970), Sunyaev and Zel'dovich (1970), Wilson and Silk (1981) and Bond and Efstathiou (1984); see also Jones and Wyse (1985).

Qualitatively, however, the result is easy to understand: observed temperature fluctuations on scales less than the thickness of the last scattering surface will be reduced by

* Given the convention that I adopted in Section 8.4.1.

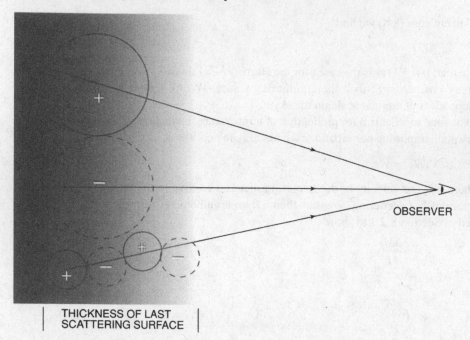

THICKNESS OF LAST
SCATTERING SURFACE

Fig. 8.10   For perturbations of characteristic scale less than the thickness of the last
scattering surface, $\Delta T/T$ is reduced by averaging along the line of sight, as shown by the
lowest line of sight.

averaging over regions of higher and lower temperature, as shown schematically in fig.
8.10. Since $\Delta z_s \approx (1/12)z_s$, we may estimate the angular scale at which this averaging
effect sets in. It is, from eqn. (8.13),

$$\theta_A \lesssim 10' \Omega^{1/2} \tag{8.17}$$

More quantitative solutions of the equation of transfer produce values of θ_A somewhat
smaller than this value, about $7' \Omega^{1/2}$. We thus lose information on density perturbations
on scales corresponding roughly to 5 Mpc (comoving) or for masses $\lesssim 10^{13} M_\odot$. Matter
at the surface of last scattering may very well be perturbed on smaller scales, but these
inhomogeneities will leave no observable trace in the CBR.*

8.5   Quantitative results

In the section immediately preceding, we discussed both the classification of density
perturbations and the imprint they leave on the CBR. We did not specifically discuss the
amplitude of $\Delta \rho / \rho$. Nor do currently available theories of inflationary cosmology
provide even order of magnitude predictions of $\Delta \rho / \rho$. In this section we will explore the
means used to calculate the amplitude of $\Delta \rho / \rho$ on the surface of last scattering, and
hence to produce numerical predictions of $\Delta T/T$.

---

* An exception to this conclusion is treated in Section 8.7.

8.5.1 Normalization

Since $\Delta\rho/\rho$ cannot be calculated from first principles, we must instead calculate it from some observable property of the Universe today. The basis of this calculation is straightforward. It relies on the fact that we have available measures of the inhomogeneity of the Universe today. Suppose we can establish that the present inhomogeneity on some scale l has an amplitude

$$(\Delta\rho/\rho)_{l,0} = x.$$

Assuming that inhomogeneity grew from a smaller amplitude density perturbation through the action of gravity alone, we may use the results of Section 8.4.2 to calculate the amplitude at any earlier epoch, in particular the epoch of last scattering at $z_s \simeq 1000$:

$$(\Delta\rho/\rho)_{l,s} = (z_s + 1)^{-1} (\Delta\rho/\rho)_{l,0} \approx 10^{-3}\, x. \tag{8.18}$$

Two constraints on the use of this simple argument must be noted. First, we are implicitly using linear perturbation theory and ignoring all but gravitational growth of the perturbation. For both reasons, our argument works only if $x \leq 1$; i.e., for small amplitude perturbations. The second constraint applies only to open ($\Omega < 1$) cosmological models. In these models gravitational growth ceases at a redshift $(z + 1) \simeq \Omega^{-1}$. From that redshift until the present, the amplitude of $\Delta\rho/\rho$ is 'frozen.' For $\Omega < 1$ models, it follows that eqn. (8.18) must be modified to

$$(\Delta\rho/\rho)_{l,s} = \Omega^{-1}(z_s + 1)^{-1} (\Delta\rho/\rho)_{l,0} \approx 10^{-3}\Omega^{-1}x. \tag{8.19}$$

Clearly, for a fixed value of x, both $(\Delta\rho/\rho)_s$ and the amplitude of the resulting CBR fluctuations will be larger in open cosmological models than in a comparable Einstein–de Sitter model. The opposite holds for closed models.

We turn now to the two normalization schemes actually used in such calculations. Both rely on the (inhomogeneous) distribution of galaxies, which is assumed to represent or 'trace' the underlying distribution of density in the Universe (an assumption that we will reexamine below). We have known for decades that galaxies are not randomly distributed in space. Their actual distribution may be described by a two-point correlation function $\xi(r)$ defined by

$$\langle \delta n(x_1)\, \delta n(x_2) \rangle = \bar{n}^2[1 + \xi(r)], \tag{8.20}$$

where $\bar{n}$ is the average density of galaxies, $r = |x_1 - x_2|$ and $\delta n(x_1) = n(x_1) - \bar{n}$ (see Peebles, 1980 for a much fuller discussion). When ξ is calculated from galaxy surveys such as those reviewed by Geller and Huchra (1988), it appears to have a power law dependence, as shown in fig. 8.11. Crudely, we see from fig. 8.11 that $\Delta\rho/\rho \approx 1$ (or $\xi = 1$) at a scale of $\sim 10h^{-1}$ Mpc. Thus the observational results summarized in fig. 8.11 provide an estimate of $(\Delta\rho/\rho)_0$ on a particular scale, as desired. In practice, we need a more precise estimate than is provided by this order of magnitude argument.

One approach is to use the observations to calculate the linear scale l on which the variance of the galaxy distribution within a sphere of radius l reaches unity. That scale has been found (Davis and Peebles, 1983b) to be $8h^{-1}$ Mpc.* Thus the predicted variance in spherical regions of the same comoving radius on the surface of last scattering was

---

* Note that the normalization depends explicitly on h, the value of Hubble's constant.

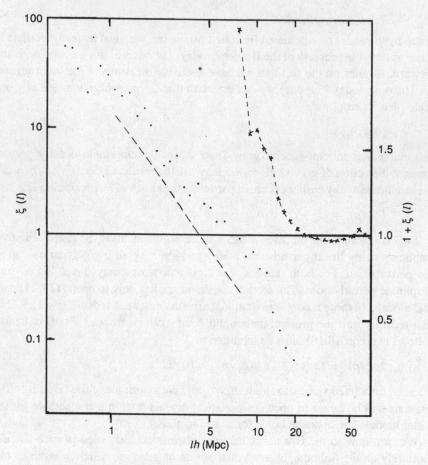

Fig. 8.11   Correlation function for galaxies in the CfA survey (from Davis and Peebles, 1983b). The dashed line is a power law of index −1.8, which provides a good fit to $\xi(l)$ over most of the range of the observations. The dots show $\xi(l)$ on a logarithmic scale; the crosses $\xi(l) + 1$ at $l > 8$ Mpc on a linear scale.

$[(z_s + 1)^{-1}]^2$ or about 10^{-6}. That value in turn can be used to establish the amplitude of the Fourier components of the density field at last scattering:

$$10^{-6} \approx \sum_k |\delta(k)|^2 \, W^2(k,l),$$

where $W(k, l)$ is the Fourier transform of the spherical window function used to calculate the variance, given by Efstathiou (1990) as

$$W(k,l) = \frac{3}{(kl)^3}[\sin kl - kl\cos kl].$$

Values of $\delta(k)$ in turn provide values of $(\Delta\rho/\rho)_{l,s}$.

Note that this method rests on a measurement of $\Delta\rho/\rho$, that is not small, but approximately unity. One would prefer to work on larger scales, where $\Delta\rho/\rho$ is $\ll 1$; unfortunately, at these larger scales, the galaxy distribution is less well determined. Large-scale surveys now underway are likely to improve our knowledge of the galaxy distribution on scales $> 8h^{-1}$ Mpc, and hence to provide a more secure normalization of $\delta(k)$ or $\Delta\rho/\rho$ on the surface of last scattering.

Another (but not independent) normalization is provided by the second moment of the galaxy correlation function,

$$J_3 = \int_0^L \xi \, l^2 \, dl$$

(Davis and Peebles, 1983b). Values of J_3 can be calculated from the observations of galaxies like those summarized in fig. 8.11, or from two-dimensional surveys. Values of J_3 can also be calculated for models of the density perturbation field at the epoch of last scattering, and then compared with the observations to provide a normalization for $\delta(k)$ or $\Delta\rho/\rho$.

This technique is less susceptible than the previous one to errors introduced by non-linearity (Groth and Peebles, 1976). It is gratifying that both provide very similar normalizations.

Both methods, however, rely explicitly on the assumption that the distribution of galaxies accurately traces the inhomogeneous distribution of *all* matter in the Universe. As we saw in Section 8.2.4, it may not: bias may increase the correlation of galaxies. If bias is present, normalizations based on the distribution of galaxies will be too high by a factor b. Correcting for bias thus means lowering our estimates of $(\Delta\rho/\rho)_s$ and of the predicted amplitude of CBR temperature fluctuations by the same factor. Values currently favored for b lie in the range $1 \lesssim b \lesssim 2.5$, so the correction factor is not large.

It is also important to bear in mind that both methods produce a normalization on quite a small distance scale, 10–20 times smaller than the horizon scale, for instance.

An entirely independent means of normalizing calculations of CBR fluctuations was opened up by the detection of $(\Delta T/T)$ fluctuations on angular scales $\gtrsim 10°$ by the COBE satellite (Section 7.7.3). Here the normalization is much more direct (and, incidentally, is unaffected by concerns about bias and non-linearity): we merely extrapolate the measured values of $\Delta T/T$ to smaller angular scales. As the discussion in Section 8.4 suggests, that extrapolation depends on the model assumed for the nature of the perturbations, on the nature of the dark matter present, if any, and on the index n. Hence a comparison of the results of observations on angular scales 0.1–10° with the COBE measurements on larger scales may constrain theories of galaxy formation or even reveal the nature or existence of dark matter (see Appendix C).

It is obviously of considerable interest to ask whether the COBE measurements produce a normalization in agreement with the results based on observations of the present distribution of galaxies. The rough answer is yes – providing welcome evidence that the fluctuations detected by the satellite are the long-sought $\Delta T/T$ fluctuations on the surface of last scattering. On the other hand, the agreement is not exact. The amplitude of the fluctuations reported by the COBE group (Smoot *et al.*, 1992) appears to be up to twice as large as expected on the basis of the correlation function of galaxies, at least for some theories of galaxy formation. The COBE results are too recent to allow us to assess whether or not this discrepancy presents a serious problem for current theories of structure formation. Pessimists worry about the factor of two; optimists marvel at the agreement between measurements of $\Delta T/T$ (or $\Delta\rho/\rho$) at an early epoch of about 3×10^5 years and observations of the present inhomogeneous structure of the Universe at $t_0 \simeq 10^{10}$ years. It is worth remarking that the discrepancy is made worse if the bias parameter b exceeds unity.

8.5.2   Typical numerical results for primary fluctuations in the CBR

We now have in hand the building blocks needed to construct predictions of the amplitude of CBR fluctuations as a function of angular scale. These building blocks include the amplitude of $\Delta\rho/\rho$ as a function of linear scale, the connections between $\Delta\rho$ and ΔT (Section 8.4), and a normalization.

Within a few years of the discovery of the CBR, Sachs and Wolfe (1967), Silk (1968) and others were predicting values of $\Delta T/T$ on a range of angular scales. In general, the predicted values were far too large.* More detailed calculations by a number of groups and individuals appeared by the early 1980s. Among the papers presenting such calculations are those of Doroshkevich *et al.* (1978), Wilson and Silk (1981), Bond and Efstathiou (1984, 1987), Vittorio and Silk (1984), Efstathiou and Bond (1987), Martinez-Gonzalez and Sanz (1989), Fukugita *et al.* (1990) and Gorski (1991). Reviews by Holtzman (1989) and Suto *et al.* (1990) summarize a wide range of models; see also the review articles by Bond (1988), Efstathiou (1990), and Gouda *et al.* (1991). These authors have considered both adiabatic and isocurvature fluctuations; a wide variety of cosmological models; a variety of assumptions about the nature (or existence) of dark matter; different values of h and Ω_b; and a range of values of the index n in eqn. (8.6).

The calculated amplitudes of CBR fluctuations for a particular model are generally presented as either a power spectrum of fluctuations or an autocorrelation function. Let us begin by writing down the predicted CBR temperature distribution in terms of spherical harmonics (as in eqn. (6.3)):

$$T(\theta,\phi) = \sum_{n,m} C_{nm} Y_n^m (\theta,\phi).$$

The expectation value of the coefficients C_{nm} averaged over m defines the *power spectrum*:

$$C_n = \langle |C_{nm}|^2 \rangle. \tag{8.21}$$

Values of C_n are calculated, multiplied by n^2 and plotted as a function of n – see fig. 8.12.

The coefficients C_n may be related to the multipole moments introduced in eqn. (6.5) as follows:

$$T_n = \left(\frac{1}{4\pi} \sum_{m=-n}^{n} C_{nm}^2 \right)^{1/2} = \left(\frac{(2n+1)C_n}{4\pi} \right)^{1/2}.$$

In particular,

$$T_2 = \left(\frac{5C_2}{4\pi} \right)^{1/2}. \tag{8.22}$$

Note that some authors employ the notation $a_2 = C_2^{1/2}$.

The power spectrum C_n may also be related to the autocorrelation function $C(\theta)$

* In large part because only baryonic matter was considered.

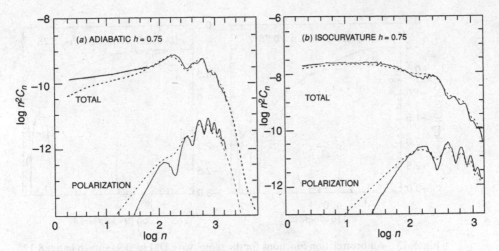

Fig. 8.12   Power spectra of temperature anisotropies in two cold dark matter models: (a) adiabatic fluctuations and (b) isocurvature fluctuations (from Bond and Efstathiou, 1987). The dashed lines are from an approximation developed in that paper. See text for discussion of the details of these curves.

defined by

$$C(\theta) = \langle \Delta T(\mathbf{x}_1) \, \Delta T(\mathbf{x}_2) \rangle \tag{8.23}$$

with $\mathbf{x}_1 \cdot \mathbf{x}_2 = \cos \theta$. The average is taken over all points separated by angle θ. The relation between the two is

$$C(\theta) = \frac{1}{4\pi} \sum_n (2n+1) C_n P_n(\cos \theta), \tag{8.24}$$

where P_n are the Legendre polynomials. The r.m.s. temperature fluctuation across the sky is given by $C^{1/2}(0)$.

Note that models for temperature fluctuations on the surface of last scattering in general produce non-zero values for T_1 and T_2, the dipole and quadrupole moments of the CBR distribution. These represent the long wavelength tail of the fluctuation spectrum, and are independent of the Doppler-induced dipole discussed in Section 8.2 or the quadrupole produced by anisotropic expansion (Section 8.3). The Doppler-induced dipole completely dominates that produced by perturbations on or near the surface of last scattering. On the other hand, the predicted quadrupole moment (produced largely by the Sachs–Wolfe effect) may be detectable. Indeed, as we shall see in Section 8.6, predicted amplitudes of T_2 are in order-of-magnitude agreement with the COBE measurement described in Section 6.7.3.

In the remainder of this section, I will single out four models for detailed discussion. The range is wide enough to reveal most qualitative properties of the great variety of calculations performed in the papers and reviews listed above.

Cold dark matter (CDM) models.   The first is a cosmological model with cold dark matter as the dominant component of density. In general, the total density is assumed to be $\Omega = 1$, with $\Omega_b \simeq 0.03$–0.1 and $\Omega_{CDM} \simeq 0.97$–0.90. Most authors have chosen to work with adiabatic fluctuations and to assume a power law for the spectrum of perturbations (often with index $n = 1$). In fig. 8.12, I display a set of typical results for the power spectrum, taken from Bond and Efstathiou (1987).

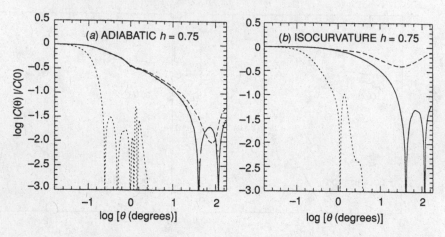

Fig. 8.13   Autocorrelation functions for the same two CDM models shown in fig. 8.12 (from Bond and Efstathiou, 1987). As is conventional $|C(\theta)|$ not $C(\theta)$ is plotted; $C(\theta)$ crosses zero whenever $|C(\theta)| = 0$ on these plots. The short dashes are $|C(\theta)|$ for the polarized component as in fig. 8.12.

The flat region at low values of n^* is that dominated by the Sachs–Wolfe effect; as expected from eqn. (8.18) for index $n = 1$, there is little or no dependence of ΔT on scale at large angular scales. At smaller scales ($n \gtrsim 100$ or $\theta \lesssim \frac{1}{2}°$), Doppler perturbations dominate. As noted in Section 8.4.6, this mechanism introduces linear polarization into the CBR fluctuations. As the graphs show, the polarization is typically 3–5%. Most of the 'bumps and wiggles' at $n \approx 100$–1000 are produced by compressions and rarefactions in oscillatory sound waves (see Section 8.4.2). Finally, note the gradual roll-off in C_n at large n in the isocurvature case. This is a direct consequence of the fact that isocurvature perturbations do not grow before they enter the horizon, unlike adiabatic ones.

In fig. 8.13, $C(\theta)$ is shown for the same models, normalized to $C(0)$. As expected, the isocurvature model has more power at large scales.

Figure 8.12 and fig. 8.13 were constructed for $\Omega_b = 0.03$, $\Omega_{CDM} = 0.97$ and $P(k) \propto k^1$. In fig. 8.14, I plot $n^2 C_n$ for CDM models with several different values of Ω_b, with $\Omega_{CDM} = 1 - \Omega_b$. For small Ω_b, say $\Omega_b \lesssim 0.2$, the power spectrum does not depend strongly on Ω_b. As Ω_b becomes dominant, however, C_n rises sharply.

We will see in Section 8.6 that the predicted amplitudes of ΔT on scales 0.1–10° for some CDM models are in good agreement with the observational results of Chapter 7. On the other hand, other astronomical observations have raised some doubts about the validity of this model (see recent discussions by Efstathiou, 1990, and Efstathiou *et al.*, 1992). Basically, the CDM models produce perturbations of too small an amplitude at large scales ($l \gtrsim 20h^{-1}$ Mpc) to explain large-scale inhomogeneities in the distribution of galaxies (see, e.g., Maddox *et al.*, 1990) or the cluster–cluster correlation reported by Bahcall and Soneira (1983). The lack of large-scale power also makes it difficult to explain the large-scale streaming velocities mentioned in Section 8.2.3, or the large value of the velocity of the local group derived from the dipole moment of the CBR itself (Section 8.2.2). The magnitude of these discrepancies is *roughly* a factor of 2–3, like the

---

* Do not confuse n, the order of the multipole, with the power law index for fluctuations.

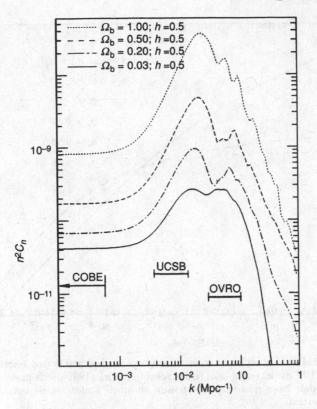

Fig. 8.14   Power spectra for a wider range of CDM models (from Vittorio and Silk, 1992, with permission). In this case $h = 0.5$. C_n is shown as a function of comoving wave number; the length scales spanned by several different observational programs are indicated. Note the strong dependence on Ω_b for values of $\Omega_b \gtrsim 0.2$.

discrepancy between the COBE measurements on scales $> 10°$ and the normalizations based on J_3 or the variance within a sphere of $8h^{-1}$ Mpc radius. Some cosmologists take these discrepancies seriously and consider the CDM models dead or at least terminally ill. Others defend the models (in part because they agree well with CBR anisotropy limits and measurements) or modify them slightly to add extra power where needed (see Section 8.8).

Hot dark matter models.   Cosmological models dominated by hot dark matter (say neutrinos of nonzero rest mass) do provide extra power (larger amplitude perturbations) at large scales. More precisely, these models, once normalized, can in principle provide adequate power at large scales without introducing too much variance, and hence unacceptably high values of $\Delta T/T$, at small scales. The cutoff shown in fig. 8.9 is responsible for this feature of these models. In figure 8.15, taken from Schaeffer *et al.* (1989), I display $\delta M/M$ for one massive neutrino hot dark matter model, and contrast it with the same quantity predicted in CDM models. Note the increased power at large masses, corresponding to large angular scales (low multipole orders). Since predictions like those shown in fig. 8.15 are generally normalized at relatively small scales, of order 10 Mpc or $0°.1$, models dominated by hot dark matter predict somewhat larger values of $\Delta T/T$ on degree scales and above (including T_2) than do comparable models dominated by cold dark matter.

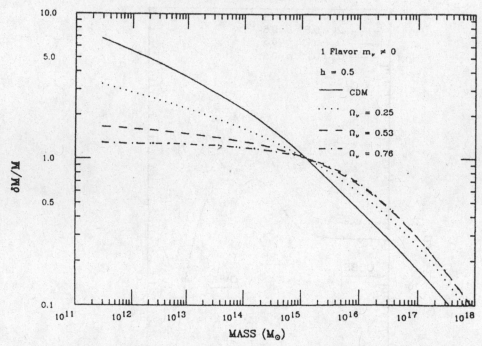

Fig. 8.15 HDM models with one massive neutrino (dashed and dotted lines) compared with a CDM model (solid line) (from Schaeffer *et al.* (1989), with permission). The HDM models have relatively less power on small scales (small masses), as explained in the text.

In the case of hot dark matter models, the observational problems are the opposite of those encountered by cold dark matter models – there is too little power on small scales to explain structure formation on scales $\lesssim 10h^{-1}$ Mpc. For instance, there is essentially no power (i.e., no perturbations) on the scale of galaxies. Hence galaxy formation requires fragmentation of large systems, and is expected to occur later than in CDM models. On the other hand, HDM models do have the advantage of possessing a potential candidate for the dark matter particle, a light neutrino of non-zero rest mass. While there have been plenty of theoretical suggestions for cold particles (axions, WIMPs, etc.), we have no experimental evidence as yet for any of them.

Baryon-only (adiabatic) models. Given the lack of direct evidence for dark matter, it makes sense to look carefully at cosmological models containing only baryons. The earliest predictions of CBR fluctuations, made before dark matter came into vogue, were naturally based on this assumption, which also has the appeal of simplicity.

Baryon models may be divided into two classes – those that accept the upper limits set on Ω_b by nucleosynthesis (see Section 1.6.4) and those that do not. In the former case, since $\Omega_b = \Omega_{TOT} \ll 1$, the Universe must be open. In the latter case, we are free to assume $\Omega_b = \Omega_{TOT} = 1$.

In either case, the amplitude of *adiabatic* fluctuations in the CBR is much larger than found in dark matter models. The explanation for this difference is simple; in baryon-only models no potential wells are formed by dark-matter perturbations for the baryons

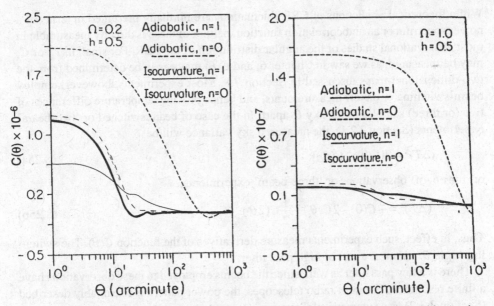

Fig. 8.16 $C(\theta)$ for baryon isocurvature models (from Gouda *et al.*, 1989, with permission). See also fig. 8.24 where these models are compared with the upper limit of Readhead *et al.* (1989), and shown to exceed it unless $\Omega \gtrsim 1$ and $h \lesssim 0.5$.

to fall into after recombination. Hence, to produce the structure that we see today, we need to 'build in' larger fluctuations on the surface of last scattering. That effect is displayed clearly in fig. 8.14; compare the curves for $\Omega_b = 0.03$ ($\Omega_{CDM} = 0.97$) and for $\Omega_b = \Omega_{TOT} = 1$. In the latter case, there is no dark matter present; hence large initial perturbations are required. If $\Omega_b = \Omega_{TOT} \ll 1$, as given by the nucleosynthesis argument, still larger adiabatic fluctuations are required since gravitational growth ceases early.

Baryon isocurvature models. One way to sidestep the prediction of large values of $\Delta T/T$ in baryon models is to assume that the perturbations are isocurvature (approximately isothermal) rather than adiabatic. Such models have been championed by Peebles (1987). Power spectra and/or autocorrelation functions for baryon isocurvature models have been constructed by Bond and Efstathiou (1987) and Gouda *et al.* (1989); fig. 8.16 is taken from the latter. As expected, we see that isocurvature models have more power on large scales than the corresponding adiabatic ones. We also see that changing the power law index n has a strong effect on $C(\theta)$, especially in the isocurvature case. An $n = 0$ spectrum has intrinsically less small-scale power than an $n = 1$, scale invariant spectrum. Since the normalization is fixed at small scales, $C(\theta)$ is 'magnified' for the $n = 0$ case. In section 8.5.3, we will see, however, that most experiments measure not $C(\theta)$ but its derivatives; hence measured $\Delta T/T$ values for $n = 0$ would not necessarily be much larger than those for $n = 1$. As we will see in Section 8.6 (see fig. 8.19 later), the predicted values of $\Delta T/T$ for $n = 0$ are at most a factor of three larger than $\Delta T/T$ for $n = 1$, and then only on scales $\gtrsim 1°$.

8.5.3 Beam geometries and filter functions

While theoretical predictions of CBR fluctuations are normally presented in terms of a power spectrum or an autocorrelation function, neither of these is directly measurable in most observational studies of the angular distribution of the CBR. For small n, T_n or C_n may be measured, as we saw in Chapter 6; and $C(\theta)$ may easily be determined from the two-dimensional maps discussed in Section 7.8. Most experiments, however, employ beam-switching or on–off measurements, and hence provide temperature differences of two (or three) samples of the sky θ_s apart. In the case of beam switched or 'two-beam' experiments (Section 7.2.1), the measured sky variance will be

$$(\Delta T)^2 = 2[C(0) - C(\theta_s)];\tag{8.25a}$$

or, for on–off observations or 'three-beam' experiments,

$$(\Delta T)^2 = \frac{3}{2}C(0) - 2C(\theta_s) + \frac{1}{2}C(2\theta_s).\tag{8.25b}$$

Thus, in effect, such experiments measure derivatives of the function $C(\theta)$. The switching angle θ_s serves as a sort of high pass filter.

There is a low pass filter as well, since the beams employed in these observations have a finite resolution. For most radio telescopes, the power pattern is reasonably described (Section 3.3.2) as a symmetrical Gaussian,

$$P(\theta) = e^{-1/2(\theta/\theta_0)^2} = \exp\left[-(4\ln 2)\left(\frac{\theta}{\theta_{1/2}}\right)^2\right],\tag{8.26a}$$

where $\theta_{1/2}$ is the full width at half power of the beams, numerically equal to $2.35\theta_0$. Since for small θ, θ and the multipole order n are related by $\theta \approx 1/n$, we may also write

$$P(n) \approx e^{-(n\theta_0)^2}\tag{8.26b}$$

with θ_0 expressed in radians. Clearly, sky variations of scale $\lesssim \theta_{1/2}$ will be washed out.

A particular experiment is thus sensitive to a restricted range of $C(\theta)$, very roughly $\theta_{1/2} < \theta < \theta_s$. These ranges are displayed in fig. 8.17 for several of the most sensitive experiments discussed in Chapter 7.

The effect of beam geometry may also be expressed in terms of a filter function, F_n, where n is again the order of the multipole. Combining (8.24), (8.25) and (8.26), we have

$$(\Delta T)^2 = \frac{1}{2\pi} \sum_n (2n+1)\, C_n\, F_n,\tag{8.27}$$

where

$$F_n = [1 - P_n(\cos \theta_s)]e^{-(n\theta_0)^2}\tag{8.28a}$$

for a two-beam experiment and

$$F_n = \left[\frac{3}{4} - P_n(\cos \theta_s) + \frac{1}{4}\, P_n(\cos 2\theta_s)\right] e^{-(n\theta_0)^2}\tag{8.28b}$$

for an 'on–off' or three-beam experiment.

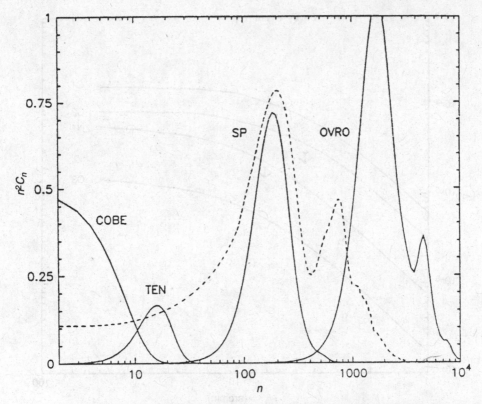

Fig. 8.17 The solid lines show filter functions for four experimental arrangements discussed in Chapter 7: COBE (Smoot *et al.*, 1992); 'TEN,' the experiment of Davies *et al.* (1987); the South Pole experiment of Meinhold and Lubin (1991); and the Owens Valley observations of Readhead *et al.* (1989). The height of each curve is a measure of the sensitivity of the experiment. The dashed line is n^2C_n for a CDM model with $\Omega_b =$ 0.1 and $h = 0.5$ (adopted from Bond, 1990, with permission).

A useful special case of (8.27) is obtained for large n (small angles):

$$(\Delta T)^2 = \frac{1}{\pi} \int n^2 \, C_n \, F_n \, d(\ln \, n). \tag{8.29}$$

Thus the variance expected in a particular observation depends on the convolution of n^2C_n found from the model predictions with the filter function F_n appropriate for that experiment. That convolution, in turn, is represented graphically as the area of n^2C_n under each of the curves in fig. 8.17. Note how well the South Pole experiment being conducted by Lubin and his colleagues matches C_n for this particular CDM model.

Incidentally, in the small angle approximation the Legendre polynomials tend to Bessel functions as

$$P_n(\cos \theta) \to J_0\left[\left(n + \frac{1}{2}\right)\theta\right].$$

Therefore the filter functions of (8.28) may be rewritten using Bessel functions as follows, for small θ or equivalently, large multipole order n:

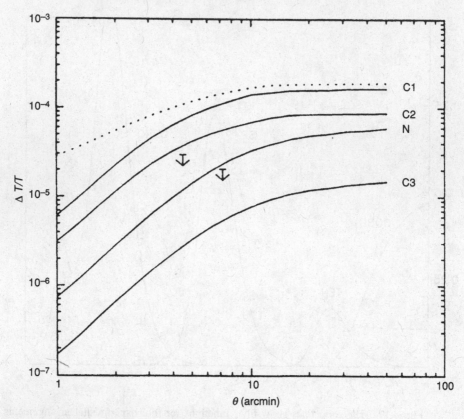

Fig. 8.18   Temperature fluctuations as a function of angular scale as calculated for the beam geometry employed by Readhead *et al.* (1989) (from Efstathiou (1990), with permission). The Readhead *et al.* upper limit is shown at $\theta \sim 7'$; also shown at $\sim 4'$ is the earlier upper limit of Uson and Wilkinson (1984) employing similar beam geometry. Models: C1 – adiabatic CDM with $\Omega = 0.2$, $h = 0.5$; C2 – adiabatic CDM with $\Omega = 0.2$, $h = 0.75$; C3 – adiabatic CDM with $\Omega = 1.0$, $h = 0.75$; N – adiabatic HDM with $\Omega = 1.0$, $h = 0.75$. In all cases, the index $n = 1$ and $\Omega_b = 0.03$.

$$F_n = [1 - J_0(n\theta_s)]e^{-(n\theta_0)^2} \tag{8.30a}$$

for a two-beam experiment, and

$$F_n = \left[\frac{3}{4} - J_0(n\theta_s) + \frac{1}{4}J_0(2n\theta_s)\right]e^{-(n\theta_0)^2} \tag{8.30b}$$

for the more usual three-beam, 'on–off,' arrangement.

It is important to keep the central result of this section in mind: the fluctuations in the CBR detected by a particular experiment will depend not only on C_n or $C(\theta)$ but on the beam geometry as well. Once the beam geometry is understood, we may convolve the appropriate filter function with $C(\theta)$ for a model to produce plots of $\Delta T/T$ as a function of θ (as shown in fig. 8.18), or a two-dimensional contour map of the fluctuations over a region of the sky (as shown in fig. 8.20 later). Alternatively, eqn.(8.29) can be evaluated numerically to provide a prediction of $\Delta T/T$ for each model and for a particular experiment.

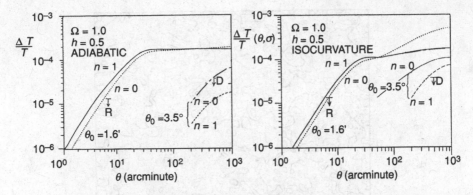

Fig. 8.19   Predicted amplitude of CBR fluctuations for pure baryon models with $\Omega = \Omega_b = 1$ (from Gouda *et al.*, 1989, with permission). In both cases, the longer curves are drawn for the experimental arrangement of Readhead *et al.* (1989); that experimental upper limit, shown as *R*, essentially eliminates all the baryon-only models for any $\Omega \lesssim 1$. The inset shorter curves are drawn for the experimental arrangement of Davies *et al.* (1987) with a much broader beam, $\theta_0 = 3°.5$. The upper limit (*D*) of Davies *et al.* (1987) eliminates the baryon isocurvature models shown here.

8.5.4 *Predicted amplitudes of primary fluctuations for three specific beam geometries*

We now present the results of such calculations for three specific experiments – COBE ($\theta_{1/2} = 7°$, $\theta_s = 60°$); the South Pole experiments now underway ($\theta_{1/2} = 0°.5$, $\theta_s = 1°$); and the experiment of Readhead *et al.* ($\theta_{1/2} = 1'.8$, $\theta_s = 7'.15$).

Fig. 8.18 shows the predicted value of $\Delta T/T$ as a function of beam switch angle for the beam geometry employed by Redhead *et al.* (1989). The rapid fall off at $\theta \lesssim 7'$ results from the averaging effect discussed in Section 8.4.7. As expected, for similar cosmological parameters ($h = 0.75$, $\Omega = 1.0$), a hot dark matter model predicts larger values of $\Delta T/T$ than a CDM model. We see also that $\Delta T/T$ increases as Ω decreases – compare eqn. (8.18) and eqn. (8.19). Finally, fig. 8.18 reveals one additional dependence that we have not commented on previously – $\Delta T/T$ depends on Hubble's constant, increasing as h decreases. This dependence is a consequence of the normalization described above.

In fig. 8.19, from Gouda *et al.* (1989), are displayed comparable results for baryon-only models. The upper limits shown are those of Readhead *et al.* (1989) at $\theta = 7.15'$ and Davies *et al.* (1987) at $\theta \simeq 500'$.

In fig. 8.20, taken from Bond (1990), a calculated contour map of the sky is shown, smoothed by the beam size of the experiment of Readhead *et al.*, with $\theta_{1/2} = 1.8'$. The contour lines are at multiples of 3×10^{-5} in $\Delta T/T$. The input model was a CDM model with $\Omega_b = 0.1$, $h = 0.5$ (line 12 of table 8.2).

While plots like Fig. 8.18 and maps like fig. 8.20 reveal interesting features of the models, the most straightforward means of testing a theoretical model is to use it to calculate the expected amplitude of $\Delta T/T$ for a given beam geometry, then compare the results with the measurement or upper limit set by the observations employing that beam geometry. As noted earlier in this section, such calculations have been performed for a wide range of models by Bond (1988), Holtzman (1989) and Suto *et al.* (1990). Holtzman

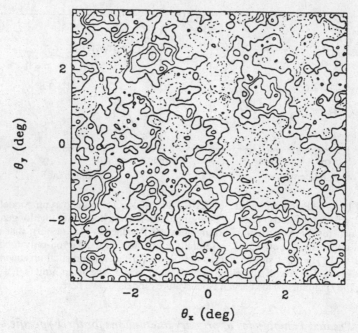

Fig. 8.20   A contour map of the microwave sky, smoothed to an angular scale $\theta_0 \sim 1'.8$. A patch of $\sim 6° \times 6°$ is shown (compare with fig. 8.26). In this case, a CDM model with $\Omega_{TOT} = 1$ and $\Omega_b = 0.1$ was used to generate the plot (from Bond, 1990, with permission).

(1989) provides results for a power law spectrum of fluctuations with index $n = 1$, allowing for a wide range of values of Ω_b, Ω_{TOT} and different types and mixtures of dark matter. He also considers models with non-zero cosmological constants (Section 8.8.3). Suto *et al.* (1990), on the other hand, fix Ω_b at approximately the value set by nucleosynthesis results (0.03), then vary Ω_{TOT}, the index n and the type of dark matter involved. Both treat adiabatic and isocurvature fluctuations. Both were written before the COBE measurements were released; the constraints imposed by these recent measurements are treated in Wright *et al.* (1992). In table 8.2, I have singled out some of these results, and others from the literature, for the three beam geometries described above. The three experiments of Smoot *et al.* (1992), Meinhold and Lubin (1991) and Readhead *et al.* (1989) are the most sensitive ones on scales of about 1'–90°. In table 8.2, I have also included some predicted values of T_2, for comparison with the COBE measurement (Section 6.7.3).

8.5.5 *Summary of general trends and dependencies in $\Delta T/T$*

Before we turn to the observational constraints on these many models for early perturbations, it seems worthwhile to me to review and summarize the qualitative dependence of $\Delta T/T$ on a number of cosmological parameters and a variety of assumptions about the nature of the density perturbations. Most of these trends are revealed in table 8.2; reasons for them are scattered throughout Sections 8.4 and 8.5.

Hubble's constant, $h = H_0/100$ km s^{-1} per megaparsec. For adiabatic perturbations, lower values of h produce larger values of $\Delta T/T$ if all other parameters of the model are kept constant. For isocurvature fluctuations, the dependence of $\Delta T/T$ on h is weak and

Table 8.2 *Predicted amplitudes for $\Delta T/T$ for a variety of models and three specific experimental arrangements. These are the OVRO experiment of Readhead et al. (1989), the South Pole experiment of Meinhold and Lubin (1991) and the quadrupole moment. Throughout, the bias factor is taken as 1. Only predictions in bold figures are compatible with the observational upper limits of measurements. The references are B88, Bond (1988); B90, Bond (1990); EB87, Efstathiou and Bond (1987); G87, Gouda et al., (1987); VS84, Vittorio and Silk (1984); W83, Wilson (1983); and WS81, Wilson and Silk (1981).*

| Reference | Model parameters | | | | Predicted $\Delta T/T$ (10^{-6}) | | |
	h	Ω	Ω_b	n	OVRO	S. Pole	T_2
			Adiabatic				
W83	0.5	0.1	0.1	1	870		
W83	0.5	0.1	0.1	0	1000		
W83	0.5	0.1	0.1	−1	720		
WS81	0.5	1	1	0	43		
WS81	0.5	1	1	−1	24		
B88	0.5	0.2	0.03	1	130		
B88	0.5	1	0.1	1	7		
B88	0.5	1	0.5	1	29		
VS84	0.5	0.4	0	1	54		
VS84	0.5	1	0	1	**13**		100
B90	0.5	1	0.01	1		**17**	
B90	0.5	1	0.1	1		**26**	9.9
B90	0.5	1	0.5	1		93	
VS84	1	0.4	0	1	20		
B90	1	1	0.03	1			4.5
VS84	1	1	0	1	7		
			Isocurvature				
G87	0.5	0.1	0.1	0	550		
B90	0.5	0.2	0.2	0	94	**21**	
B90	0.5	0.2	0.2	−1	145	71	
G87	0.5	0.4	0.4	1	140		
G87	0.5	1	1	1	19		
B90	0.5	1	1	0	89	78	
B90	0.5	1	1	−1	87	112	
EB87	0.5	1	1	2	53		
G87	1	0.1	0.1	1	800		
EB87	1	0.2	0.2	0	100		
EB87	1	0.2	0.2	−1	170		
EB87	1	1	1	1	40		

of the opposite sign. In both cases, the dependence on h results from a combination of the dependence of $P(k)$ on h^{-1} and the scale-dependence of the normalization described in Section 8.5.1.

Total density, Ω. Smaller Ω implies larger $\Delta T/T$, roughly as Ω^{-1} – see eqn. (8.19).

Baryon density, Ω_b. For fixed Ω, generally $\Omega = 1$, $\Delta T/T$ increases as Ω_b increases and $\Omega_{DM} = 1 - \Omega_b$ decreases. The increase in $\Delta T/T$ is small (proportional to $\Omega_b^{1/3}$) until Ω_b attains a value of $\gtrsim 0.2$ (Vittorio and Silk, 1992).

Power law index, n in eqn. (8.6). Here the dependence is complicated because of the details of normalization. The dependence of $\Delta T/T$ on n itself depends on the nature of the perturbations and the nature of the dark matter, if any, involved. As n increases, the amplitude of Doppler fluctuations rises relative to T_2 (there is more power on small scales).

Hot versus cold dark matter.   For a fixed set of cosmological parameters, HDM models in general predict somewhat larger values of $\Delta T/T$ than CDM models. This difference becomes more pronounced at larger angular scales, where HDM models have more power.

8.6   Observational constraints on models for density perturbations

Let us now compare the predicted values of $\Delta T/T$ with the available measurements and upper limits of Chapter 7 (especially table 7.2). We will soon discover that many of the models discussed predict values of $\Delta T/T$ substantially larger (or, in a few cases, smaller) than the observations permit. Hence whole classes of models may be dismissed. For instance, the results of the COBE measurements (Smoot *et al.*, 1992) establish limits on the index of power law density perturbations, $0.5 \leq n \leq 1.6$, thus eliminating many models for structure formation. We will begin our comparison of models and data by looking at several special cases (generally those singled out for discussion in Section 8.5) and then go on to summarize the features of models that do survive this observational test.

8.6.1   Baryon-only models

As table 8.2 shows, adiabatic, pure baryon models with $\Omega_b = \Omega_{TOT} < 1$ exceed the upper limit of Readhead *et al.* (1989) by a wide margin for any reasonable assumption about the power law index in eqn. (8.6). An adiabatic model with $\Omega_b = 1$ and $n \simeq -0.75$ might just squeak by, as Suto *et al.* (1990) show, but such a low value of n is ruled out by COBE.

Isocurvature models allow a lower value of $\Delta T/T$. Even these are in trouble with the OVRO experiment, however, unless $h \simeq 0.5$ and $\Omega_b \simeq 1$, i.e. well above the value suggested by nucleosynthesis (Section 1.6.4).* Furthermore, the set of parameters allowed by the OVRO results of Readhead *et al.* produces (Suto *et al.* 1990) values of $\Delta T/T$ on degree scales that substantially exceed the upper limit established several years ago by Davies *et al.* (1987). No baryon model is consistent with both the OVRO results (Readhead *et al.*, 1989) and the COBE results (Wright *et al.*, 1992).

Unless we introduce new physical arguments (some discussed in Section 8.7), we must give up the simplest models for structure formation, those involving only baryonic matter.

8.6.2   'Whirl' models

In the late 1960s Ozernoi and his colleagues (Chibisov and Ozernoi, 1969; Ozernoi, 1974) suggested that galaxy formation might be ascribed to large-scale vortices or 'whirls'. Such a theory offers a nice explanation for the observed rotation of galaxies. However, since such vortical motion dies away as the Universe expands, quite large values of the velocity field must be assumed on the surface of last scattering at $z \simeq 1000$. Large values of v in turn produce, via the Doppler effect, large values of $\Delta T/T$. As Anile and his colleagues showed in 1976, even the rough upper limits on $\Delta T/T$ available 15 years ago were sufficient to eliminate this model.

---

* Or unless there is reionization, which erases small scale fluctuations (see Section 8.7 below); reionization is a feature of Peebles' (1987) baryon isocurvature models.

8.6.3 Hot dark matter models

As the review by Suto *et al.* (1990) shows, the upper limit on $\Delta T/T$ set several years ago by Davies *et al.* (1987) essentially eliminates all isocurvature HDM models unless both Ω and n exceed unity. The COBE limits are even more stringent. We thus turn to adiabatic models.

A range of adiabatic HDM models is still viable if we consider *only* the CBR constraints, especially those models with large Ω and h and the index n in the range -0.5 to 1.0 More specifically, the plots of Suto *et al.* (1990) can be used to show that $\Omega \gtrsim 0.7$ and $n > 0$ for $h = 0.5$, and $\Omega \gtrsim 0.4$ and $n > -0.5$ for $h = 1.0$. Now let us consider the COBE results, not available to Suto *et al.* If we fix both n and $\Omega = 1$ for consistency with the inflationary picture, we then find that h must lie in the completely acceptable range $0.6 \lesssim h \leq 1.1$ by comparing* T_2 computed from the models with the COBE value of $13 \pm 4 \,\mu\text{K}$.

8.6.4 Cold dark matter models

A range of adiabatic CDM models also survives the confrontation with CBR observations. Indeed, as expected, the somewhat lower predicted values of $\Delta T/T$ allow more latitude in the selection of parameters in the models.

Let us begin our treatment of the CDM case by fixing the power law index at $n = 1$. For $h = 1.0$, both open and closed models with $0.2 \leq \Omega \leq 1.0$ are viable, provided Ω_b is selected appropriately (see Wright *et al.*, 1992, for the importance of the COBE results). For h this large and $\Omega = 1$, a value of Ω_b 2–5 times the value predicted by the nucleosynthesis argument of Section 1.6.4 is required to match the value of T_2 determined by COBE. Lower values of Ω_b predict too small a value for T_2. If $h = 0.5$, consistency with the observations is obtained if $\Omega \simeq 1$ and $\Omega_b \lesssim 0.3$, since somewhat larger values of T_2 are expected for a lower value of h (see Section 8.5.5).

Allowing n to vary does not change the situation much. As Suto *et al.* (1990) show, if n falls below zero, predictions of the models begin to violate the upper limit of Davies *et al.* (1987). Although Wright *et al.* (1992) did not consider models with index $n < 1$, such models would have a harder time meeting the COBE constraints than would comparable models with $n = 1$.

8.6.5 General properties of 'surviving' models

We will look primarily at the constraints imposed by CBR measurements, but also at some other astrophysical constraints. The interaction of such constraints is an area of rapidly increasing interest in the field.

No baryon-only model with $b = 1$ survives. Hence we must either assume substantial bias in galaxy formation or assume that some form of dark matter exists. If we accept the nucleosynthesis limit on Ω_b, even bias cannot 'save' baryon models (but see Section 8.7 and 8.8 below).

If we choose hot dark matter, agreement with the CBR observations pushes us towards high values of h and Ω and $n \geq -0.5$. High values of h and Ω both result in a

---

* I base my calculations of the range of h on the assumptions that $0.01 \leq \Omega_b \leq 0.1$ and that $\Omega_{HDM} = 1 - \Omega_b$; I also take the bias factor $b = 1$. The limits on h follow from an interpolation of the 'one massive neutrino' models given by Holtzmann (1989). Note that h could be as small as about 0.35 if $b = 2$.

short time scale for expansion (see Sections 1.2 and 1.3.3). That in turn makes the problem of galaxy formation in HDM models more severe; in effect, there is less time for the large-scale structure to fragment to produce galaxies. White *et al.* (1983) and Lilje (1990) among others have performed N-body simulations of this process; the models did not reproduce the observed large-scale distribution of galaxies well. Forcing a match with the observed power on large scales would result in redshifts of galaxy formation $\lesssim 1$. Formation at so recent an epoch is hard to reconcile with counts of faint galaxies, the existence of QSOs at $z > 4$, and the lack of success in finding primeval galaxies in direct searches for systems forming their first generation of stars (see Koo, 1986, for a review of the last point).

Let us now turn to cold dark matter models. How do they survive the confrontation with the CBR measurements and upper limits? If $\Omega = 1$, the observational constraints force us to low values of both h and Ω_b, e.g. $\Omega_b < 0.3$ for $h = 0.5$. Thus a model with $\Omega = 1$ and $h = 0.5$ is consistent with both the CBR observations and the value of Ω_b derived from observations of light element abundances. Agreement with both sets of observations can also be obtained in open cosmological models with $\Omega_b < 0.1$ if $h = 1$. Here the major observational constraint is the upper limit on small-scale anisotropy set by Readhead *et al.* (1989). If h is as small as 0.5, that experiment rules out open, CDM, adiabatic models.

It is tempting to conclude that CDM models are favored by these observations. As it happens, however, there is another set of observations which casts substantial doubt on them. It is the report of significant 'excess power' in the observed two-point correlation function of galaxies (Maddox *et al.*, 1990). On scales greater than $8h^{-1}$ Mpc (say 3–10 times larger), galaxies are more highly correlated than expected from predictions based on CDM models – there is 'excess power' on large scales. In principle, such an excess could be accommodated or explained if the power spectrum $P(k)$ of eqn. (8.6) had a different index. Our freedom to lower the index n, however, is limited by CBR observations, especially the COBE results. Taken together, the two sets of observations sharply limit the range of acceptable CDM models if we maintain the assumption $\Omega = 1$ (Wright *et al.*, 1992). Some (small) increase in large-scale structure relative to that on scales $\lesssim 8h^{-1}$ Mpc is achieved if we move to an open CDM model. Finally, to 'save the phenomena,' we may adopt a model with a mixture of hot and cold dark matter (in addition to baryons). Discussion of such 'mixed' models is deferred to Section 8.8. First, let us examine one possible way to evade some of the CBR constraints, by erasing CBR anisotropies on some angular scales.

8.7   Reionization and secondary fluctuations

All the calculations presented in the previous sections are based on the assumption that CBR photons reach us unimpeded and unscattered from the epoch of recombination at $z_s \simeq 1000$. The fluctuations imprinted at $z_s \simeq 1000$ are called *primary* fluctuations. It is possible, however, that the baryonic component of matter in the Universe was reionized at a subsequent epoch with redshift $< z_s$. Once reionized, the intergalactic component of the baryonic matter almost certainly remained reionized, as the Gunn–Peterson test (1965) shows.

Reionization produces free electrons, which Thomson scatter the CBR photons as they propagate from the surface at $z_s = 1000$ to our detectors. Since Thomson scattering

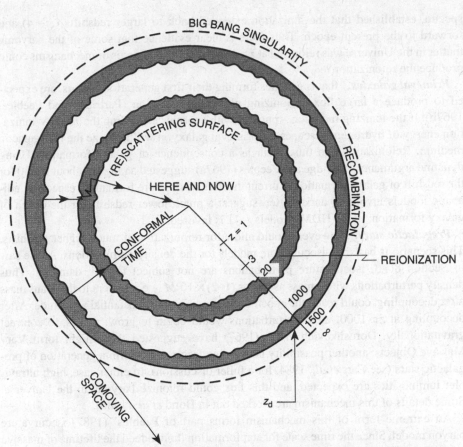

Fig. 8.21   Photons traveling towards us from the epoch of recombination will be scattered if the Universe is reionized. Thus a new, *(re)scattering* surface is established (at $z \sim 20$ in this sketch). If the matter on that surface is inhomogeneous, secondary fluctuations are induced in the CBR.

is inherently frequency-independent, this scattering has essentially no effect on the CBR spectrum.* On the other hand, if the scattering optical depth is substantial, the primary fluctuations can be erased by redistribution of the photon directions, as shown schematically in fig. 8.21.

Most of the astrophysical mechanisms capable of reionizing the baryon content of the Universe will themselves produce perturbations in it. Hence the (re)scattering surface itself will be inhomogeneous, and may imprint fluctuations in the CBR. These are called *secondary* fluctuations. As we shall see, the amplitude of these secondary fluctuations may have quite a different angle dependence than the angle dependence that we investigated in Section 8.4.6 for primary fluctuations.

8.7.1 *Mechanisms for reionization*

In the same year the CBR was discovered, Gunn and Peterson (1965) showed that the intergalactic medium at $z \simeq 2$ is highly ionized. Later work, also based on quasar

---

* As the precision of COBE spectral measurements improves, we may be able to constrain reionization models by looking for the small inverse Compton distortion (Section 5.2.4).

spectra, established that the ionization extended back to larger redshifts ($z \sim 4$) and forward to the present epoch. Thus there is clear evidence that some of the baryonic matter in the Universe was reionized at $z > 4$. What sorts of sources or mechanisms could produce the reionization?

Primeval galaxies, that is galaxies forming their first generation of stars, are expected to produce a large flux of ionizing ultraviolet radiation (Partridge and Peebles, 1967b). If the ionizing radiation – photons with energies exceeding the 13.6 eV ionization energy of hydrogen – escapes the primeval galaxy, it will reionize the intergalactic medium. Reionization can thus occur as a consequence of galaxy formation. Using dynamic arguments, Partridge and Peebles (1967a) suggested a range of about 10–30 for the redshift of galaxy formation. Current theories of structure formation, especially adiabatic models involving dark matter, suggest a much lower redshift for the epoch of galaxy formation; in the HDM models $z \lesssim 1$ is favored.

Pregalactic stars, however, would allow for reionization at much higher redshifts. This scenario is based on isocurvature models for the density perturbations. As we saw in Section 8.4.3, isocurvature perturbations are not subject to Silk damping. Thus density perturbations with a mass as low as $(1-2) \times 10^5 M_\odot$, equivalent to the Jeans mass after decoupling, could reach the epoch of decoupling with substantial amplitude. After decoupling at $z \simeq 1000$, these perturbations would begin to grow; that is, to contract gravitationally. Doroshkevich *et al.* (1967) have suggested they might form Very Massive Objects; another possibility is that they fragment to form a generation of pregalactic stars (see Carr *et al.*, 1984, for a fuller discussion). In either case, high ultraviolet luminosities are expected, and this flux could reionize baryons in the Universe. Some details of this mechanism are worked out in Bond *et al.* (1986).

An extreme form of this mechanism forms part of Peebles' (1987) isocurvature baryon model. Since the time scale for star formation (and indeed the lifetime of massive stars) is roughly comparable with the expansion time scale at $z \simeq 1000$, Peebles assumes that reionization occurred almost immediately after the recombination epoch at $z \simeq 1000$. He also assumes a steep *initial* power spectrum of fluctuations $P(k) \propto k^{3-4}$ in the notation adopted here.

We need to keep in mind that this mechanism is present only in isocurvature models. Adiabatic perturbations of subgalactic mass are heavily damped, and hence star formation occurs only as a consequence of galaxy formation.

'Explosions', which generate shock waves, could also be responsible for reionization of the Universe. The energy to drive these explosions is supplied by a kind of chain reaction of supernovae. This scenario was introduced by Ostriker and Cowie (1981) and Ikeuchi (1981).

While reionization is a consequence of the explosive model, this model's main contribution is a new mechanism for the formation of large-scale structure. Up to now, we have assumed that only gravitational forces were at work in the formation of galaxies and larger structures. The 'explosions' introduced by Ikeuchi and by Ostriker and Cowie open up the possibility of non-gravitational compression of matter. Growth of perturbations is no longer required to follow the $\Delta\rho/\rho \propto (z+1)^{-1}$ law developed in Section 8.4.2; in particular $\Delta\rho/\rho$ could be increased more rapidly by these non-gravitational forces.

The decay of unstable particles offers a fourth and more exotic mechanism for reionizing the baryonic content of the Universe. Clearly, for this mechanism to be

Table 8.3 *Reionization must occur at a redshift* $\geq z_{ri}$ *for the Thomson scattering optical depth (eqn. (8.31)) to exceed unity.*

h	Ω	Ω_b	z_{ri} for $\tau = 1$
0.5	0.2	0.03	68
0.5	0.2	0.2	27
0.5	1.0	0.06	96
0.5	1.0	0.2	43
0.5	1.0	1.0	14
1.0	0.2	0.03	49
1.0	0.2	0.2	19
1.0	1.0	0.06	60
1.0	1.0	0.2	27
1.0	1.0	1.0	8.6

effective, the half-life of the particle must exceed about 3×10^5 years, so that decay follows recombination at $z \simeq 1000$. Thus the decay will not affect the CBR spectrum (see Section 5.3.2). In that same section, we established that the mass of the decaying particle had to exceed 27 eV to produce ionizing photons. It is worth adding that there is no plausible candidate with these properties in any theory of elementary particles.

With the exception of this last mechanism, all the suggested scenarios for reionization are directly linked to the formation of structure in the Universe. We will explore some consequences of that linkage in the next two subsections.

8.7.2   Erasure of primary fluctuations in the CBR

Reionization at any redshift reintroduces free electrons, which Thomson scatter the CBR photons. To erase the primary fluctuations imprinted at $z_s \simeq 1000$, however, the optical depth of the reionized matter must exceed unity. The Thomson scattering optical depth may be written down as follows, if we assume that reionization occurred at redshift z_{ri}, and that the matter stayed reionized until $z = 0$:

$$\tau = \int_{t(z_{ri})}^{t_0} \sigma_T \, n_e c \, dt = \frac{0.88c \, \sigma_T \, H_0 \, \Omega_b}{4\pi \, Gm_p \, \Omega^2} [2 - 3\Omega + (1+\Omega z)^{1/2} \, (\Omega z + 3\Omega - 2)], \qquad (8.31)$$

where σ_T is the Thomson scattering cross-section, and I have included a factor of 0.88 to allow for 24% by mass of ^{4}He (see Gunn and Peterson, 1965; Efstathiou, 1990). We will invert eqn. (8.31) to find the value of z_{ri} that ensures $\tau = 1$. That value will depend on the cosmological parameters h and Ω and, more sensitively, on the baryon density Ω_b. Some sample results are presented in table 8.3.

Even in the most extreme case, the Thomson scattering optical depth remains below 1 unless reionization occurs prior to an epoch corresponding to $z \simeq 8$–9.* If $\Omega_b \lesssim 0.06$, as suggested by the nucleosynthesis arguments of Section 1.6.4, z_{ri} must exceed about 60. Is reionization at such early epochs reasonable? In the case of isocurvature fluctuations, which produce a pregalactic generation of stars, the answer is yes, though the energy demands are high (see Lacey and Field, 1988). On the other hand, in adiabatic

---

* Recall that we have no direct evidence that matter is ionized beyond $z \simeq 5$.

dark matter models, structure forms late, and it is much harder to see how to arrange for reionization at $z \gtrsim 60$. We thus conclude that erasure of primary fluctuations is unlikely in the case of adiabatic models.

Nor does the erasure extend to all angular scales even if $\tau > 1$. The argument is a simple variant of the discussion of Section 8.4.4. As fig. 8.21 shows, there is a surface of last (re)scattering in the reionized matter. The redshift of the surface will be close to the values tabulated in the last column of table 8.3. As in section 8.4.4, we ask what horizon scale is at this redshift. For large z, Weinberg (1972) shows

$$\theta_H \approx z^{-1/2} \text{ rad} \approx 2° \left(\frac{1000}{z} \right)^{1/2}. \tag{8.32}$$

Since Thomson scattering is a causal process, it cannot occur on scales $> \theta_H$ as given in eqn. (8.32). Therefore, fluctuations on scales $\gtrsim 10°$ cannot be erased even if the baryonic content of the Universe is reionized.

Since the measurements reported by the COBE team (Smoot *et al.*, 1992) are all at angular scales $\geq 10° \simeq \theta_H$, they are not affected by reionization. It follows that any model for density perturbations rejected by the COBE measurements cannot be 'saved' by arguing that the primary fluctuations were erased by reionization and rescattering. Many baryon isocurvature models fall into this category (see Section 8.7.4 below, however). In baryon isocurvature models, reionization is quite plausible. On the other hand, reionization cannot erase the large amplitude fluctuations that the model predicts on scales $\gtrsim 10°$, and these are in general in conflict with the COBE measurements. Thus, in the model where reionization is most likely to occur naturally, it has little effect on the observational constraints. In adiabatic dark matter models, the opposite is true. Erasure of primary fluctuations could indeed ease the observational constraints imposed at small angular scales (by the observations of Readhead *et al.* (1989) in particular); but in these models, reionization is likely to occur too late to be effective. Thus, in spite of the hopes of proponents of some models for galaxy formation, reionization has no *major* impact on the constraints on models laid out in Section 8.6.

Thus far, we have considered only scattering by free electrons. However, as La (1989) pointed out, neutral hydrogen atoms can also scatter photons. This Rayleigh scattering has a much smaller cross-section than Thomson scattering, and is strongly wavelength dependent (as noted in Section 5.2.5). It will erase fluctuations only at $\lambda \lesssim 1$ mm. Again, this mechanism has no effect on the constraints discussed in Section 8.6.

8.7.3 Secondary fluctuations in the CBR

With the exception of particle decay, all the mechanisms for reionization discussed above are likely to involve large-scale inhomogeneities in the matter and/or velocity distribution. Hence the surface of (re)scattering itself is also likely to be inhomogeneous, and will thus imprint fluctuations on the CBR. These so-called *secondary* fluctuations have received less attention to date than primary fluctuations (but see Vishniac, 1987, and Bond *et al.*, 1991). Theoretical work on secondary fluctuations may be less fully developed because the connection between fluctuations in the CBR and the spectrum of primordial density perturbations is much less direct. Complicated, non-linear, astrophysical processes intervene, unlike the case for primary fluctuations, where $\Delta \rho / \rho$ is small. Nevertheless, observations of, or upper limits on, the amplitude of secondary fluctuations may provide useful information about the formation of both galaxies and larger scale structure in the Universe.

Let us begin our study of secondary fluctuations by recalling the argument of Section 8.4.7 on the thickness of the scattering surface: eqn. (8.31) may be used to show that the thickness of the (re)scattering surface produced by reionization is substantial. For instance, for one of the models listed in table 8.3 (with $h = 0.5$, $\Omega = 1.0$ and $\Omega_b = 0.2$), it may be shown that the range of redshift corresponding to an increase of τ from 1/2 to 1 is $\Delta z \simeq 16$. Thus for rescattering $\Delta z/z \simeq 0.5$, not about 0.1 as found for the epoch of decoupling at $z_s \simeq 1000$. The consequence is that CBR fluctuations produced by the (linear) mechanisms described in Section 8.4.6 are almost completely washed out on all angular scales below the horizon scale, $\theta_H \simeq 10°$. The case of linear fluctuations is examined in more detail by Kaiser (1984).

All is not lost, however. There are mechanisms that produce fluctuations, that are *not* subject to such averaging or 'washing out'. One is the Sunyaev–Zel'dovich (SZ) effect described in Section 5.2.4. Another is a second-order source of CBR fluctuations: anisotropy introduced by the product of $\Delta\rho/\rho$ and large-scale, coherent inhomogeneities in the velocity field (Ostriker and Vishniac, 1986; Vishniac, 1987). There are other second-order (and higher) terms as well, which can produce CBR fluctuations on angular scales $\ll \theta_H$.

Since the amplitude of the SZ signal is proportional to the product of the electron number density n_e and the electron temperature T_e, fluctuations in the CBR will be introduced only if T_e is high, and there are large perturbations in n_e. Under what circumstances would we expect these conditions? One special case is the intergalactic plasma in clusters of galaxies, both nearby clusters and a background of more distant clusters – this case is treated separately in Section 8.9. Clusters of galaxies, however, are relaxed, highly non-linear dynamical structures. We also need to consider SZ fluctuations introduced by more generally distributed hot plasma, whether formed in collapsing Zel'dovich pancakes or by chain-reaction explosions of supernovae. These two scenarios have been investigated by Szalay *et al.* (1983) and Hogan (1984), respectively. Hogan, for instance, finds

$$\frac{\Delta T}{T} \approx 2 \times 10^{-4} h \left(\frac{z_{ri} + 1}{6} \right)^{11/4}$$

for an Einstein–de Sitter model with a baryon density $\Omega_b = 0.03$ in the form of ionized gas. I have ignored a factor of order unity depending on the exact beam geometry; the result given above holds for angles $\lesssim 5'$. More precise calculations are discussed below.

Now let us consider a particular case of second-order fluctuations, the 'Vishniac effect' from his 1987 paper, in which the observed fluctuations are proportional to $(\Delta\rho/\rho)v$. Under what conditions will this product be substantial? Clearly, both large amplitude density perturbations and large-scale coherent velocity fields are needed; so is a large scattering optical depth. Just these conditions may be present if the Universe is reionized early by one or more of the mechanisms discussed in Section 8.7.1. As we have seen, reionization at high redshift is much more likely in an isocurvature model than in adiabatic models.

Both Vishniac (1987) and Efstathiou and Bond (1987) have made quantitative predictions of the autocorrelation function $C(\theta)$ for secondary fluctuations arising from second-order and SZ mechanisms (see table 8.4). Figure 8.22, from Vishniac's paper, is typical of the results. Comparing that figure with fig. 8.13 reveals the most interesting qualitative difference between primary and secondary fluctuations – a change in the dependence of $\Delta T/T$ on angular scale. As we have already noted, the horizon scale θ_H is larger for the latter. Of more interest, however, is the power at small angular scales, $\lesssim 7'$, where primary fluctuations are sharply cut off, as explained in Section 8.4.7. Secondary fluctuations, associated with the non-linear processes of galaxy formation, can appear at smaller angular scales. Indeed, this difference provides a useful test for secondary fluctuations – if CBR fluctuations are detected on scales $< 7'$, they must be secondary. Such observations would then imply an epoch of reionization.

Finally, since anisotropic velocity fields are involved in the Vishniac effect, the resulting CBR fluctuations will be (partially) linearly polarized. Also, because the horizon scale is larger, polarized fluctuations arising from a surface of (re)scattering can be present at larger angular scales than can the polarized components of primary fluctuations; we thus have another potential test of reionization models.

In addition to fluctuations imprinted on the CBR, photons may be added to it by high-redshift galaxies. In this case, since the angular scale of an individual galaxy at high z is of order $10''$, CBR fluctuations on subarcminute scales are expected. The mechanism involved in production of microwave photons is emission by warm dust. We have treated this mechanism as a source of foreground noise in Section 7.3. It is strongly wavelength-dependent, as suggested by fig. 7.12. Thus the amplitude of the resulting fluctuations depends sensitively on the temperature of the reemitting dust, the redshift of the sources and the observing wavelength. A wide network of models has been assessed by Bond *et al.* (1991). That paper also treats in detail the statistical properties of fluctuations produced by these point-like sources. We turn in Section 8.7.4 to the confrontation of these models with the observations of Chapter 7. It is worth remarking, however, that most of the experimental work described in Chapter 7 was performed at wavelengths too long to provide a real test of the dust models.

8.7.4 *Observational tests of reionization models*

This section will treat two different classes of tests of models involving reionization. It is convenient to delineate the two classes by angular scale. First, we need to consider the effect of observational constraints on scales $> 7'$. We will be concerned in this case with the erasure of primary fluctuations as well as the addition of secondary fluctuations. Second, we need to consider phenomena unique to reionization models, especially CBR fluctuations on scales $< 7'$, where no primary fluctuations are expected. In the first case, our central question will be whether reionization can rescue models that would otherwise have been ruled out by observational upper limits on $\Delta T/T$. In the second case, the question is whether observations of the CBR provide any direct evidence for an epoch of reionization at redshifts of order 10–100.

Can reionization smooth away primary fluctuations without introducing even larger secondary fluctuations, and thus save a range of models for structure formation? In general, no model ruled out by the COBE DMR results can be saved, since reionization cannot erase fluctuations on the angular scales investigated by that instrument. Thus the class of models most susceptible to 'rescue' by reionization includes those consistent with large-scale limits on $\Delta T/T$, but with amplitudes of primary fluctuations too large to fit the Readhead *et al.* (1989) constraint. One such example is the adiabatic, baryon-only model with $\Omega_b = \Omega = 1$, which predicts values of $\Delta T/T$ exceeding the OVRO upper limit by factors between 1.5 and 2, depending on n (see table 8.2). If we allow for reionization, a range of values of n from roughly –0.7 to –1.7 becomes acceptable, as shown in table 8.4. For open models with $\Omega = \Omega_b \lesssim 0.2$, an even wider range of adiabatic, baryon models can be 'rescued' – again, see table 8.4. Recall, however, that adiabatic fluctuations are unlikely to produce structure early enough to ensure reionization. As I have noted above, the models most easily 'saved' by invoking reionization are those in which it is unlikely to occur.

In the case of CDM models, to avoid overproducing secondary fluctuations, we require late reionization – see the lower part of table 8.4, based on the models of Bond *et al.* (1991). In particular, to avoid overproducing secondary fluctuations or introducing distortions in the spectrum of the CBR (Section 5.4.2), we must accept late reionization at $z \lesssim 10$ (e.g., their models 7, 8 or 11). The energy density of reionizing radiation is limited to $\Omega \lesssim 10^{-6}$. In other words, models that would reionize the intergalactic gas early enough to erase the primary fluctuations are also likely to produce secondary fluctuations and spectral distortions exceeding the observational constraints.

In hot dark matter models, the formation of Zel'dovich 'pancakes' is expected; these in turn produce CBR fluctuations through the SZ effect (see Szalay *et al.*, 1983). The redshift at which reionization occurs is too low in these models to erase primary fluctuations. Not only that, large secondary fluctuations are produced. These exceed the COBE limits unless the pancakes form at $z \lesssim 3$, and then only if the pancake scale is $\lesssim 50$ Mpc (see Szalay *et al.*, 1983). These are not unreasonable values.

Do CBR observations provide any direct evidence for reionization, specifically of reionization at sufficiently large redshift to ensure erasure of primary fluctuations? Here, observations on angular scales well below 7' are of particular value; hence we will concentrate on the results summarized in Section 7.8.

Upper limits on ΔT on scales 6"–60" set by VLA observations are able to rule out Hogan's (1984) model, but subsequent theoretical work has suggested that his predictions of $\Delta T/T$ are too large in any case. These same results sharply constrain early models for the evolution of ionized intergalactic matter in clusters, such as that presented by Korolev *et al.* (1986). That conclusion was drawn by P. M. Hartnett in an unpublished thesis. Much more interesting constraints should soon be available when new VLA measurements underway at 3.6 cm wavelength become available.

At present, the VLA limits do not constrain the reionization models calculated by Vishniac (1987), as fig. 8.22 shows. We need an order-of-magnitude improvement in sensitivity (not impossible, as it happens). It is also worth remarking that Vishniac did not take into account all the second-order sources of anisotropy in the CBR. Some of these may have the effect of raising the predicted value of $\Delta T/T$ for such models. It is to be hoped that further, more detailed calculations on such secondary fluctuations will be performed soon.*

Constraints on reemission anisotropy are even weaker, essentially because all but one of the observations are at wavelengths $\gtrsim 1$ cm, where $\Delta T/T$ is expected to be very small.

---

* Note added in proof: recent theoretical work suggests only the source identified by Vishniac will play a major role.

Table 8.4 *Secondary CBR fluctuations. Top half (from Efstathiou and Bond, 1987, with permission): baryon-only models including either no recombination at z = 1000 or recombination followed by reionization at z $\simeq$ 200, as specified in the third column. In the fourth and fifth columns, the predicted values of $\Delta T/T$ are given for the experimental arrangements of Readhead et al. (1989) and Davies et al. (1987), respectively. The corresponding experimental upper limits are 17 and 35 in the units employed. As in Table 8.2, numbers in bold are consistent with the experimental upper limits. The figures in parentheses in the fourth and fifth columns give the (approximate) factors by which $\Delta T/T$ is reduced (or increased) by reionization. Bottom half (from Bond et al., 1991, with permission): predicted values of $\Delta T/T$ from reemission models for the experimental arrangements of Kreysa and Chini (1989) and Readhead et al. (1989). The corresponding experimental upper limits are 260 and 17 in the units employed. Numbers in bold are consistent with these limits.*

			$\Delta T/T$ (10^{-6})	
$\Omega = \Omega_b$	n^*	Ionization history	OVRO	Davies *et al.*
0.1	4	No recombination	**6.2**	**3.6**
0.1	3	No recombination	**13**	35
0.1	2	No recombination	140	350
0.2	4	No recombination	**11** (0.18)	**3.8** (1.4)
0.2	3	No recombination	**14** (0.17)	**31** (2.0)
0.2	2	No recombination	64 (0.58)	240 (2.0)
0.2	4	Ionization at $z = 200$	**10** (0.16)	**1.4** (0.5)
0.2	3	Ionization at $z = 200$	**9.4** (0.11)	**11** (0.7)
0.2	2	Ionization at $z = 200$	30 (0.27)	91 (0.76)
1.0	4	No recombination	34 (0.9)	**4.1** (0.61)
1.0	3	No recombination	**12** (0.43)	**14** (1.0)
1.0	2	No recombination	**4.7** (0.2)	48 (1.0)

			$\Delta T/T$ (10^{-6})**	
Ω_{dust}	Ω_{rad}	Ionization history	IRAM	OVRO
10^{-5}	10^{-6}	Ionization at $z \simeq 50$	2000	59
10^{-5}	10^{-6}	Ionization at $z \simeq 5$	**140**	**4.7**
10^{-5}	10^{-7}	Ionization at $z \simeq 50$	**235**	**7.4**
10^{-5}	10^{-7}	Ionization at $z \simeq 10$	**89**	**2.9**
10^{-6}	10^{-6}	Ionization at $z \simeq 50$	580	21
10^{-6}	10^{-6}	Ionization at $z \simeq 5$	**31**	**1.8**
10^{-6}	10^{-7}	Ionization at $z \simeq 10$	**7**	**0.8**

* Using the convention adopted in Section 8.4.1.
** Approximately corrected to bias $b = 1$.

The one exception, the work of Kreysa and Chini (1989) at $\lambda = 1.3$ mm, limits $\Delta T/T$ only to $\leqslant 2.6 \times 10^{-4}$. In general, this upper limit provides less restrictive constraints than spectral measurements on models which include reemission from dusty galaxies. As we saw in Section 5.4.2, the limits on CBR spectral distortions established by the FIRAS instrument on COBE serve to rule out many such models. Of the many models presented by Bond *et al.* (1991), fewer than half are consistent with the COBE spectral measurements. These are, in general, models involving late reionization. Only a single model permitted by the spectral measurements (model 9 of Bond *et al.*, 1991) is (barely) called into question by the upper limit of Kreysa and Chini (1989). To really test dust reemission models, we will need an order-of-magnitude improvement in the sensitivity

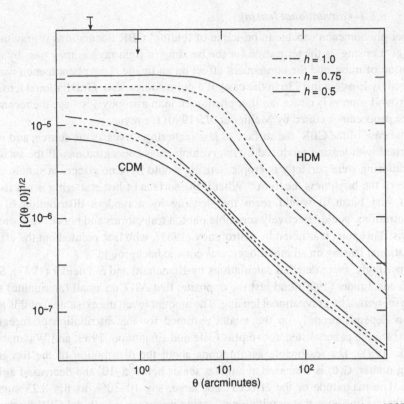

Fig. 8.22   Predicted amplitudes of secondary fluctuations (from Vishniac, 1987, with permission). The upper curves are drawn for a model with neutrinos of non-zero mass (HDM); the lower curves for a conventional CDM model. The two upper limits are from the interferometric observations described in Section 7.8. The sensitivity of such measurements is expected to improve by a factor ~5 in the next few years.

of short wavelength searches for CBR fluctuations. Although its launch may lie as much as a decade in the future, the Space Infrared Telescope Facility (SIRTF) planned by NASA will be ideally suited to test such models of secondary fluctuations, and hence to impose interesting constraints on theories of galaxy formation.

8.8   Ways around the CBR limits, including alternative theories of structure formation

In this section, we turn to modifications of or additions to standard models for structure formation as discussed in Sections 8.4–8.7 above. It is reasonable to ask why such modifications or additions seem attractive or even plausible. In some cases, there exist fundamental physical arguments for new possibilities. Cosmic strings or textures (Section 8.8.4) are examples. In other cases, I would suggest, the changes were proposed principally to meet constraints imposed by the observations, including the CBR results and the observation of 'excess power' in the two-point correlation function of galaxies. Models incorporating a period of reionization, which can 'hide' fluctuations on scales ≲ 10° were so motivated. The introduction of bias into theories of galaxy formation had a somewhat similar history.

8.8.1 Gravitational lensing

Another phenomenon capable in principle of 'hiding' CBR fluctuations is gravitational lensing. 'Lensing' is the term used for the bending of light rays as they pass by a concentration of mass, and the consequent effect on an image. This phenomenon was first explored by Einstein in 1936 in the case of a point source lens. Gravitational lensing of background sources is now a familiar phenomenon in astrophysics – see the recent conference proceedings edited by Mellier *et al.* (1990) for reviews.

In the case of the CBR, the surface of last scattering serves as the source, and we are concerned with lensing produced by intervening mass concentrations. If the surface of last scattering were perfectly isotropic, lensing would have no effect – a simple consequence of the brightness theorem.* What if the surface of last scattering is not isotropic? At first blush, it would seem that lensing by a random distribution of mass concentrations would effectively scramble photon trajectories and hence wash out fluctuations. This point was noted by Mitrofanov (1981), who first pointed out the effect of gravitational lensing on an inhomogeneous cosmic background.

Surprisingly, more detailed calculations by Blanchard and Schneider (1987), Sasaki (1989) and Linder (1990) find just the opposite: that $\Delta T/T$ on small (arcminute) scales can be *increased* by gravitational lensing. The amount (even the existence) of this amplification depends strongly on the model assumed for the distribution of foreground matter (see the papers listed above plus Cole and Efstathiou, 1989, and Watanabe and Tomita, 1991). For reasonable assumptions about the distribution of the foreground lensing matter, $C(\theta)$ is increased at angular scales below 5–10′ and decreased at larger scales. The magnitude of the effect is not large, say 10–30%, as fig. 8.23 suggests. Nevertheless, the hope that gravitational lensing might serve to 'hide' CBR fluctuations is not realized.†

8.8.2 Modifications to the spectrum of fluctuations

If we cannot erase CBR fluctuations, can we alter the angular spectrum of CBR fluctuations in such a way as to 'save the phenomena'? For instance, plots of $C(\theta)$ are calibrated at an angle corresponding to a linear scale of about $8h^{-1}$ Mpc, only about 1% of the angular scale resolved by COBE. By violating the assumption underlying eqn. (8.6) (that the fluctuations have a power law spectrum), can we modify $C(\theta)$ so as to meet the various observational constraints? In particular, can we ensure adequate power on large scales without producing too large a value of $\Delta T/T$ on arcminute scales? In principle, of course, the answer is yes, and one such model was introduced by Bardeen *et al.* in 1987. 'Extra power' was added at an appropriate range of wave number k to produce a better match to the CBR constraints and the galaxy correlation function. In the absence of a physical explanation for the extra power, such models strike me as *ad hoc*, like the addition of another epicycle or two to save the geocentric model of the solar system, and there are other constraints as shown in Section 5.4.4.

---

* I am excluding here effects that depend on the gravitational redshift of photons as they pass through a density perturbation, as opposed to alterations in the photon trajectories. For the former, see Rees and Sciama (1968) and, for a more thorough analysis, Dyer (1976). I am also ignoring the effects of a moving lens; these are treated in Sections 8.8.4 and 8.9.2 below.
† One paper (Kashlinsky, 1988) does make the claim that gravitational lensing can wash out CBR fluctuations – but it contains an error, as noted by Watanabe and Tomita (1991).

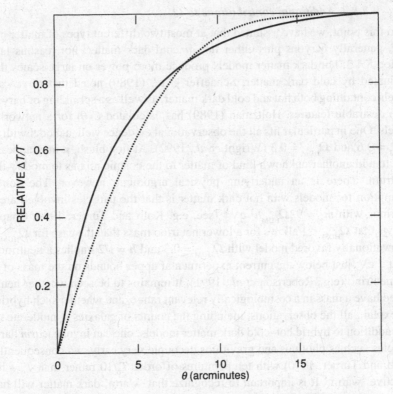

Fig. 8.23   The effect of gravitational lensing on CBR fluctuations: the dotted line is the original spectrum and the solid line the spectrum as modified by lensing (from Watanabe and Tomita, 1991, with permission).

Another way of modifying the *observed* spectrum of CBR fluctuations, in my opinion, holds more promise (it effectively modifies the transfer function, not the power spectrum $P(k)$). At its base is a mixture of tensor and scalar perturbations (see Section 8.4.1). As we noted in that section, virtually all calculations of the connection between $P(k)$ and observed $\Delta T(\theta)$ have been based on the assumption that scalar modes dominate. Suppose they do not, and that tensor modes contribute equally or even dominate (some contribution is expected if the index n is less than 1). Both scalar and tensor modes produce Sachs–Wolfe fluctuations, but only the former induce gravitational motion of matter. Hence only scalar perturbations produce the Doppler fluctuations that dominate $\Delta T/T$ at small angular scales. A mixture of scalar and tensor fluctuations would thus produce lower values of ΔT on scales $\lesssim 1°$ than expected from an extrapolation of the COBE observations on scales $\gtrsim 10°$. At this time, this possibility has barely been investigated. Nor is it clear that tensor modes can produce fluctuations of amplitude comparable with those produced by scalar modes (Starobinskii, 1985). To my mind, we need additional careful calculations of the amplitude of tensor fluctuations, just as we do of secondary fluctuations (some recent work along these lines is listed in Appendix C).

8.8.3 Additional contributions to Ω

Up to this point, we have assumed that at most two different types of matter contribute to Ω, generally baryons plus either hot *or* cold dark matter. For reasons laid out in Section 8.4.3, hot dark matter models produce more power on large scales than those dominated by cold dark matter. Schaeffer *et al.* (1989) noted that *three-component* models containing both hot and cold dark matter, as well as a sprinkling of baryons, have some desirable features. Holtzman (1989) has calculated $C(\theta)$ for a network of such models. One in particular fits all the observational evidence well, a model with $\Omega_b = 0.1$, $\Omega_{CDM} = 0.6$ and $\Omega_{HDM} = 0.3$ (Wright *et al.*, 1992). At first blush, it may appear as arbitrary to add another unknown kind of matter to the cosmic mix as to modify the power spectrum. There is an underlying physical argument, however. The conventional assumption for models with hot dark matter is that the particles involved are massive neutrinos with $m_\nu = 92\Omega_{HDM}h^2$ eV* (see, e.g. Kolb and Turner, 1990, Chapter 5). It follows that $\Omega_{HDM} < 1$ allows for a lower neutrino mass than the case for $\Omega_{HDM} \approx 1$. The observationally favored model with $\Omega_{HDM} = 0.3$ and $h = 1/2$ implies a neutrino mass of about 7 eV, just below the current experimental upper bounds on the mass of the electron neutrino (e.g., Robertson *et al.*, 1991). It remains to be seen whether neutrinos do indeed have a mass in a cosmologically relevant range, and whether such hybrid models can explain all the observations, including the counts of galaxies at moderate redshifts.

In addition to hybrid hot–cold dark matter models, one can invoke *warm* dark matter. Particles such as photinos and gravitinos decouple very early and consequently end up (Kolb and Turner, 1990) with temperatures of order $T_0/10$ rather than T_0 – hence the adjective 'warm.' It is important to recognize that 'warm' dark matter will have quite different dynamical and clustering properties than a mixture of hot and cold dark matter. I know of no calculations of CBR anisotropies based on 'warm' dark matter as the dominant contributor to Ω.

Finally, let us consider the possibility that the cosmological constant (Section 1.3.6) is not equal to zero. What effect does a non-zero Λ have on predictions of CBR fluctuations? There are two issues to consider. First, a non-zero value of Λ affects the propagation of photons in the expanding Universe (Blanchard, 1984). Second, as we have seen in Section 1.3.6, a non-zero Λ results from a non-zero energy density of the vacuum. Hence the cosmological constant can induce space curvature just as 'ordinary' matter does. It is for that reason that we treat a non-zero value of Λ as an additional contributor to Ω. If we are free to adjust Λ, we may maintain the flat space case $k = 0$ ($\Omega_{tot} = 1$) favored in inflationary models, and yet have $\Omega_{matter} < 1$.

Now let us turn to the consequences for the amplitude and spectrum of CBR fluctuations as discussed by Blanchard (1984), Kofman and Starobinsky (1985), Vittorio and Silk (1985) and Sugiyama *et al.* (1990). The last of these presents a wide range of $\Lambda \neq 0$ models. That same paper also illustrates nicely one of two processes which cause $\Lambda \neq 0$ models to have lower amplitudes of $\Delta T/T$ than comparable $\Lambda = 0$ models: the effect of Λ on photon propagation. For a fixed value of Ω_{matter}, a perturbation of proper scale l subtends a larger and larger angle as Ω_Λ and hence Ω_{TOT} increases. Consequently, for a given *angular* scale, smaller proper distances are spanned in a $\Lambda \neq 0$ model. Hence the lines in fig. 8.18 are shifted to the right, reducing $\Delta T/T$.

---

* The connection between m_ν and Ω arises from the assumption that the neutrinos form a thermal sea with $T = (4/11)^{1/3} T_0$ – see Cowsik and McClellan (1972).

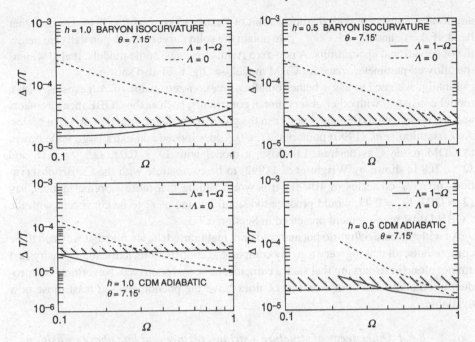

Fig. 8.24   Predicted fluctuations for several models with $\Lambda = 0$ (dotted lines) and $\Lambda = 1 - \Omega$ (solid lines) drawn for the observational geometry of Readhead *et al.* (1989), with $\theta_s = 7.15$. The observational upper limit is shown shaded (adapted from Sugiyama *et al.*, 1990, with permission). In the case of baryon-only models, a non-zero Λ can 'save' some models. It also lowers the predicted amplitude of $\Delta T/T$ for CDM models, as shown in the lower panels.

The second effect of a non-zero value of Λ is to alter the normalization described in Section 8.5.1. In that section, we assumed that perturbations grew in amplitude as $\Delta \rho/\rho \propto (z + 1)^{-1}$ until the present for an $\Omega = 1$ model, or until a redshift $(z + 1) \approx 1/\Omega$ for $\Omega < 1$. The situation is different in an expanding model dominated by a cosmological constant term. In the case of a spatially flat $k = 0$ model with non-zero Λ, growth ceases at a redshift given by

$$(z + 1) \approx (1/\Omega_{matter} - 1)^{1/3},$$

rather than $(z + 1) \approx 1/\Omega_{matter}$ for the corresponding open model with $\Lambda = 0$. By comparing the two formulas, it is easy to see that $\Lambda \neq 0$ models allow for a longer period of growth. Hence we may start with smaller amplitude fluctuations on the surface of last scattering. In turn, a smaller value of $(\Delta \rho/\rho)_{l,s}$ implies lower amplitude temperature fluctuations in the CBR. Thus the curves in fig. 8.18 are shifted down as well as to the right. If $\Lambda \neq 0$, lower values of $\Delta T/T$ are expected on all angular scales for a fixed value of Ω_{matter}.

As this brief discussion may suggest, both processes have a particularly large effect on predictions of $\Delta T/T$ on the small angular scales observed by Uson and Wilkinson (1984) or Readhead *et al.* (1989). That is shown effectively in fig. 8.24, adapted from Sugiyama *et al.* (1990). The dotted lines in the upper panels are predictions for baryon isocurvature models with $\Omega \leq 1$ and $\Lambda = 0$. Only if $\Omega = 1$ and $h \leq 0.5$ are the predicted amplitudes consistent with the upper limit of Readhead *et al.* at $\theta = 7'.15$. On the other hand, if $\Lambda \neq 0$, and $\Omega_\Lambda = 1 - \Omega_{matter}$, we obtain the solid curves, which generally lie below the observational upper limits. A non-zero Λ thus 'rescues' some models. It also widens the allowed parameter space for CDM models, as fig. 8.24 also shows.

Finally, we need to ask whether building in a non-zero value of Λ leaves us with a model consistent with other observational constraints, such as the COBE measurements and the observations of excess power in the galaxy correlation function on large scales. As Efstathiou *et al.* (1990) point out, $\Omega_\Lambda \simeq 0.8$ does indeed add extra large scale power to CDM models, as desired. Likewise, a model with $\Omega_b = 0.02$, $\Omega_{CDM} = 0.18$ and $\Omega_\Lambda = 0.8$ is shown by Wright *et al.* (1992) to be compatible with the COBE observations of $\Delta T/T$ on scales of 10–90°. It is worth adding that more extreme models, say $\Omega_b = 0.03$, $\Omega_\Lambda = 0.97$, would produce too small a value of T_2 to be consistent with the COBE DMR measurement presented in Section 6.7.3.

As with the case of hybrid hot and cold dark matter models, we find that we need three components, all making very roughly comparable contributions to the total density Ω. I am not alone in remarking that such a coincidence seems contrived. Nevertheless, introducing an additional component to Ω does 'save the phenomena,' at least those now available.

8.8.4 *Other seeds of structure – strings, texture and late phase transitions*

An even more fundamental modification to theories of galaxy formation is to drop the assumption that the gravitational growth of quantum fluctuations alone is responsible for the growth of structure. In Section 8.7, we examined one such possibility, the 'explosive' scenario of Ikeuchi (1981) and Ostriker and Cowie (1981). Here we examine another possibility, one with far-reaching implications for both CBR observations and fundamental theory. It is that large-scale structure is 'seeded,' or more accurately 'induced,' by relics of the phase transition described in Section 1.5.2. A detailed description of the origin and topological properties of such relics lies beyond the scope of this book (see Chapter 7 of Kolb and Turner, 1990; or Vilenkin, 1985; for useful reviews). It is appropriate to devote some attention to classification of the relics, however. Then I will describe in more detail how they affect the CBR.

In conventional inflationary models, the symmetry breaking phase transition occurs very early, at an epoch of very roughly 10^{-32} s (see Kolb and Turner, 1990; or Section 1.5). Recall that quantum fluctuations induced in these epochs are the ultimate source of density perturbations in the standard model. Hill *et al.* (1989), however, following Wasserman (1986), have suggested that a phase transition may occur at a much later epoch, one corresponding to $z < 10^3$. In this case, the surface of last scattering at $z_s \simeq 10^3$ could be entirely isotropic, and all the observed structure in the present Universe would arise later from fluctuations in a 'late' phase transition. As a consequence, the CBR would also be highly isotropic, at least on small scales. On larger scales, $\theta \gtrsim 1°$, the Sachs–Wolfe (1967) effect will still produce anisotropies. The *amplitude* of these anisotropies, however, cannot be found using the methods of Section 8.5, since non-gravitational forces may be involved. Late phase transitions appear to produce no novel features in the CBR, and their status is controversial from the point of view of high energy physics.

Let us turn, therefore, to the relics of a phase transition at $\lesssim 10^{-32}$ s. These may be classified by their dimensionality – magnetic monopoles are remnants of the high energy, false vacuum of dimension zero. Cosmic strings are thin, almost one-dimensional 'tubes' of false vacuum, and domain walls are two-dimensional. 'Texture' (Turok, 1989) arises from three-dimensional topological knots. The effects of these various possible relics on the formation of structure in the Universe have been investigated by, among others, Kolb and Turner (1990); by Kibble (1976), Vilenkin (1980), Turok and Brandenberger (1986), Bertschinger (1987) and Stebbins *et al.* (1987); by Stebbins and Turner (1989); and by Bennett and Rhie (1990) and Turok and Spergel (1990); respectively. As the relative number of references just cited suggests, most investigations to date have focused on cosmic strings. Physical and astrophysical evidence (summarized well in Chapter 7 of Kolb and Turner, 1990) has established strong constraints on the number density of monopoles. They are unlikely to contribute substantially to structure formation. Domain walls left from an early phase transition are so massive as to produce huge anisotropies in the Universe; hence we know walls do not exist. Until very recently, little theoretical work has been done on the effect of texture on either the CBR or the formation of structure in the Universe. In the remainder of this section we will concentrate on cosmic strings.

A cosmic string can be parameterized by its mass per unit length, μ which in turn depends on the square of the symmetry-breaking energy scale:

$$\mu \approx 10^{22} \left(\frac{E}{10^{16}\,\text{GeV}} \right)^2 \text{g cm}^{-1}.$$

Strings may be present either as infinite one-dimensional structures in rapid motion or as closed loops. Both can serve to 'seed' structure formation. Loops act as gravitating mass concentrations of mass $M = \mu\pi d$, where d is the characteristic diameter of the loop. On the basis of numerical simulations, it is now argued that string loops are less important in the formation of large-scale structure than the wakes formed by strings moving perpendicular to their extension (see e.g. Stebbins *et al.*, 1987; Bouchet and Bennett, 1990).

A moving cosmic string will also produce a sharp, step-like discontinuity in the observed temperature of the CBR (Bouchet *et al.*, 1988). The amplitude of the 'step' shown in fig. 8.25 is at most

$$\Delta T/T = 8\pi G\mu(v/c^3)\gamma, \tag{8.33}$$

where $\gamma = [1-(v/c)^2]^{-1/2}$ and v is the velocity of the string perpendicular to the line of sight (Stebbins, 1988). Because of their high tension, strings move at relativistic velocity; hence a typical value for their velocity perpendicular to the line of sight may be taken to be $c/\sqrt{3}$. The absence of visible, large-scale discontinuities in the COBE map (fig. 6.8) at a level of about 10^{-4} in $\Delta T/T$ may then be used to show $\mu \lesssim 10^{23}$ g cm^{-1}, if cosmic strings are present. That limit in turn places limits on the symmetry-breaking energy scale. Once again, we have discovered an interesting link between fundamental physics and observable properties of the CBR.

With its poor angular resolution, COBE is not ideally suited to searching for sharp temperature discontinuities as sketched in fig. 8.25, especially if there are a number of such steps within the antenna pattern. Let us thus ask generally what pattern a stochastic background of cosmic strings (and loops) would produce in the CBR. Just such calculations were performed by Bouchet *et al.* (1988); their results are shown as fig. 8.26.

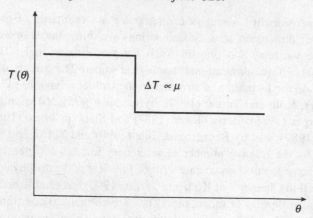

Fig. 8.25   The 'step' introduced into $T(\theta)$ for a cosmic string moving transverse to the line of sight.

These simulations confirm an earlier result of Kaiser and Stebbins (1984): the statistics of the CBR fluctuations produced by this mechanism are no longer Gaussian. The presence of large-scale order is visible in fig. 8.26, and is responsible for the non-Gaussian distribution of values of $\Delta T/T$. This novel statistical property of string-induced anisotropies has some interesting consequences. Clearly, the discovery of a manifestly non-Gaussian distribution of CBR fluctuations would provide a unique 'fingerprint' for strings (or texture). In addition, it is worth pointing out that many of the observational upper limits described in Chapter 7 are presented in terms of an r.m.s. value, or some other statistic which implicitly assumes that values of $\Delta T/T$ are distributed in a Gaussian fashion. Hence the upper limits presented in table 7.2 do not apply directly to CBR anisotropies produced by strings. Finally, as an inspection of fig. 8.26 will suggest, a network of cosmic strings will produce occasional, non-Gaussian 'hotspots' with a much larger value of $\Delta T/T$ than typical.

All of these properties of string-induced CBR fluctuations suggest that the ideal experiment to search for them would combine high angular resolution and wide sky coverage. The aperture synthesis experiments described in Section 7.8 provide the former but not the latter. Current limits on $\Delta T/T$ from such measurements (table 7.3) establish limits on μ roughly comparable with those derived above from the COBE maps. Filled aperture searches, such as that of Readhead *et al.*. (1989), while more sensitive, have lower angular resolution and sample less of the sky. I will end this section with the obvious remark that a high resolution map of a large solid angle will require a very large number of resolution elements, and hence long integrating times.

8.8.5 Summary

We began Section 8.8 with the remark that many of the mechanisms sketched in it were devised to 'save the phenomena.' How successful has that enterprise been? In particular, do these processes allow lower values of $\Delta T/T$ and hence prevent conflicts with the upper limits described in Chapter 7?

In some cases – for instance, when we take account of bias or of gravitational lensing – relatively modest changes in the amplitude of $\Delta T/T$ result. Hence any model 'saved'

Fig. 8.26 Pattern of fluctuations introduced by a random distribution of cosmic strings (from Bouchet *et al.*, 1988, with permission). Notice the sharp discontinuities and occasional 'hot spots' in this $7° \times 7°$ image (compare with fig. 8.20).

by such mechanisms is certain soon to be tested as observers push down limits on $\Delta T/T$ (crudely, upper limits on $\Delta T/T$ have been dropping by a factor of ten each 10–12 years – Partridge, 1990). Recent progress in such searches is assessed in Appendix C.

In other cases, modifications to the standard model can be engineered to save the phenomena, but seem *ad hoc* to me. Modifications of the power spectrum of perturbations (eqn. (8.6)), to my way of thinking, fall into this category. While complicating the mix of cosmic matter by adding a third component is less *ad hoc*, it is also unsatisfying philosophically. It is bad enough that the baryonic content of the Universe (which is all that astronomers have so far detected) is but a few percent of the total: now we must confront the possibility that we have 'missed' two *different* components of matter, both more important dynamically than the baryons. Before adopting that view, I would like to see some evidence directly favoring such models, not just arguments that they can be made to be consistent with the observations we now have. Nevertheless, a network of such models is consistent with the CBR observations, as a recent paper by Wright *et al.* (1992) shows. These are shown in fig. 8.27, adapted from that paper.

In interesting contrast is the cosmic string model. It introduces novel physics, makes novel predictions about the pattern of CBR anisotropy, and opens up the possibility of a direct measurement of the symmetry-breaking energy scale. Unfortunately the observers have not yet achieved the requisite sensitivity, sky coverage and angular resolution to test this model.

Finally, there are models, or processes, which have not been examined in the same detail as the standard CDM or baryon-only models. Here, I would mention textures, warm dark matter and tensor fluctuations as examples.

I would like not to end on this inconclusive note. Perhaps the most remarkable thing about this collection of processes and models is that they should be needed at all. The need to save the phenomena is a testament to the growing power of the observational results on the CBR, the distribution of galaxies and others, to winnow out the range of models for the formation of large-scale structure in the Universe.

8.9   Implications of observations of the Sunyaev–Zel'dovich effect for astrophysics and cosmology

To complete this chapter, we turn to a more specialized topic, the implications of searches for the Sunyaev–Zel'dovich (SZ) effect in low redshift objects. The results of searches for the SZ effect in clusters of galaxies are summarized in Section 7.10. The characteristic temperature decrement in the CBR produced by inverse Compton scattering by a hot plasma has been detected in several clusters. In addition, SZ signals produced by a random foreground of clusters of galaxies will be present in any search for primary CBR fluctuations. As we will see, both the detection of the SZ effect in a few well-studied clusters and upper limits on the amplitude of fluctuations produced by a random foreground of clusters have provided useful data on the properties and evolution of clusters of galaxies.

Before turning to these results, however, let us draw one other conclusion from the detection of SZ signals, one of direct relevance for the topic of this book. The detection of an SZ signal from the distant cluster 0016 + 16 clearly establishes that the CBR photons originate beyond the cluster, that is at redshifts greater than the cluster redshift of 0.54. Thus any local origin of the CBR is ruled out.

8.9.1   *Properties of intergalactic plasma in clusters*

The individual galaxies of a cluster are known to make up only a small fraction of the cluster mass (see, e.g., Trimble, 1987). From X-ray observations of clusters (Pravdo *et al.*, 1979; Forman and Jones, 1982; Birkinshaw *et al.*, 1991) we have known for some time that some of the remaining mass of clusters is in the form of hot ($T_e \simeq 10^8$ K $\simeq 10$ keV), ionized gas. Observations of the SZ signal allow us to measure the product $n_e T_e$ through a cluster; if we have available spatially resolved observations such as those shown in fig. 7.26, we can also obtain the radial profile of the gas.

Recall from eqn. (7.23) that the observed SZ profile for an isothermal cluster will be proportional to

$$[1 + (\theta/\theta_c)^2]^{1/2-3\tau/2},$$

where the parameter τ is a measure of the relative scale heights of the cluster galaxies and the intergalactic gas. Since both the galaxies and the gas occupy the same gravitational potential well (probably dominated by dark matter), we expect the same scale

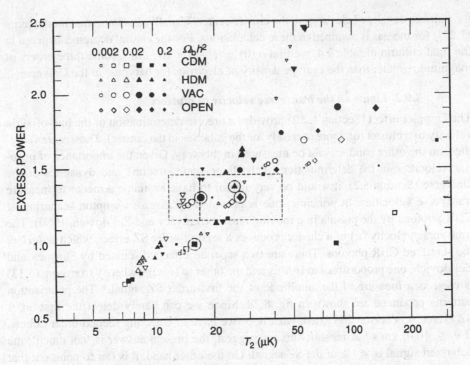

Fig. 8.27   A summary figure from Wright *et al.* (1992) based on the recent COBE DMR observations (Sections 6.7.3 and 7.7.3). The COBE measurements constrain T_2 (and higher order multipoles of $T(\theta)$). Other astrophysical observations, as we have seen in Section 8.6.5, require 'excess power.' Hence only the limited number of models enclosed in the dashed-line box are compatible with all the observations. Among the 'survivors' are: ▲, a hot dark matter model with one massive neutrino and $\Omega_b h^2 = 0.02$; ◆, an open cosmological model with $\Omega_{CDM} = 0.18$ and $\Omega_b = 0.02$; ●, a flat model with $\Lambda \neq 0$ and $\Omega_{CDM} = 0.18$, $\Omega_b = 0.02$. For further details, see Wright *et al.* (1992).

height unless one of the two components has relaxed or cooled (or, conversely, unless one of the two has been heated). That the scale heights are not radically different (so τ is of order unity) may be demonstrated by converting the velocity dispersion of galaxies in a cluster into a kinetic temperature, then comparing it with the plasma temperature derived from X-ray observations. For $\langle v \rangle^{1/2} = 800$–$1200$ km s^{-1}, typical of the massive clusters observed in SZ searches, the kinetic temperature is 5–8 keV, close to (or at most a factor of two below) the measured plasma temperatures of about 10 keV $= 1.1 \times 10^8$ K.

As fig. 7.26 and the discussion in Section 7.10 suggest, radial profiles derived from SZ measurements by themselves are not yet accurate enough to set useful constraints on τ. When combined with the X-ray images of clusters now becoming available from the Einstein Observatory and ROSAT, however, the SZ observations can help constrain both τ and θ_c. For Abell 665, for instance, Birkinshaw *et al.* (1991) find $0.6 \leq \tau \leq 0.8$. Since $\tau = 1$ is excluded, we conclude that there has been some energy equipartition, 'cooling' the galaxies relative to the gas.

If τ, θ_c and the temperature T_e can be found from some combination of X-ray and SZ observations, the value of the central SZ decrement may then be used to calculate $n_e(0)$, the number density of electrons at the cluster center. Usually, an isothermal model (eqn. (7.22)) for the gas is assumed in these calculations. For the central decrements given in the final column of table 7.4, we find $n_e(0) \simeq (1-3) \times 10^{-3}$ cm^{-3}, some three orders of magnitude greater than the average density of electrons (or baryons) in the Universe.

8.9.2 Limits on the transverse velocity of clusters

The Doppler effect (Section 1.2.2) provides a precise determination of the line-of-sight velocity of a cluster (or, more precisely, of the galaxies in the cluster). *Transverse* velocities, on the other hand, cannot be measured in this way. Given the importance of peculiar velocities in the determination of the large-scale structure and dynamics of the Universe (Section 8.2), it would be very useful to have available a means to measure transverse velocities. In principle, one is provided by inverse Compton scattering of CBR photons by the plasma in a moving cluster (Sunyaev and Zel'dovich, 1980). The transverse velocity (v_t) of a cluster produces a second-order SZ effect, which *polarizes* the scattered CBR photons. There are two separate effects discussed by Sunyaev and Zel'dovich, one proportional to $(v_t/c)^2 y$ and the other to $(v_t/c)y^2$, where y from eqn. (5.13) is used as a measure of the amplitude of the first-order SZ signal.* The polarization patterns produced are shown in fig. 8.28. Since we can barely detect the first-order SZ effect, it is reasonable to ask what hope we have of detecting second-order effects. If $v_t \lesssim 1000$ km s^{-1} as redshift surveys suggest, the present answer is 'not much'; the polarized signal is $\ll 1\%$ of the SZ signal. On the other hand, it is fair to point out that measurements of small polarizations are easier to make than measurements of small changes in absolute power (some sources of systematic error play a smaller role – see Section 7.9).

In addition, as Birkinshaw and Gull pointed out in 1983, a moving cluster is also a moving gravitational lens. A moving lens will produce an antisymmetric temperature change about an axis perpendicular to v_t (fig. 8.28). The amplitude of the gravitationally induced signal is given as

$$\Delta T = 0.04(v_t/1000 \text{ km s}^{-1}) \, (M/10^{15} M_\odot) \, (r_c/100 \text{ kpc})^{-1} \text{ mK},$$

where M and r_c are the cluster mass and core radius, respectively. In 1987, using appropriate values of M and r_c for Abell 2218, I made use of our VLA observations of that cluster to set an upper limit of 70 000 km s^{-1} on its transverse velocity. For conventional cosmologies, such a weak upper limit on v_t is of little value (but see Nel, 1987, for a discussion of models that permit larger values). The more sensitive measurements on 0016 + 16 and Abell 665 (fig. 7.26; Birkinshaw, 1991) permit us to set more stringent upper limits, at least on the east–west component of v_t in these clusters. From fig. 7.26, I judge any asymmetry in the north–south scan of 665 to be $\lesssim 0.2$ mK. Appropriate values for M and r_c for this cluster are about $3 \times 10^{15} h^{-1} M_\odot$ and $250 h^{-1}$ kpc, respectively (Dressler, 1978). We thus find $(v_t)_{\text{E-W}} < 5000$ km s^{-1}. A roughly comparable limit may be obtained from 0016 + 16. Improved SZ observations, which should become available over the next few years, may permit us to establish interesting limits on, or even measurements of, v_t, as well as the peculiar velocity in the line of sight.

---

* There is also a first-order 'kinematic' SZ effect proportional to the peculiar velocity in the line of sight (Birkinshaw, 1991).

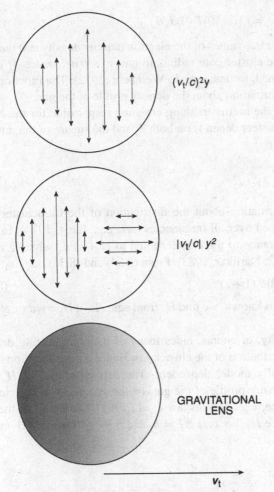

Fig. 8.28   Patterns of linear polarization (top two diagrams) induced by the transverse motion of a cluster of galaxies through the CBR. There are two separate terms, as noted in Section 8.9.2. The lowest figure shows schematically the variation in T produced by a moving gravitational lens – a cluster with a velocity as shown.

8.9.3 *Using the SZ effect to determine H_0*

As first pointed out in 1978 both by Gunn and by Silk and White, observations of the SZ effect may allow us to determine Hubble's constant, H_0, by providing a direct measurement of the distance to the cluster studied. By 'direct' here, I mean a measurement independent of all the intermediate rungs of the 'distance ladder' used to determine cosmological distances (Rowan-Robinson, 1985). The basis for the measurement is a comparison of the SZ signal from a cluster with its X-ray flux. Qualitatively, it is easy to see how an estimate of the cluster's distance arises from such a comparison: ΔT is independent of distance, and the X-ray flux S_x is proportional to $1/d^2$.

More quantitatively, the SZ 'dip' observed along a line of sight through the cluster center may be written (from eqn. (7.20)):

$$\Delta T(0) = f_1(l)n_e(0)T_e(0)r_c = f_1(l)n_e(0)T_e(0)d_A\theta_c, \tag{8.34}$$

where $n_e(0)$ and $T_e(0)$ are the central values of the electron number density and temperature, respectively, and r_c is the cluster core radius. In turn $r_c = \theta_c d_A$, where d_A is the angular-diameter distance, defined, for instance, in Weinberg (1972). The function $f_1(l)$ contains both constants and information about the density profile of the gas.

As we saw in Section 3.7.1, the bremsstrahlung emission responsible for the X-ray emission from the hot gas in clusters depends on both T_e and the square of the number density, n_e^2. Hence we may write

$$S_x \propto f_2(l)n_e^2(0)\frac{(r_c)^3}{d_L^2} f_3(T_e). \tag{8.35}$$

Once again, $f_2(l)$ contains information about the distribution of the intracluster gas; $f_3(T_e) \propto T_e^{1/2}$ if the flux is summed over all frequencies (see, e.g., Lang, 1974). In eqn. (8.35), d_L is the luminosity distance to the cluster, equal to $(1 + z)^2 d_A$ where z is the cluster redshift (Weinberg, 1972; Narlikar, 1983). From (8.34) and (8.35), we find

$$d_L \propto \Delta T^2(0) S_x^{-1} T_e^{-3/2}(0)\theta_c(1 + z)^{-2}. \tag{8.36}$$

Once the distance to the cluster is known, we find H_0 from eqn. (1.3); $H_0 = v/d = zc/d$ for small z.

Writing d_L as a proportionality, of course, hides many of the complicating details. These involve primarily the distribution of the cluster gas. Hence a determination of H_0 using this method is intrinsically model dependent. The derived values of H_0 will depend, for instance, on the density profile of the gas, on the symmetry of the cluster, and on clumping of the gas (see, e.g., Birkinshaw *et al.*, 1991). Clumping of the gas affects estimates of d_L and hence H_0 because $\Delta T \propto n_e$ and $S_x \propto n_e^2$. Define a clumping parameter C by

$$C = \frac{\overline{n_e^2}}{(\overline{n_e})^2},$$

where the averages are taken over the cluster. It is then easy to see that small-scale clumping will cause us to overestimate H_0 by a factor C (i.e., the true value of H_0 will be smaller than that found from eqn. (8.36) assuming $C = 1$).

In the past decade, several attempts have been made to apply this technique to the few clusters for which we have both X-ray measurements (of S_x and T_e) and a reliable SZ signal. Boynton *et al.* (1982), for instance, found $H_0 \approx 54$ km s^{-1} per megaparsec on the basis of early (and erroneous) values of $\Delta T(0)$ and T_e in Abell 2218. McHardy *et al.* (1990) returned to Abell 2218 and found $H_0 = 24^{+13}_{-10}$ in the same units. However, the amplitude of the SZ signal from 2218 is in some dispute (Section 7.10.2), as is the cluster core radius.

Observations of Abell 665 are more secure; $\Delta T = 0.43 \pm 0.05$ mK and $\theta_c = 1'.6$ (Section 7.10.2; Birkinshaw *et al.*, 1991). It was therefore selected by Birkinshaw and his colleagues, Hughes and Arnaud, for further study using two X-ray satellites, Ginga and the Einstein Observatory. The former provided an X-ray spectrum, from which they found $T_e = 8.2 \pm 0.5$ keV for the cluster. The latter provided an X-ray image. When combined with the SZ observations, the X-ray data yielded $0.6 \leqslant \tau \leqslant 0.8$ and $\theta_c = 1'.6 \pm 0'.4$ at 95% confidence.

Birkinshaw *et al.* (1991) then calculated H_0, finding 40–50 km s^{-1} per megaparsec with an error of ±12 in the same units. They also discussed at length various sources of error, in particular those that might have caused them to *underestimate H_0*.* One possibility is that Abell 665 is prolate, with its long axis lying approximately along the line of sight. It is also worth remarking, as they do, that the major uncertainty in their work is the value of $\Delta T(0)$, not the X-ray parameters. Nevertheless, it is encouraging that this technique has provided a value of H_0 roughly equal to and roughly as accurate as values obtained from the classical astronomical techniques described by Rowan-Robinson (1985). Present and planned X-ray satellite observatories will produce better measurements of S_x, θ_c and perhaps τ; I hope observers will continue to improve the microwave measurements. The ability of aperture synthesis to map a target cluster makes it a promising technique for such observations (see Appendix C for an update on such attempts).

8.9.4 Fluctuations produced by a random distribution of clusters

Now let us turn to the effect on the CBR of many clusters lying along the line of sight. Each cluster will produce a small temperature decrement in the CBR (for observations at $\nu < 220$ GHz). These will add up to produce random fluctuations in the intensity of the sky at radio wavelengths. These 'foreground' fluctuations will add to – or even mask – primary or secondary fluctuations discussed in Sections 8.6 and 8.7.

The amplitude of this random foreground of SZ signals has been estimated by Korolev *et al.* (1986), Cole and Kaiser (1988) and Schaeffer and Silk (1988). The predicted levels of SZ fluctuation are quite model-dependent. Korolev *et al.*, for instance, adopt an isothermal model and assume no evolution in the intrinsic properties of clusters. They then assume that the telescope used to observe CBR fluctuations does not resolve the individual clusters, and go on to calculate the amplitude of flux density fluctuations that a random foreground of clusters would produce. In these calculations, the authors take into account cosmological effects, such as the increase in angular diameter of a source of fixed proper diameter beyond a critical redshift (for $\Omega = 1$, that redshift is $z = 1.25$; see, e.g., Narlikar 1983, Chapter 4). Schaeffer and Silk (1988), on the other hand, base their calculations on a specific model for the evolution of clusters. Evolution of the properties of the cluster gas has the effect of making the SZ signal from a cluster redshift-dependent, $\Delta T \propto (z + 1)^\gamma$, with γ close to unity. The amplitude of the resulting fluctuations will thus depend on the redshift (and hence epoch) when clusters first formed. One sample calculation includes clusters out to $z \simeq 10$ (with a mean redshift of about 3). In this case, on a 1′ scale, $\Delta T/T \simeq 4 \times 10^{-5}$ from the effect of foreground clusters. The amplitude of $\Delta T/T$ on this scale is thus roughly comparable with the maximum amplitude expected (on a larger angular scale) for primary fluctuations in the CBR.

The predicted $\Delta T/T$ is also very close to the upper limits already established by the VLA observations described in Section 7.8. Indeed, the new 3.6 cm observations described in Section 7.8 should be able to provide a clear test of the model. Even if the particular model of Schaeffer and Silk does not survive confrontation with the observations, we should not abandon the notion that SZ fluctuations at some level will contribute to fluctuations in the microwave sky. Schaeffer and Silk assumed a specific

---

* Note that clumping of the cluster gas would have the opposite effect.

model for cluster evolution, for instance; assuming a different model with a more rapid, more recent growth of n_e or T_e would allow for smaller SZ fluctuations.

These constraints on the evolution of clusters of galaxies may serve as useful adjuncts to the more general constraints on the formation and evolution of large-scale structure in the Universe imposed by the upper limits on primary fluctuations in the CBR discussed earlier in this chapter.

References

Aaronson, M., Huchra, J., Mould, J., and Schechter, P. L. 1982, *Ap. J.*, **258**, 64.

Aaronson, M., Bothun, G., Mould, J., Huchra, J., Schommer, R. A., and Cornell, M. E. 1986, *Ap. J.*, **302**, 536.

Anile, A. M., Danese, L., De Zotti, G., and Motta, S. 1976, *Ap. J. (Lett.)*, **205**, L59.

Bahcall, N. A., and Soneira, R. M. 1983, *Ap. J.*, **270**, 20.

Bardeen, J. 1986, in *Inner Space/Outer Space*, eds. E. W. Kolb, M. S. Turner, D. Lindley, K. Olive, and D. Seckel, University of Chicago Press, Chicago.

Bardeen, J. 1980, *Phys. Rev.* D, **22**, 1882.

Bardeen, J., Bond, J. R., and Efstathiou, G. 1987, *Ap. J.*, **321**, 28.

Barrow, J. D. 1976, *Monthly Not. Roy. Astron. Soc.*, **175**, 359.

Barrow, J. D. 1984, *Monthly Not. Roy. Astron. Soc.*, **211**, 221.

Barrow, J. D., Juszkiewicz, R., and Sonoda, D. H. 1983, *Nature*, **305**, 397.

Barrow, J. D., Juszkiewicz, R., and Sonoda, D. H. 1985, *Monthly Not. Roy. Astron. Soc.*, **213**, 917.

Basko, M. M., and Polnarev, A. G. 1980, *Monthly Not. Roy. Astron. Soc.*, **191**, 207.

Bennett, D. P., and Rhie, S. H. 1990, *Phys. Rev. Lett.*, **65**, 1709.

Bertschinger, E. 1987, *Ap. J.*, **316**, 489.

Bertschinger, E., Dekel, A., Faber, S. M., Dressler, A., and Burstein, D. 1990, *Ap. J.*, **364**, 370.

Bianchi, L. 1898, *Mem. Soc. Ital.*, **11**, 267.

Birkinshaw, M. 1991, in *Physical Cosmology*, eds. A. Blanchard *et al.*, Editions Frontiers, Gif-sur-Yvette.

Birkinshaw, M., and Gull, S. F. 1983, *Nature*, **302**, 315.

Birkinshaw, M., Hughes, J. P., and Arnaud, K. A. 1991, *Ap. J.*, **379**, 466.

Blanchard, A. 1984, *Astron. Astrophys.*, **132**, 359.

Blanchard, A., and Schneider, J. 1987, *Astron. Astrophys.*, **184**, 1.

Bond, J. R. 1988, in *The Early Universe*, eds. W. G. Unruh and G. W. Semenoff, D. Reidel, Dordrecht.

Bond, J. R. 1990, in *Frontiers in Physics - From Colliders to Cosmology*, eds. B. Campbell and F. Khanna, World Scientific, Singapore.

Bond, J. R., and Efstathiou, G. 1984, *Ap. J. (Lett.)*, **285**, L45.

Bond, J. R., and Efstathiou, G. 1987, *Monthly Not. Roy. Astron. Soc.*, **226**, 655.

Bond, J. R., and Szalay, A. S. 1983, *Ap. J.*, **274**, 443.

Bond, J. R., Carr, B. J., and Hogan, C. 1986, *Ap. J.*, **306**, 428.

Bond, J. R., Carr, B. J., and Hogan, C. 1991, *Ap. J.*, **367**, 420.

Bouchet, F. R., and Bennett, D. P. 1990, *Ap. J. (Lett.)*, **354**, L41.

Bouchet, F. R., Bennett, D. P., and Stebbins, A. 1988, *Nature*, **335**, 410.

Boynton, P. E., Radford, S. J. E., Schommer, R. A., and Murray, S. S. 1982, *Ap. J.*, **257**, 473.

Brans, C. H. 1975, *Ap. J.*, **197**, 1.

Burke, W. L. 1975, *Ap. J.*, **196**, 329.

Carr, B. J., Bond, J. R., and Arnett, W. D. 1984, *Ap. J.*, **277**, 445.

Chibisov, G. V., and Ozernoi, L. M. 1969, *Astrophys. Lett.*, **3**, 189.

Cole, S., and Efstathiou, G. 1989, *Monthly Not. Roy. Astron. Soc.*, **239**, 195.

Cole, S., and Kaiser, N. 1988, *Monthly Not. Roy. Astron. Soc.*, **233**, 637.

Collins, C. B., and Hawking, S. W. 1973, *Monthly Not. Roy. Astron. Soc.*, **162**, 307.

Collins, C. B., Joseph, R. D., and Robertson, N. A. 1986, *Nature*, **320**, 506.

Cowsik, R., and McClellan, J. 1972, *Phys. Rev. Lett.*, **29**, 669; also 1973, *Ap. J.*, **180**, 7.

Dautcourt, G. 1969, *Monthly Not. Roy. Astron. Soc.*, **144**, 255.

Davies, R. D. 1983, in *I.A.U. Symposium 104*, eds. G. Abell and G. Chincarini, D. Reidel, Dordrecht.

Davies, R. D., Lasenby, A. N., Watson, R. A., Daintree, E. J., Hopkins, J., Beckman, J., Sanchez-Almeida, J., and Rebolo, R., 1987, *Nature*, **326**, 462.

Davis, M., and Peebles, P. J. E. 1983a, *Ann. Rev. Astron. Astrophys.*, **21**, 109.

Davis, M., and Peebles, P. J. E. 1983b, *Ap. J.*, **267**, 465.

Dekel, A., Bertschinger, E., and Faber, S. M. 1990, *Ap. J.*, **364**, 349.

de Vaucouleurs, G., de Vaucouleurs, A., and Corwin, H. G. 1976, *Second Reference Catalogue of Bright Galaxies*, University of Texas Press, Austin, Texas.

Doroshkevich, A. G., Zel'dovich, Ya. B., and Novikov, I. D. 1967, *Soviet Astron. J.*, **11**, 233.

Doroshkevich, A. G., Zel'dovich, Ya. B., and Sunyaev, R. A. 1978, *Soviet Astron. J.*, **22**, 523.

Dressler, A. 1978, *Ap. J.*, **226**, 55.

Dressler, A. 1988, *Ap. J.*, **329**, 519.

Dyer, C. C. 1976, *Monthly Not. Roy. Astron. Soc.*, **175**, 429.

Efstathiou, G. 1990, in *Physics of the Early Universe*, eds. J. A. Peacock, A. F. Heavens and A. T. Davies, SUSSP, Edinburgh.

Efstathiou, G., and Bond, J. R. 1986, *Monthly Not. Roy. Astron. Soc.*, **218**, 103.

Efstathiou, G., and Bond, J. R. 1987, *Monthly Not. Roy. Astron. Soc.*, **227**, 33.

Efstathiou, G., and Silk, J. 1983, *Fundamentals Cosmic Phys.*, **9**, 1.

Efstathiou, G., Sutherland, W. J., and Maddox, S. J. 1990, *Nature*, **348**, 705.

Efstathiou, G., Bond, J. R., and White, S. D. M., 1992, *Monthly Not. Roy. Astron. Soc.* **258**, 1.

Einstein, A. 1936, *Science*, **84**, 506.

Fabbri, R., Lucchin, F., and Matarrese, S. 1987, *Ap. J.*, **315**, 1.

Fall, S. M., and Jones, B. J. T. 1976, *Nature*, **262**, 457.

Forman, W., and Jones, C. 1982, *Ann. Rev. Astron. Astrophys.*, **20**, 547.

Fukugita, M., Sugiyama, N., and Umemura, M. 1990, *Ap. J.*, **358**, 22.

Geller, M. J., and Huchra, J. P. 1988, in *Large Scale Motions in the Universe*, eds. V. C. Rubin and G. V. Coyne, Princeton University Press, Princeton, New Jersey.

Gorski, K. M. 1991, *Ap. J. (Lett.)*, **370**, L5.

Gouda, N., Sasaki, M., and Suto, Y. 1987, *Ap. J. (Lett.)*, **321**, L1.

Gouda, N., Sasaki, M., and Suto, Y. 1989, *Ap. J.*, **341**, 557.

Gouda, N., Sasaki, M., and Sugiyama, N. 1991, *Prog. Theor. Phys.*, **85**, 1023.

Grishchuk, L. P., and Zel'dovich, Ya. B. 1978, *Soviet Astron. J.*, **22**, 125.

Groth, E. J., and Peebles, P. J. E. 1976, *Astron. Astrophys.*, **53**, 131.

Gunn, J. E. 1978, in *Observational Cosmology*, eds. A. Maeder *et al.*, Geneva Observatory.

Gunn, J. E., and Peterson, B. A. 1965, *Ap. J.*, **142**, 1633.

Harmon, R. T., Lahav, O., and Meurs, E. J. A. 1987, *Monthly Not. Roy. Astron. Soc.*, **228**, 5.

Hart, L. and Davies, R. 1982, *Nature*, **297**, 191.

Hartnett, P. M. 1991, BA thesis, Haverford College.

Hawking, S. W. 1969, *Monthly Not. Roy. Astron. Soc.*, **142**, 129.

Hill, C. T., Schramm, D. N., and Fry, J. 1989, *Comments Nuclear Particle Phys.*, **19**, 25.

Hogan, C. J. 1984, *Ap. J. (Lett.)*, **284**, L1.

Holtzman, J. A. 1989, *Ap. J. Suppl.*, **71**, 1.

Ikeuchi, S. 1981, *Publ. Astron. Soc. Japan*, **33**, 211.

Jeans, J. 1902, *Phil. Trans. Roy. Soc. A*, **199**, 49.

Jones, B. J. T., and Wyse, R. F. G. 1985, *Astron. Astrophys.*, **149**, 144.

Kaiser, N. 1984, *Ap. J.*, **282**, 374.

Kaiser, N. 1986, in *Inner Space/Outer Space*, eds. E. W. Kolb, M. S. Turner, D. Lindley, K. Olive, and D. Seckel, University of Chicago Press, Chicago.

Kaiser, N. 1991, in *After the First Three Minutes*, eds. S. S. Holt, C. L. Bennett and V. Trimble, American Institute of Physics, New York.

Kaiser, N., and Silk, J. 1986, *Nature*, **324**, 529.

Kaiser, N. and Stebbins, A. 1984, *Nature*, **310**, 391.

Kaiser, N., Efstathiou, G., Ellis, R., Frenk, C., Lawrence, A., Rowan-Robinson, M., and Sauders, W. 1991, *Monthly Not. Roy. Astron. Soc.*, **252**, 1.

Kashlinsky, A. 1988, *Ap. J. (Lett.)*, **331**, L1.

Kibble, T. W. B. 1976, *J. Phys. A*, **9**, 1387.

Kofman, L. A., and Starobinsky, A. A. 1985, *Soviet Astron. J. Lett.*, **11**, 271.
Kolb, E. B., and Turner, M. S. 1990, *The Early Universe*, Addison-Wesley, New York.
Koo, D. C. 1986, in *Spectral Evolution of Galaxies*, eds. C. Chiosi and A. Renzini, D. Reidel, Dordrecht.
Korolev, V. A., Sunyaev, R. A. and Yakubtsev, L. A. 1986, *Soviet Astron. J. Lett.*, **12**, 141.
Kreysa, E., and Chini, R. 1989, in *Third ESO/CERN Symposium, Astronomy, Cosmology and Fundamental Physics*, eds. M. Caffo *et al.*, Kluwer, Dordrecht.
La, D. 1989, *Ap. J.*, **341**, 575.
Lacey, C. G., and Field, G. B. 1988, *Ap. J. (Lett.)*, **330**, L1.
Lahav, O. 1987, *Monthly Not. Roy. Astron. Soc.*, **225**, 213.
Lahav, O., and Partridge, R. B. 1988, *Monthly Not. Roy. Astron. Soc.*, **235**, 1p.
Lang, K. R. 1974, *Astrophysical Formulae*, Springer-Verlag, New York.
Lifschitz, E. M. 1946, *Zh. Eksp. Teor. Fiz.*, **16**, 587.
Lilje, P. B. 1990, *Ap. J.*, **351**, 1.
Linder, E. V. 1990, *Monthly Not. Roy. Astron. Soc.*, **243**, 353.
Lubin, P. M., Epstein, G. L., and Smoot, G. F. 1983, *Phys. Rev. Lett.*, **50**, 616.
Lynden-Bell, D., and Lahav, O. 1989, in *Large Scale Motions in the Universe*, Princeton University Press, Princeton, New Jersey.
Lynden-Bell, D., Faber, S. M., Burstein, D., Davies, R. L., Dressler, A., Terlevich, R. J., and Wegner, G. 1988, *Ap. J.*, **326**, 19.
Lynden-Bell, D., Lahav, O., and Burstein, D. 1989, *Monthly Not. Roy. Astron. Soc.*, **241**, 325.
MacCallum, M. A. H. 1985, in *Observational and Theoretical Aspects of Relativistic Astrophysics and Cosmology*, eds. J. L. Sanz and L. J. Goicoechea, World Scientific, Singapore.
Maddox, S. J., Sutherland, W. J., Efstathiou, G., Loveday, J. 1990, *Monthly Not. Roy. Astron. Soc.*, **243**, 692.
Martinez-Gonzalez, E., and Sanz, J. L. 1989, *Ap. J.*, **347**, 11.
McHardy, I. M., Stewart, G. C., Edge, A. C., Cooke, B., Yamashita, K., and Hatsukade, I. 1990, *Monthly Not. Roy. Astron. Soc.*, **242**, 215.
Meiksin, A., and Davis, M. 1986, *A. J.*, **91**, 191.
Meinhold, P., and Lubin, P. 1991, *Ap. J. (Lett.)*, **370**, L11.
Mellier, Y., Fort, B., and Soucail, G., eds 1990, *Gravitational Lensing*, Springer-Verlag, Berlin.
Misner, C. W. 1968, *Ap. J.*, **151**, 431.
Mitrofanov, I. G. 1981, *Soviet Astron. J. Lett.*, **7**, 39.
Mukhanov, V. F., and Chibisov, G. V. 1984, *Soviet Astron. J. Lett.*, **10**, 374.
Narlikar, J. V. 1983, *Introduction to Cosmology*, Jones and Bartlett, Boston.
Negroponte, J., and Silk, J. 1980, *Phys. Rev. Lett.*, **44**, 1433.
Nel, S. D. 1987, in *Theory and Observational Limits in Cosmology*, ed. W. R. Stoeger, Specola Vaticana, Vatican City.
Novikov, I. D. 1968, *Soviet Astron. J.*, **12**, 427.
Ostriker, J. P., and Cowie, L. L. 1981, *Ap. J. (Lett.)*, **243**, L127.
Ostriker, J. P., and Vishniac, E. T. 1986, *Ap. J. (Lett.)*, **306**, L51.
Ozernoi, L. M. 1974, in *I.A.U. Symposium 63*, ed. M. S. Longair, D. Reidel, Dordrecht.
Paczynski, B., and Piran, T. 1990, *Ap. J.*, **364**, 341.
Partridge, R. B. 1990, in *The Cosmic Microwave Background: 25 Years Later*, eds. N. Mandolesi and N. Vittorio, Kluwer, Dordrecht.
Partridge, R. B., and Peebles, P. J. E. 1967a, *Ap. J.*, **147**, 868.
Partridge, R. B., and Peebles, P. J. E. 1967b, *Ap. J.*, **148**, 377.
Peebles, P. J. E. 1971, *Physical Cosmology*, Princeton University Press, Princeton, New Jersey.
Peebles, P. J. E. 1976, *Ap. J.*, **205**, 318.
Peebles, P. J. E. 1980, *The Large Scale Structure of the Universe*, Princeton University Press, Princeton, New Jersey.
Peebles, P. J. E. 1981, *Ap. J.*, **248**, 885.
Peebles, P. J. E. 1982, *Ap. J.*, **258**, 413.
Peebles, P. J. E. 1984, *Ap. J.*, **277**, 470.
Peebles, P. J. E. 1987, *Ap. J. (Lett.)*, **315**, L73.
Peebles, P. J. E., and Wilkinson, D. T. 1968, *Phys. Rev.*, **174**, 2168.

Peebles, P. J. E., and Yu, J. T. 1970, *Ap. J.*, **162**, 815.

Pravdo, S. H., Boldt, E. A., Marshall, F. E., McKee, J., Mushotzky, R. F., Smith, B. W., and Reichert, G. 1979, *Ap. J.*, **234**, 1.

Readhead, A. C. S., Lawrence, C. R., Myers, S. T., Sargent, W. L. W., Hardebeck, H. E., and Moffet, A. T. 1989, *Ap. J.*, **346**, 566.

Rees, M. J., and Sciama, D. W. 1968, *Nature*, **217**, 511.

Robertson, R. G. H., Bowles, T. J., Stephenson, G. J., Wark, D. L., and Wilkerson, J. F. 1991, *Phys. Rev. Lett.*, **67**, 957.

Rothman, T., and Matzner, R. 1985, *Phys. Rev.* D, **30**, 1649.

Rowan-Robinson, M. 1985, *The Cosmological Distance Ladder*, W. H. Freeman, San Francisco.

Rubin, V. C., and Coyne, G. V., eds 1989, *Large Scale Motions in the Universe*, Princeton University Press, Princeton, New Jersey.

Rubin, V. C., Ford, W. K. Jr, Thonnard, N., Roberts, M. S., and Graham, J. A. 1976a, *A. J.*, **81**, 687.

Rubin, V. C., Thonnard, N., Ford, W. K. Jr, and Roberts, M. S. 1976b, *A. J.*, **81**, 719.

Sachs, R. K., and Wolfe, A. M. 1967, *Ap. J.*, **147**, 73.

Sasaki, M. 1989, *Monthly Not. Roy. Astron. Soc.*, **240**, 415.

Scaramella, R., Vettolani, G., and Zamorani, G. 1991, *Ap. J. (Lett.)*, **376**, L1.

Schaeffer, R., and Silk, J. 1988, *Ap. J.*, **333**, 509.

Schaeffer, R. K., Shafi, Q., and Stecker, F. W. 1989, *Ap. J.*, **347**, 575.

Schechter, P. L. 1977, *A. J.*, **82**, 569.

Silk, J., 1968, *Ap. J.*, **151**, 459.

Silk, J., and Vittorio, N. 1991, in *Confrontation between Theories and Observations in Cosmology*, eds. J. Audouze and F. Melchiorri, North Holland, Amsterdam.

Silk, J., and White, S. D. M. 1978, *Ap. J. (Lett.)*, **226**, L103.

Smoot, G. F., *et al.* 1992, *Ap. J. (Lett.)*, **396**, L1.

Starobinskii, A. A. 1985, *Soviet Astron. J. Lett.*, **11**, 133.

Stebbins, A. 1988, *Ap. J.*, **327**, 584.

Stebbins, A., and Turner, M. S. 1989, *Ap. J. (Lett.)*, **339**, L13.

Stebbins, A., Veeraraghavan, S., Brandenberger, R., Silk, J., and Turok, N. 1987, *Ap. J.*, **322**, 1.

Strauss, M. A., and Davis, M. 1989, in *Large Scale Motions in the Universe*, Princeton University Press, Princeton, New Jersey.

Strauss, M. A., Yahil, A., Davis, M., Huchra, J. P., and Fisher, K. 1992, *Ap. J.*, **397**, 395.

Sugiyama, N., Gouda, N., and Sasaki, M. 1990, *Ap. J.*, **365**, 432.

Sunyaev, R. A. 1971, *Astron. Astrophys.*, **12**, 190.

Sunyaev, R. A., and Zel'dovich, Ya. B. 1970, *Astrophys. Space Sci.*, **7**, 3.

Sunyaev, R. A., and Zel'dovich, Ya. B. 1980, *Monthly Not. Roy. Astron. Soc.*, **190**, 413.

Suto, Y., Gouda, N., and Sugiyama, N. 1990, *Ap. J. Suppl.*, **74**, 665.

Szalay, A. S., Bond, J. R., and Silk, J. 1983, in *Formation and Evolution of Galaxies and Large Scale Structure in the Universe*, eds. S. J. Audouze and Tran Thanh Van, D. Reidel, Dordrecht.

Thorne, K. S. 1967, *Ap. J.*, **148**, 51.

Tolman, B. W. 1985, *Ap. J.*, **290**, 1.

Trimble, V. 1987, *Ann. Rev. Astron. Astrophys.*, **25**, 425.

Turok, N. 1989, *Phys. Rev. Lett.*, **63**, 2625.

Turok, N., and Brandenberger, R. 1986, *Phys. Rev.* D, **33**, 2175.

Turok, N., and Spergel, D. N. 1990, *Phys. Rev. Lett.*, **64**, 2736.

Uson, J. M., and Wilkinson, D. T. 1984, *Ap. J.*, **283**, 471; see also Uson, J. M., and Wilkinson, D. T. 1984, *Nature*, **312**, 427.

Vilenkin, A. 1980, *Phys. Rev. Lett.*, **46**, 1169 and 1496.

Vilenkin, A. 1985, *Phys. Rep.*, **121**, 263.

Vishniac, E. T. 1987, *Ap. J.*, **322**, 597.

Vittorio, N., and Silk, J. 1984, *Ap. J. (Lett.)*, **285**, L39.

Vittorio, N., and Silk, J. 1985, *Ap. J. (Lett.)*, **297**, L1.

Vittorio, N., and Silk, J. 1992, *Ap. J. (Lett.)*, **385**, L9.

Wasserman, I. 1986, *Phys. Rev. Lett.*, **57**, 2234.

Watanabe, K., and Tomita, K. 1991, *Ap. J.*, **370**, 481.
Weinberg, S. 1972, *Gravitation and Cosmology*, John Wiley, New York.
White, S. D. M., Frenck, C. S., and Davis, M. 1983, *Ap. J. (Lett.)*, **274**, L1.
Wilson, M. L. 1983, *Ap. J.*, **273**, 2.
Wilson, M. L., and Silk, J. 1981, *Ap. J.*, **243**, 14.
Wright *et al.* 1992, *Ap. J. (Lett.)*, **396**, L13.
Yahil, A. 1985, in *The Virgo Cluster of Galaxies*, eds. O. G. Richter and B. Binggeli, ESO,
    Garching bei München.
Yahil, A., Tammann, G. A., and Sandage, A. 1977, *Ap. J.*, **217**, 903.
Yahil, A., Walker, D., and Rowan-Robinson, M. 1986, *Ap. J. (Lett.)*, **301**, L1.
Yahil, A., Strauss, M. A., Davis, M., and Huchra, J. P. 1991, *Ap. J.*, **372**, 380.

Appendix A

A measurement of excess antenna temperature at 4080 Mc/s*

Measurements of the effective zenith noise temperature of the 20 foot horn-reflector antenna (Crawford, Hogg, and Hunt 1961) at the Crawford Hill Laboratory, Holmdel, New Jersey, at 4080 Mc/s have yielded a value about 3.5 K higher than expected. This excess temperature is, within the limits of our observations, isotropic, unpolarized, and free from seasonal variations (July, 1964–April, 1965). A possible explanation for the observed excess noise temperature is the one given by Dicke, Peebles, Roll, and Wilkinson (1965) in a companion letter in this issue.

The total antenna temperature measured at the zenith is 6.7 K of which 2.3 K is due to atmospheric absorption. The calculated contribution due to Ohmic losses in the antenna and back-lobe response is 0.9 K.

The radiometer used in this investigation has been described elsewhere (Penzias and Wilson 1965). It employs a traveling-wave maser, a low-loss (0.027 db) comparison switch, and a liquid helium-cooled reference termination (Penzias 1965). Measurements were made by switching manually between the antenna input and the reference termination. The antenna, reference termination, and radiometer were well matched so that a round-trip return loss of more than 55 db existed throughout the measurement; thus errors in the measurement of the effective temperature due to impedance mismatch can be neglected. The estimated error in the measured value of the total antenna temperature is 0.3 K and comes largely from uncertainty in the absolute calibration of the reference termination.

The contribution to the antenna temperature due to atmospheric absorption was obtained by recording the variation in antenna temperature with elevation angle and employing the secant law. The result, 2.3 ± 0.3 K, is in good agreement with published values (Hogg 1959; DeGrasse, Hogg, Ohm, and Scovil 1959; Ohm 1961).

The contribution to the antenna temperature from Ohmic losses is computed to be 0.8 ± 0.4 K. In this calculation we have divided the antenna into three parts: (1) two non-uniform tapers approximately 1 m in total length which transform between the 2⅛ inch round output waveguide and the 6 inch-square antenna throat opening; (2) a double-choke rotary joint located between these two tapers; (3) the antenna itself. Care was taken to clean and align joints between these parts so that they would not significantly increase the loss in the structure. Appropriate tests were made for leakage and loss in the rotary joint with negative results.

*Reprinted with permission from the *Astrophysical Journal*, 1965.

The possibility of losses in the antenna horn due to imperfections in its seams was eliminated by means of a taping test. Taping all the seams in the section near the throat and most of the others with aluminum tape caused no observable change in antenna temperature.

The backlobe response to ground radiation is taken to be less than 0.1 K for two reasons. (1) Measurements of the response of the antenna to a small transmitter located on the ground in its vicinity indicate that the average back-lobe level is more than 30 db below isotropic response. The horn-reflector antenna was pointed to the zenith for these measurements, and complete rotations in azimuth were made with the transmitter in each of ten locations using horizontal and vertical transmitted polarization from each position. (2) Measurements on smaller horn-reflector antennas at these laboratories, using pulsed measuring sets on flat antenna ranges, have consistently shown a back-lobe level of 30 db below isotropic response. Our larger antenna would be expected to have an even lower back-lobe level.

From a combination of the above, we compute the remaining unaccounted-for antenna temperature to be 3.5 ± 1.0 K at 4080 Mc/s. In connection with this result it should be noted that DeGrasse *et al.* (1959) and Ohm (1961) give total system temperatures at 5650 Mc/s and 2390 Mc/s, respectively. From these it is possible to infer upper limits to the background temperatures at these frequencies. These limits are, in both cases, of the same general magnitude as our value.

We are grateful to R. H. Dicke and his associates for fruitful discussions of their results prior to publication. We also wish to acknowledge with thanks the useful comments and advice of A. B. Crawford, D. C. Hogg, and E. A. Ohm in connection with the problems associated with this measurement.

Note added in proof. The highest frequency at which the background temperature of the sky had been measured previously was 404 Mc/s (Pauliny-Toth and Shakeshaft 1962), where a minimum temperature of 16 K was observed. Combining this value with our result, we find that the average spectrum of the background radiation over this frequency range can be no steeper than $\lambda^{0.7}$. This clearly eliminates the possibility that the radiation we observe is due to radio sources of types known to exist, since in this event, the spectrum would have to be very much steeper.

A. A. PENZIAS
R. W. WILSON

MAY 13, 1965
BELL TELEPHONE LABORATORIES, INC.
CRAWFORD HILL, HOLMDEL, NEW JERSEY

References

Crawford, A. B., Hogg, D. C., and Hunt, L. E. 1961, *Bell System Tech. J.*, **40**, 1095.
DeGrasse, R. W., Hogg, D. C., Ohm, E. A., and Scovil, H. E. D. 1959, Ultra-low Noise Receiving System for Satellite or Space Communication, *Proc. National Electronics Conf.* Vol. 15, 370.
Dicke, R. H., Peebles, P. J. E., Roll, P. G., and Wilkinson, D. T. 1965, *Ap. J.*, **142**, 414.
Hogg, D. C. 1959, *J. Appl. Phys.*, **30**, 1417.
Ohm, E. A. 1961, *Bell System Tech. J.*, **40**, 1065.
Pauliny-Toth, I. I. K., and Shakeshaft, J. R. 1962, *M.N.*, **124**, 61.
Penzias, A. A. 1965, *Rev. Sci. Instrum.*, **36**, 68.
Penzias, A. A., and Wilson, R. W. 1965, *Ap. J.* (in press).

Appendix B
Cosmic black-body radiation*†

One of the basic problems of cosmology is the singularity characteristic of the familiar cosmological solutions of Einstein's field equations. Also puzzling is the presence of matter in excess over antimatter in the universe, for baryons and leptons are thought to be conserved. Thus, in the framework of conventional theory we cannot understand the origin of matter or of the universe. We can distinguish three main attempts to deal with these problems.

1   The assumption of continuous creation (Bondi and Gold 1948; Hoyle 1948), which avoids the singularity by postulating a universe expanding for all time and a continuous but slow creation of new matter in the universe.

2   The assumption (Wheeler 1964) that the creation of new matter is intimately related to the existence of the singularity, and that the resolution of both paradoxes may be found in a proper quantum mechanical treatment of Einstein's field equations.

3   The assumption that the singularity results from a mathematical over-idealization, the requirement of strict isotropy or uniformity, and that it would not occur in the real world (Wheeler 1958; Lifshitz and Khalatnikov 1963).

If this third premise is accepted tentatively as a working hypothesis, it carries with it a possible resolution of the second paradox, for the matter we see about us now may represent the same baryon content of the previous expansion of a closed universe, oscillating for all time. This relieves us of the necessity of understanding the origin of matter at any finite time in the past. In this picture it is essential to suppose that at the time of maximum collapse the temperature of the universe would exceed 10^{10} K, in order that the ashes of the previous cycle would have been reprocessed back to the hydrogen required for the stars in the next cycle.

Even without this hypothesis it is of interest to inquire about the temperature of the universe in these earlier times. From this broader viewpoint we need not limit the discussion to closed oscillating models. Even if the universe had a singular origin it might have been extremely hot in the early stages.

Could the universe have been filled with black-body radiation from this possible high-temperature state? If so, it is important to notice that as the universe expands the cos-

* This research was supported in part by the National Science Foundation and the Office of Naval Research of the U.S. Navy.
† Reprinted with permission from the *Astrophysical Journal*, 1965.

357

mological redshift would serve to adiabatically cool the radiation, while preserving the thermal character. The radiation temperature would vary inversely as the expansion parameter (radius) of the universe.

The presence of thermal radiation remaining from the fireball is to be expected if we can trace the expansion of the universe back to a time when the temperature was of the order of 10^{10} K ($\simeq m_e c^2$). In this state, we would expect to find that the electron abundance had increased very substantially, due to thermal electron-pair production, to a density characteristic of the temperature only. One readily verifies that, whatever the previous history of the universe, the photon absorption length would have been short with this high electron density, and the radiation content of the universe would have promptly adjusted to a thermal equilibrium distribution due to pair-creation and annihilation processes. This adjustment requires a time interval short compared with the characteristic expansion time of the universe, whether the cosmology is general relativity or the more rapidly evolving Brans–Dicke theory (Brans and Dicke 1961).

The above equilibrium argument may be applied also to the neutrino abundance. In the epoch where $T > 10^{10}$ K, the very high thermal electron and photon abundance would be sufficient to assure an equilibrium thermal abundance of electron-type neutrinos, assuming the presence of neutrino–antineutrino pair-production processes. This means that a strictly thermal neutrino and antineutrino distribution, in thermal equilibrium with the radiation, would have issued from the highly contracted phase. Conceivably, even gravitational radiation could be in thermal equilibrium.

Without some knowledge of the density of matter in the primordial fireball we cannot predict the present radiation temperature. However, a rough upper limit is provided by the observation that black-body radiation at a temperature of 40 K provides an energy density of 2×10^{-29} g cm^{-3}, very roughly the maximum total energy density compatible with the observed Hubble constant and acceleration parameter. Evidently, it would be of considerable interest to attempt to detect this primeval thermal radiation directly.

Two of us (P.G.R. and D.T.W.) have constructed a radiometer and receiving horn capable of an absolute measure of thermal radiation at a wavelength of 3 cm. The choice of wavelength was dictated by two considerations, that at much shorter wavelengths atmospheric absorption would be troublesome, while at longer wavelengths galactic and extragalactic emission would be appreciable. Extrapolating from the observed background radiation at longer wavelengths ($\simeq 100$ cm) according to the power-law spectra characteristic of synchrotron radiation or bremsstrahlung, we can conclude that the total background at 3 cm due to the Galaxy and the extragalactic sources should not exceed 5×10^{-3} K when averaged over all directions. Radiation from stars at 3 cm is $< 10^{-9}$ K. The contribution to the background due to the atmosphere is expected to be approximately 3.5 K, and this can be accurately measured by tipping the antenna (Dicke, Beringer, Kyhl, and Vane 1946).

While we have not yet obtained results with our instrument, we recently learned that Penzias and Wilson (1965) of the Bell Telephone Laboratories have observed background radiation at 7.3 cm wavelength. In attempting to eliminate (or account for) every contribution to the noise seen at the output of their receiver, they ended with a residual of 3.5 ± 1 K. Apparently this could only be due to radiation of unknown origin entering the antenna.

It is evident that more measurements are needed to determine a spectrum, and we expect to continue our work at 3 cm. We also expect to go to a wavelength of 1 cm. We

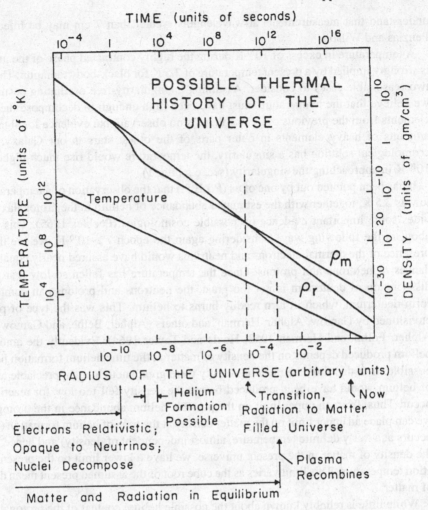

Fig. 1A Possible thermal history of the Universe. The figure shows the previous thermal history of the Universe assuming a homogeneous istropic general-relativity cosmological model (no scalar field) with present matter density 2×10^{-29} gm/cm^3 and present thermal radiation temperature 3.5 K. The bottom horizontal scale may be considered simply the proper distance between two chosen fiducial co-moving galaxies (*points*). The top horizontal scale is the proper world time. The line marked 'temperature' refers to the temperature of the thermal radiation. Matter remains in thermal equilibrium with the radiation until the plasma recombines, at the time indicated. Thereafter further expansion cools matter not gravitationally bound faster than the radiation. The mass density in radiation is ρ_r. At present ρ_r is substantially below the mass density in matter, ρ_m, but, in the early Universe ρ_r exceeded ρ_m. We have indicated the time when the Universe exhibited a transition from the characteristics of a radiation-filled model to those of a matter-filled model.

Looking back in time, as the temperature approaches 1010 K the electrons become relativistic, and thermal electron-pair creation sharply increases the matter density. At temperatures somewhat greater than 1010 K these electrons should be so abundant as to assure a thermal neutrino abundance and a thermal neutron-proton abundance ratio. A temperature of this order would be required also to understand that measurements at wavelengths greater than 7 cm may be filled in by Penzias and Wilson.

A temperature in excess of 10^{10} K during the highly contracted phase of the universe is strongly implied by a present temperature of 3.5 K for black-body radiation. There are two reasonable cases to consider. Assuming a singularity-free oscillating cosmology, we believe that the temperature must have been high enough to decompose the heavy elements from the previous cycle, for there is no observational evidence for significant amounts of heavy elements in outer parts of the oldest stars in our Galaxy. If the cosmological solution has a singularity, the temperature would rise much higher than 10^{10} K in approaching the singularity (see, e.g., fig. 1).

It has been pointed out by one of us (P.J.E.P.) that the observation of a temperature as low as 3.5 K, together with the estimated abundance of helium in the protogalaxy, provides some important evidence on possible cosmologies (Peebles 1965). This comes about in the following way. Considering again the epoch $T \gg 10^{10}$ K, we see that the presence of the thermal electrons and neutrinos would have assured nearly equal abundances of neutrons and protons. Once the temperature has fallen so low that photodissociation of deuterium is not too great, the neutrons and protons can combine to form deuterium, which in turn readily burns to helium. This was the type of process envisioned by Gamow, Alpher, Herman, and others (Alpher, Bethe, and Gamow 1948; Alpher, Follin, and Herman 1953; Hoyle and Tayler 1964). Evidently the amount of helium produced depends on the density of matter at the time helium formation became possible. If at this time the nucleon density were great enough, an appreciable amount of helium would have been produced before the density fell too low for reactions to occur. Thus, from an upper limit on the possible helium abundance in the protogalaxy we can place an upper limit on the matter density at the time of helium formation (which occurs at a fairly definite temperature, almost independent of density) and hence, given the density of matter in the present universe, we have a lower limit on the present radiation temperature. This limit varies as the cube root of the assumed present mean density of matter.

While little is reliably known about the possible helium content of the protogalaxy, a reasonable upper bound consistent with present abundance observations is 25% helium by mass. With this limit, and assuming that general relativity is valid, then if the present radiation temperature were 3.5 K, we conclude that the matter density in the universe could not exceed 3×10^{-32} g cm^{-3}. (See Peebles 1965 for a detailed development of the factors determining this value.) This is a factor of 20 below the estimated average density from matter in galaxies (Oort 1958), but the estimate probably is not reliable enough to rule out this low density.

(*Fig. 1A contd.*)
decompose the nuclei from the previous cycle in an oscillating Universe. Notice that the nucleons are non-relativistic here.

The thermal neutrons decay at the right-hand limit of the indicated region of helium formation. There is a left-hand limit on this region because at higher temperatures photodissociation removes the deuterium necessary to form helium. The difficulty with this model is that most of the matter would end up in helium.

Conclusions

While all the data are not yet in hand we propose to present here the possible conclusions to be drawn if we tentatively assume that the measurements of Penzias and Wilson (1965) do indicate black-body radiation at 3.5 K. We also assume that the universe can be considered to be isotropic and uniform, and that the present energy density in gravitational radiation is a small part of the whole. Wheeler (1958) has remarked that gravitational radiation could be important.

For the purpose of obtaining definite numerical results we take the present Hubble redshift age to be 10^{10} years.

Assuming the validity of Einstein's field equations, the above discussion and numerical values impose severe restrictions on the cosmological problem. The possible conclusions are conveniently discussed under two headings, the assumption of a universe with either an open or a closed space.

Open universe. From the present observations we cannot exclude the possibility that the total density of matter in the universe is substantially below the minimum value 2×10^{-29} g cm^{-3} required for a closed universe. Assuming general relativity is valid, we have concluded from the discussion of the connection between helium production and the present radiation temperature that the present density of material in the universe must be $\lesssim 3 \times 10^{-32}$ g cm^{-3}, a factor of 600 smaller than the limit for a closed universe. The thermal-radiation energy density is even smaller, and from the above arguments we expect the same to be true of neutrinos.

Apparently, with the assumption of general relativity and a primordial temperature consistent with the present 3.5 K, we are forced to adopt an open space, with very low density. This rules out the possibility of an oscillating universe. Furthermore, as Einstein (1950) remarked, this result is distinctly non-Machian, in the sense that, with such a low mass density, we cannot reasonably assume that the local inertial properties of space are determined by the presence of matter, rather than by some absolute property of space.

Closed universe. This could be the type of oscillating universe visualized in the introductory remarks, or it could be a universe expanding from a singular state. In the framework of the present discussion the required mass density in excess of 2×10^{-29} g cm^{-3} could not be due to thermal radiation, or to neutrinos, and it must be presumed that it is due to ordinary matter, perhaps intergalactic gas uniformly distributed or else in large clouds (small protogalaxies) that have not yet generated stars (see fig. 1).

With this large matter content, the limit placed on the radiation temperature by the low helium content of the solar system is very severe. The present black-body temperature would be expected to exceed 30 K (Peebles 1965). One way that we have found reasonably capable of decreasing this lower bound to 3.5 K is to introduce a zero-mass scalar field into the cosmology. It is convenient to do this without invalidating the Einstein field equation, and the form of the theory for which the scalar interaction appears as an ordinary matter interaction (Dicke 1962) has been employed. The cosmological equation (Brans and Dicke 1961) was originally integrated for a cold universe only, but a recent investigation of the solutions for a hot universe indicates that with the scalar field the universe would have expanded through the temperature range $T \simeq 10^9$ K so fast that essentially no helium would have been formed. The reason for this is that the static part of the scalar field contributes a pressure just equal to the scalar-field energy density. By contrast, the pressure due to incoherent electromagnetic radiation or to relativistic particles is one third of the energy density. Thus, if we traced back to a highly contracted universe, we would find that the scalar-field energy density exceeded all other contributions, and that this fast increasing scalar-field energy caused the universe to expand through the highly contracted phase much more rapidly than would be the case if the scalar field vanished. The essential element is that the pressure approaches the energy density, rather than one third of the energy density. Any other interaction which would cause this, such as the model given by Zel'dovich (1962), would also prevent appreciable helium production in the highly contracted universe.

Returning to the problem stated in the first paragraph, we conclude that it is possible to save baryon conservation in a reasonable way if the universe is closed and oscillating. To avoid a catastrophic helium production, either the present matter density should be $< 3 \times 10^{-32}$ g cm^{-3}, or there should exist some form of energy content with very high pressure, such as the zero-mass scalar, capable of speeding the universe through the period of helium formation. To have a closed space, an energy density of 2×10^{-29} gm cm^{-3} is needed. Without a zero-mass scalar, or some other "hard" interaction, the energy could not be in the form of ordinary matter and may be presumed to be gravitational radiation (Wheeler 1958).

One other possibility for closing the universe, with matter providing the energy content of the universe, is the assumption that the universe contains a net electron-type neutrino abundance (in excess of antineutrinos) greatly larger than the nucleon abundance. In this case, if the neutrino abundance were so great that these neutrinos are degenerate, the degeneracy would have forced a negligible equilibrium neutron abundance in the early, highly contracted universe, thus removing the possibility of nuclear reactions leading to helium formation. However, the required ratio of lepton to baryon number must be $> 10^9$.

We deeply appreciate the helpfulness of Drs Penzias and Wilson of the Bell Telephone Laboratories, Crawford Hill, Holmdel, New Jersey, in discussing with us the result of their measurements and in showing us their receiving system. We are also grateful for several helpful suggestions of Professor J. A. Wheeler.

R. H. DICKE
P. J. E. PEEBLES
P. G. ROLL
D. T. WILKINSON

MAY 7, 1965
PALMER PHYSICAL LABORATORY
    PRINCETON, NEW JERSEY

References

Alpher, R. A., Bethe, H. A., and Gamow, G. 1948, *Phys. Rev.*, **73**, 803.
Alpher, R. A., Follin, J. W., and Herman, R. C. 1953, *Phys. Rev.*, **92**, 1347.
Bondi, H., and Gold, T. 1948, *M.N.*, **108**, 252.
Brans, C., and Dicke, R. H. 1961, *Phys. Rev.*, **124**, 925.
Dicke, R. H. 1962, *Phys. Rev.*, **125**, 2163.
Dicke, R. H., Beringer, R., Kyhl, R. L., and Vane, A. B. 1946, *Phys. Rev.*, **70**, 340.
Einstein, A., 1950, *The Meaning of Relativity* (3d ed.; Princeton, New Jersey: Princeton
    University Press), p. 107.
Hoyle, F. 1948, *M.N.*, **108**, 372.

Hoyle, F., and Tayler, R. J. 1964, *Nature*, **203**, 1108.

Liftshitz, E. M., and Khalatnikov, I. M. 1963, *Adv. Phys.*, **12**, 185.

Oort, J. H. 1958, *La Structure et l'évolution de l'universe* (11th Solvay Conf. Brussels: editions Stoops), p. 163.

Peebles, P. J. E. 1965, *Phys. Rev.* (in press).

Penzias, A. A., and Wilson, R. W. 1965, private communication.

Wheeler, J. A., 1958, *La Structure et l'évolution de l'universe* (11th Solvay Conf. Brussels: editions Stoops), p. 112.

Wheeler, J. A., 1964, in *Relativity, Groups and Topology*, eds C. DeWitt and B. DeWitt (New York: Gordon & Breach).

Zel'dovich, Ya. B. 1962, *Soviet Phys. – J.E.T.P.*, **14**, 1143.

Appendix C

Recent results

The publishers have kindly agreed to allow me to include a brief appendix in order to update some of the observational results contained in Chapters 4, 6 and 7. As noted in the preface, references in the main body of the text are complete to early 1992; here I provide summaries of some of the results reported in 1992, 1993 and early 1994. In the space of a few pages, I cannot hope to cover all the work in this field, in which hundreds of papers were published in that time interval. Instead, I will emphasize the most crucial observational results, and provide a few references to theoretical papers which establish new results or provide good reviews of the field.

1 Spectrum

The COBE team (Mather *et al.*, Fixsen *et al.*, and Wright *et al.*, all 1994) have reported new and more complete results on the short-wavelength spectrum of the CBR. From a direct measurement of the spectrum using the FIRAS instrument (Section 4.9), Mather *et al.* (1994) determine $T_0 = 2.726 \pm 0.010$ K. Their observations allow them to place limits on distortions of the spectrum; in the notation introduced in Chapter 5, these limits are $|y| < 2.5 \times 10^{-5}$ and $|\mu| < 3.3 \times 10^{-4}$. These values, like the error in T_0, are given at the 95% confidence level. Careful measurements of the spectrum of the dipole anisotropy also allow a calculation of T_0 (see Section 4.11). Kogut *et al.* (1993) used this technique to obtain $T_0 = 2.75 \pm 0.05$ K.

A redetermination of T_0 at $\lambda = 20$–21 cm is reported by Bensadoun *et al.* (1993); other work by this group at shorter wavelengths is treated in Section 4.6.4. This 1993 paper summarizes previous 20 cm results and new observations made at two good locations. The result obtained, $T_0 = 2.26 \pm 0.19$ K, lies 2.5σ below the COBE value at $\lambda \leqslant 1$ cm, and below most other results summarized in Table 4.6. Bensadoun and his colleagues discuss some possible causes for this discrepancy, but find none of them fully convincing.

Finally, Roth and her colleagues (1993) have reexamined local excitation mechanisms and produced refined values of T_0 from measurements of CN line equivalent widths (Section 4.10). They find $T_0 = 2.729^{+0.023}_{-0.031}$ K at $\lambda = 2.64$ mm, and $T_0 = 2.656 \pm 0.057$ K at $\lambda = 1.32$ mm, both generally lower than the results shown in Table 4.5, but in excellent agreement with the newest COBE result, $T_0 = 2.726 \pm 0.010$ K (Mather *et al.*, 1994).

2 Large angular scale anisotropy

The COBE team has now analyzed a second year of data obtained with the DMR

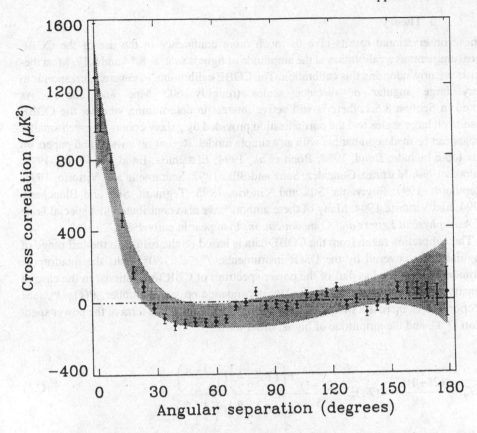

Fig. C.1 The cross correlation of two COBE–DMR maps, one at 53 GHz, the other at 90 GHz. Two years of data are represented (from Bennett *et al.*, 1994, with permission). The dipole moment has been subtracted. The shaded region is $C(\theta)$ for a theoretical CDM model with $n = 1.6$ and $(T_2)_{PS} = 12.4$ μK. Cosmic variance is included in the model; note its increase at large θ.

instrument (see Section 6.7.2). The addition of more data and a more careful assessment of systematic errors have combined to produce a much higher level of statistical certainty in the detection of large-scale fluctuations in the CBR. The tentative values for T_1 and T_2 obtained from this analysis (Bennett *et al.*, 1994) are fully consistent with earlier work summarized in Table 6.2 and with more recent results published by Kogut *et al.* (1993): T_1=3.365±0.027 mK, with a maximum in the direction l^{II}=264°.4±0°.3, b^{II}=48°.4±0°.5. Unpublished reports by the COBE–DMR team on CBR fluctuations on scales $\geqslant 10°$ have confirmed the earlier work reported in Section 6.7.4 and have refined the value for $\Delta T/T$. In fig. C.1, I display the cross correlation signal obtained from the DMR maps at two frequencies, clearly showing a significant signal. In addition, very important confirming work has been carried out using both balloon-borne (Ganga *et al.*, 1993) and ground-based (Hancock *et al.*, 1994) instruments. These observations show a positive correlation with the COBE sky maps. Thus we now have much more confidence than we did in mid-1992 that real fluctuations in the microwave sky have been detected on scales of 10° and above.

3 Theory

These observational results give us much more confidence in the use of the COBE measurements as a calibration of the amplitude of figures such as 8.14 and 8.17. Most theorists are now adopting this calibration. The COBE calibration, of course, corresponds to very large angular or distance scales (roughly 600 Mpc and above). As noted in Section 8.5.1, there is still active interest in determining whether the COBE results on large scales and the normalization provided by galaxy counts on much smaller scales can be made compatible with any simple model. Recent reviews of and papers on this topic include: Bond, 1994; Bond *et al.*, 1994; Efstathiou, Bond and White, 1992; Gorski, 1993; Martinez-Gonzalez, Sanz and Silk, 1992; Scaramella and Vittorio, 1993; Steinhardt, 1994; Sugiyama, Silk and Vittorio, 1993; Tegmark, Silk and Blanchard, 1994; and Vittorio, 1994. Many of these authors have also contributed to a special issue of *Astrophysical Letters and Communications* to appear in early 1995.

The calibration taken from the COBE data is based on the results in the full range of angular scale covered by the DMR instruments, $7° \lesssim \theta < 180°$. Thus the quadrupole moment* is included as part of the power spectrum of CBR fluctuations. In the case of density perturbations with a power law dependence on wave number, $P(k) \propto k^n$, (see Section 8.4.1), there is a simple relation between the quadrupole term of the power spectrum $(T_2)_{PS}$ and the amplitude of higher order terms:

$$\Delta T_l^2 = \frac{(2l+1)C_l}{4\pi} = (T_2)_{PS}^2 \frac{(2l+1)}{5} \frac{\Gamma\left(l + \frac{n-1}{2}\right)\Gamma\left(\frac{9-n}{2}\right)}{\Gamma\left(l + \frac{5-n}{2}\right)\Gamma\left(\frac{3+n}{2}\right)}, \tag{C.1}$$

where I have used l for the order of the multipole to avoid confusion with the power-law index n. This relation is taken from Bond and Efstathiou (1987). When eqn (C.1) is fitted to the COBE data (Bennett *et al.*, 1994, and references therein), values of $(T_2)_{PS} \sim 12$–18 μK result, with the larger value favored if n is constrained to 1.0. Since the amplitude for higher order moments is fixed by $(T_2)_{PS}$ for a given n, $(T_2)_{PS}$ in effect becomes the calibration constant derived from the COBE measurements.

It is interesting that $(T_2)_{PS}$ found as above is somewhat larger than the value of T_2 derived by fitting eqn. (6.2b) to the observations directly (the intrinsic quadrupole moment of the observations). That is, T_2 is smaller than expected from the amplitude of higher order moments. This may be a result of *cosmic variance*, the scatter expected between different realizations of the same power-law spectrum of fluctuations with random phases. Cosmic variance is naturally larger for lower multipole moments. In our Universe, one particular realization, T_2 may be smaller than average.

In the past two years, interest in the possibility of tensor as well as scalar perturbations (Section 8.4.1) has continued high. The addition of tensor perturbations will modify the power spectrum of density perturbations. In particular, Davis *et al.* (1992) have shown that there is a simple relation between the index n for the scalar perturbations and the relative amplitude of tensor to scalar perturbations:

---

* Any intrinsic dipole moment is swamped by the dipole induced by the motion of the observer (Section 8.2).

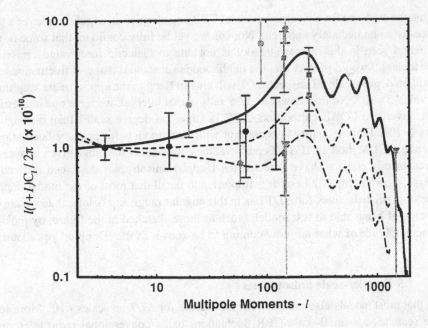

Fig. C.2 Recent measurements of CBR fluctuations on a range of scales (points with associated error bars) compared with a set of theoretical models. The point at $l \approx 3$ is the COBE–DMR measurement. The wavy structure in the theoretical curves at $100 \leqslant l \leqslant 1000$ represents the 'Doppler' or 'acoustic' peaks. (Courtesy Paul Steinhardt.)

$$\frac{(T_2)_{\text{tensor}}}{(T_2)_{\text{scalar}}} = 7\,(1-n)$$

Note that $n = 1$ implies no tensor component.

4 Fluctuations on degree scales

Returning to the observational side, I would say the greatest advances since 1992 have been made in measurements of CBR fluctuations on scales of roughly 1° (actually, from 0°.2 to several degrees). On these angular scales, it is possible to make observations from the ground, if care is taken to select a good site (the South Pole is rapidly coming to be recognized as a premier site). Results in this range of angular scale have been reported by Piccirillo and Calisse (1993); Schuster *et al.* (1993); Tucker *et al.* (1993); Wollack *et al.* (1993); and Dragovan *et al.* (1994). In addition, groups based (primarily) at Rome, the University of California and MIT have been pursuing anisotropy observations from balloons (Gunderson *et al.*, 1993; Meinhold *et al.*, 19943; Cheng *et al.*, 1994; de Bernardis *et al.*, 1994).

Most of these groups are now reporting *detections* of CBR fluctuations, often at the 2σ–3σ level. That is in sharp contrast to the situation in 1992, where most results were being reported as upper limits. Indeed, it is interesting that in some cases data initially used to derive upper limits are now being reinterpreted to provide actual measurements of $\Delta T/T$. Figure C.2 shows some of these measurements and a few of the remaining upper limits. Without the theoretical curves to guide the eye, I think it would be fair to say that the evidence for a clear detection of CBR anisotropies on angular scales of a few degrees is no immediately apparent. Nor can we yet be fully confident that some of the fluctuation seen in the microwave sky is not due to Galactic foreground emission. Nevertheless, I would predict that the likelihood is substantial that real fluctuations are present on degree scales. I suspect $\Delta T/T$ will end up lying in the approximate amplitude interval $2–5 \times 10^{-5}$. No matter where $\Delta T/T$ falls in that interval, we have one interesting result: given the COBE calibration, $\Delta T/T$ is larger on degree scales than on angular scales $\geqslant 10°$. That inequality is consistent with most models for primary fluctuations, particularly CDM models. If $\Delta T/T$ ends up at the upper end of the range given above, we can reasonably claim to have detected the Doppler anisotropies described in Section 8.4.6, as shown in fig. C.2 here. It is important to recall that most of the model-dependence of predicted values for $\Delta T/T$ lies in this angular range and below; thus we are on the verge of being able to test models, such as those sketched in the figure, by looking for the amplitude of what are now coming to be known as the 'Doppler' or 'acoustic' peaks.

5 Smaller-scale anisotropies

Note that most models also predict quite low values for $\Delta T/T$ on scales $<10'$. More sensitive searches for small-scale CBR fluctuations using conventional radio telescopes (Myers *et al.*, 1993) and interferometric techniques (Fomalont *et al.*, 1993; Radford, 1993; Subrahmanyan *et al.*, 1993) continue to provide tighter and tighter upper limits on, but no detections of, fluctuations. Much of the material discussed in Chapter 7 here has been thoroughly reviewed by Readhead and Lawrence in an article which appeared late in 1992.

I will end this brief review by noting the success of the Cambridge group in using interferometers to study the Sunyaev–Zel'dovich (SZ) effect in nearby clusters of galaxies (Jones *et al.*, 1993). They have now observed several clusters, and have convincingly detected the SZ effect in several of them. In one case (Jones, 1994), they have used their observations to derive a value for Hubble's constant, using the technique described in Section 8.9.3. That value was still tentative when this appendix was prepared, but it is likely to be $H_0 \leqslant 50$ km s^{-1} per megaparsec.

Herbig *et al.* (1994) working on a nearby well-studied cluster (Coma) have determined a somewhat higher value for H_0, 74 ± 29 in the same units. Also recently, Wilbanks *et al.* (1994) have reported the first significant detection of the SZ effect at millimeter wavelengths. The wavelength used, 2.2 mm, is still longer than the wavelength at which the SZ signal changes sign.

Finally, I emphasize that the list of references given below is a *very partial* sample of those appearing in the interval mid-1992–mid-1994. The sampling of theoretical papers is particularly sparse.

References

Bennett, C. L., *et al.* 1994, *Ap. J.*, in press.
Bensadoun, M., *et al.* 1993, *Ap. J.*, **409**, 1.
Bond, J. R. 1994, in *Cosmology and Large Scale Structure* (Les Houches Lectures, 1993), ed. R. Schaeffer, Elsevier Sci. Publ., Amsterdam.
Bond, J. R., and Efstathiou, G. 1987, *Monthly Not. Roy. Astron. Soc.*, **226**, 655.

Bond, J. R., *et al.* 1994, *Phys. Rev. Lett.*, **72**, 13.

Cheng, E. S., *et al.* 1994, *Ap. J. (Lett.)*, **422**, L37.

Davis, R., Steinhardt, P. J., and Turner, M. S. 1992, *Phys. Rev. Lett.*, **69**, 1856.

de Bernardis, P., *et al.* 1994, *Ap. J. (Lett.)*, **422**, L33.

Dragovan, M., *et al.* 1994, *Ap. J. (Lett.)*, **427**, L67.

Efstathiou, G., Bond, J. R., and White, S. D. M. 1992, *Monthly Not. Roy. Astron. Soc.*, **258**, 1 p.

Fixsen, D. J., *et al.* 1994, *Ap. J.*, **420**, 445.

Fomalont, E. B., Partridge, R. B., Lowenthal, J. D., and Windhorst, R. A. 1993, *Ap. J.*, **404**, 8.

Ganga, K., Cheng, E., Meyer, S., and Page, L. 1993, *Ap. J. (Lett.)*, **410**, L57.

Gorski, K. M. 1993, *Ap. J. (Lett.)*, **410**, L65.

Gunderson, J. O., *et al.* 1993, *Ap. J. (Lett.)*, **413**, L1.

Hancock, S., *et al.* 1994, *Nature*, **367**, 333.

Herbig, T., Lawrence, C. R., Readhead, A. C. S., and Gulkis, S. 1994, *Ap. J. (Lett.)*, in press.

Jones, M. 1994, *Astrophys. Letters and Commun.*, in press.

Jones, M., *et al.* 1993, *Nature*, **365**, 320.

Kogut, A., *et al.* 1993, *Ap. J.*, **419**, 1.

Martinez-Gonzalez, E., Sanz, J. L., and Silk, J. 1992, *Phys. Rev. D*, **46**, 4193.

Mather, J. C., *et al.* 1994, *Ap. J.*, **420**, 439.

Meinhold, P. R., *et al.* 1993, *Ap. J.*, **406**, 12.

Myers, S. T., Readhead, A. C. S., and Lawrence, C. R. 1993, *Ap. J.*, **405**, 8.

Piccirillo, L., and Calisse, P. 1993, *Ap. J.*, **411**, 529.

Radford, S. J. E. 1993, *Ap. J. (Lett.)*, **404**, L33.

Readhead, A. C. S., and Lawrence, C. R. 1992, *Ann. Rev. Astron. and Astrophys.*, **30**, 653.

Roth, K. C., Meyer, D. M., and Hawkins, I. 1993, *Ap. J. (Lett.)*, **413**, L67.

Scaramella, R., and Vittorio, N. 1993, *Ap. J.*, **411**, 1.

Schuster, J., *et al.* 1993, *Ap. J. (Lett.)*, **412**, L47.

Steinhardt, P. J. 1994, in *Evolution of the Universe and Its Observational Quest*, ed. K. Sato, Univ. Academic Press, Inc., Tokyo.

Subrahmanyan, R., *et al.*, 1993, *Monthly Not. Roy. Astron. Soc.*, **263**, 416.

Sugiyama, N., Silk, J., and Vittorio, N. 1993, *Ap. J. (Lett.)*, **419**, L1.

Tegmark, M., Silk, J., and Blanchard, A. 1994, *Ap. J.*, **420**, 484.

Tucker, G. S., *et al.* 1993, *Ap. J. (Lett.)*, **419**, L45.

Vittorio, N. 1994, in *Transaction of the IAU XXIIA: Reports on Astronomy*, ed. J. Bergeron, Kluwer Acad. Publ., Dordrecht.

Wilbanks, T. M., *et al.* 1994, *Ap. J. (Lett.)*, **427**, L75.

Wollack, E. J., *et al.* 1993, *Ap. J. (Lett.)*, **419**, L49.

Wright, E. L., *et al.* 1994, *Ap. J.*, **420**, 450.

Index

absorbing–emitting cloud, 91–2
adiabatic perturbations, 19, 297ff, 328, 338
age (of Universe), t_0, 5–6
airborne observations, 130
aliasing, 88, 261, 264
Alpher, R. A., 43–4, 50
anisotropic cosmologies, 290–5
anisotropy (of CBR – summaries), 201, 208, 260,
274–5, 343, 364–7
antenna, horn, 68–70
antenna, corrugated, 69–70
antenna pattern (*see* beam pattern)
antenna temperature, 62–3, 129, 202–3, 231
aperture synthesis, 85–90, 249–50, 260–8, 270
atmospheric emission, 115–23, 194–5, 218
atmospheric emission fluctuations, 217–21, 223
atmospheric temperature, T_{atm} (measures of), 120–2,
125, 127, 129–37
autocorrelation function (of CBR fluctuations,
$C(\theta)$), 312–24, 331–2, 336

backlobes, 109–11, 221–2
balloon-borne experiments, 56, 137–44, 198–204,
247, 258–9
bandwidth, 73–5
bandwidth smearing (fringe washing), 82, 89–90
baryogenesis, 24–5
baryon density, Ω_b, 27–8, 163–70, 180–1, 184, 323
baryon (isocurvature) models, 317, 324, 339
baryon number, 21, 24–5
baseline, 81–8
Bayesian statistics, 241–3
beam pattern, $P(\theta, \phi)$, 64–7, 70, 84, 318–21
beam solid angle, 63–7
beam switching, 68–9, 190–1, 214–20, 318
beam width, $\theta_{1/2}$, 66–7, 318
Bell Telephone Laboratories, 46–51, 59, 356
Bianchi types, 291–3
bias (*and* bias factor, b), 34, 289
bolometers, 56, 72, 77–80, 246–7
Bond, J. R., 161, 175–7, 280, 296, 301, 305, 307,
312, 334, 366
Bose–Einstein spectrum, 163–5, 169, 181–2
Boughn, S. P., 200–1, 235–44
bremsstrahlung, 92–4, 96, 114, 163–6, 196, 225–30
brightness temperature, 61, 91

calibration, 108–9, 138–9, 147–8, 365
Cartesian coordinates (on sky), 67, 85–8
Cassegrain telescope, 67–8
causal connection, 19–20, 303–4
causal horizon (*see* particle horizon)
CH, CH⁺ (interstellar), 154–5
chemical potential, μ, 163–5, 169, 181–3
chronometric cosmology, 36–7
clumping factor, 349
clusters of galaxies, 3–4, 271–4
COBE (Cosmic Background Explorer) satellite,
146–9, 205–9, 260, 321–5, 330–4, 341, 345,
364–5
COBE calibration of $\Delta T/T$, 311, 365
cold dark matter (CDM) models, 297, 302, 305,
313–15, 325
cold load calibrator, 103–4, 106–8
comoving coordinates, 8
Compton effect (inverse), 94–5, 170–3 (*see also*
Sunyaev–Zel'dovich effect)
Compton scattering, 162–3, 168
'Cool' Big Bang models, 37–8
correlation function, $\xi(r)$, 309–11, 326
corrugations (*see* antenna, corrugated)
cosmic rays, 55, 183–5
cosmic variance, 366
cosmological constant, Λ, 12–14, 337–40
cosmological equations, 8–10, 12–14
cosmological models, 7–14, 289–93
cosmological principle, 7, 37
critical density, ρ_c, 11
curvature, k, 8, 10ff, 22
cyanogen, CN (interstellar), 45–6, 149–54, 364

damping (of density purturbations), 29, 301–4
'dark matter', 1–4, 14, 32–4, 302–6, 313–17
density (cosmic), ρ_0, 8–11, 14
density parameter, Ω, 11, 14, 165, 287–9, 323, 338
density perturbations, 296
    scalar, 296ff
    tensor, 296, 336, 366
    vortices, 296, 324
deuterium (primordial), 26–7
Dicke, R. H., 32, 42, 45–50, 362
Dicke switching, 47, 75–7
dipole moment, T_1, 155, 188–9, 201 (table), 208,
211, 313, 364